CDMA: Access and Switching

CDMA: Access and Switching

For Terrestrial and Satellite Networks

Diakoumis Gerakoulis
AT&T Labs-Research, USA

Evaggelos Geraniotis
University of Maryland, USA

JOHN WILEY & SONS
Chichester · New York · Brisbane · Toronto · Singapore

John Wiley & Sons, Ltd
Baffins Lane, Chichester,
West Sussex, PO19 1UD, England

National 01243 779777
Internationl (+44) 1243 779777

e-mail (for orders and customer service enquiries): cs-books@wiley.co.uk

Visit our Home Page on http://www.wiley.co.uk or http://www.wiley.com

Other Wiley Editorial Offices

John Wiley & Sons, Inc., 605 Third Avenue,
New York, NY 10158-0012, USA

Wiley-VCH Velag GmbH
Pappelallee 3, D-69669 Weinheim, Germany

Jacaranda Wiley Ltd, 33 Park Road, Milton,
Queensland 4064, Australia

John Wiley & Son (Asia) Pte Ltd, 2 Clementi Loop #02-01,
Jin Xing Distripark, Singapore 12980

John Wiley & Sons (Canada) Ltd, 22 Worcester Road,
Rexdale, Ontario, M9W 1L1, Canada

Library of Congress Cataloging-in-Publication Data

Gerakoulis, Diakoumis P.
CDMA: access and switching for terrestrial and satellite networks/Diakoumis Gerakoulis, Evaggelos Geraniotis.
p.cm.
Includes bibliographical references and index.
ISBN 0-471-49184-5 (alk.paper)
1. Code division multiple access. 2. Artificial satellites in telecommunication. I. Geraniotis, Evaggelos. II. Title.

TK5103.452.G47 2001
621.3845—dc21

00-47751

British Library Cataloguing in Publication Data

A catalogue record for this book is available from the British Library

ISBN 0-471-49184-5

Produced from files supplied by the author, processed by Laser Words, Madras, India.
Printed and bound in Great Britain by Antony Rowe, Chippenham.
This book is printed on acid-free paper responsibly manufactured from sustainable forestry in which at least two tress are planted for each one used for paper production.

Contents

Preface

Code division may be considered as a generalized method for access and switching in communication networks. Such an approach may be viewed as a user encoding process where the choices of the code and spreading types can create a large set of access and switching techniques in which the traditional ones are included. In this sense, code division can provide a unified approach for multiple access and switching in communications.

In *CDMA: Access and Switching* we introduce new concepts and applications, and present innovative designs for Code Division Multiple Access (CDMA). Each new application is assessed and evaluated, while each innovative design is followed by rigorous performance analysis. Code division is applied for both link access and node routing (switching) of the user data. Thus, a *Switched CDMA* network may be formed for the interconnection of end users. A switched CDMA network can be either terrestrial or satellite, wireless or cable.

More specifically, in Chapter 1 we present generalized CDMA as a unifying approach to multiple access communications. We introduce the processes of spread-spectrum and spread-time, and derive each traditional access method from the generalized one.

In Chapter 2 we present spreading sequences of three main categories: Orthogonal, Pseudo-Orthogonal and Quasi-Orthogonal. We also include orthogonal Hadamard code construction methods, complex and polyphase orthogonal designs and the timing jitter properties of sequences.

In Chapter 3 we present the concept of switched CDMA networks. One such network is the satellite switched CDMA (SS/CDMA), for which we describe the network architecture and system operation.

Chapters 4 to 9 cover all aspects of a switched CDMA network, with particular focus on the SS/CDMA. Chapter 4 introduces the method of code division switching for routing calls and packets at the exchange node of the network. Code Division Switch (CDS) architectures are presented, evaluated and compared with traditional switching techniques. Chapter 5 presents a demand assignment system for joint access and switching in the SS/CDMA network, and provides its throughput performance based on optimum and random CDS scheduling algorithms. Chapter 6 describes a spectrally-efficient CDMA (SE-CDMA) for link access in the SS/CDMA network. The SE-CDMA has an innovative orthogonal CDMA design, for which we provide detailed interference analysis and bit error rate performance. Chapter 7 presents a spread-spectrum random access protocol for the SS/CDMA network access, and provides the orthogonal code synchronization mechanism for the SE-CDMA uplink transmission. It also includes the design and performance of an innovative tracking control loop. Chapter 8 presents two techniques for carrier recovery in CDMA: symbol-aided and pilot-aided demodulation. These techniques are evaluated analytically, and their bit error rate performance is compared with coherent and differentially coherent demodulation methods. Chapter 9 deals with the phenomenon of non-linear amplification of synchronous CDMA signals. This phenomenon appears at the satellite amplifier for downlink transmission. Performance analysis provides the optimum value of the input 'back-off' required for linear amplification.

While some of the previous chapters focus on the orthogonal CDMA approach, Chapter 10 considers the 'pseudo-orthogonal' CDMA alternative, and examines two basic methods for optimizing it: adaptive power control and multiuser detection. The power control is based on adaptive quantized and loop-filtering feedback. Also, a survey of multiuser detection methods and a novel iterative multiuser detection technique are presented and evaluated.

The authors would like to thank everyone who contributed to the development of this book. Special thanks to R.R. Miller at AT&T Labs Research for his support on this project. We also wish to acknowledge the contributions of the following: Wai-Chung Chan in the SS/CDMA throughput Chapter 5; Hesham El Gamal in the iterative multiuser detector of Chapter 10; Saeed Ghassemzadeh in the construction of Hadamard matrices of Chapter 2; Mohamed M. Khairy in the symbol-aided demodulation of Chapter 8; Pen C. Li on the effects of non-linearities in Chapter 9; and last but not least, Hsuan-Jung Su with the tracking loop performance in Chapter 7 and with the adaptive power control in Chapter 10.

Finally, *CDMA: Access and Switching* will be a valuable companion to many system designers who are interested in new applications of CDMA, and also to academic researchers, since it opens up new research areas in the field of multiple access and switching, such as the Generalized CDMA and Code Division Switching.

Diakoumis Gerakoulis
Evaggelos Geraniotis

1

The Generalized CDMA

1.1 Introduction

One of the basic concepts in communication is the idea of allowing several transmitters to send information simultaneously over a communication channel. This concept is described by the terms *multiple access* and *multiplexing.* The term *multiple access* is used when the transmitting sources are not co-located, but operate autonomously as a *multipoint-to-point* network, while when the transmitting sources are co-located, as in a *point-to-multipoint* network, we use the term *multiplexing.* There are several techniques for providing multiple access and multiplexing, which belong to one of two basic categories: the *orthogonal* and the *pseudo-orthogonal* (PO) division multiple accesses. In orthogonal multiple access the communication channel is divided into sub-channels or user channels which are mutually *orthogonal*, i.e. are not interfering with each other. In pseudo-orthogonal multiple access, on the other hand, there is interference between user channels since they are not perfectly orthogonal to each other. The traditional Time Division and Frequency Division Multiple Access methods (TDMA and FDMA), as well as the Orthogonal Code Division Multiple Access (O-CDMA), are orthogonal multiple accesses, while the conventional asynchronous CDMA is a pseudo-orthogonal multiple access.

Orthogonal division multiple access is achieved by assigning an orthogonal code or sequence to each accessing user (orthogonal code-sequences are presented in Chapter 2). Orthogonal sequences provide complete isolation between user channels. However, they require synchronization so that all transmissions arrive at the receiver at a given reference time (global synchronization). Pseudo-orthogonal multiple accesses, such as the asynchronous CDMA, are implemented with pseudo-random noise codes or sequences (PN-sequences) which suppress the other user interference only by the so-called *spreading factor* or *processing gain.* The pseudo-orthogonal approach, however, does not require global synchronization.

The capacity (i.e. the maximum number of accessing users) of an orthogonal multiple access is fixed, and is equal to the length or the size of the orthogonal code, which is also equal to the spreading factor. In pseudo-orthogonal multiple access, on the other hand, the capacity is not fixed but is limited by the interference between users. Such a system is said to have a 'soft' capacity limit, since excess users may be allowed access at the expense of increased interference to all users. In general, the capacity in Pseudo-Orthogonal (PO) or Asynchronous (A) CDMA is less than the spreading factor. In order to enhance capacity, PO-CDMA sytems utilize multiple access interference cancellation techniques known as *multiuser detectors* (see Chapter 10). Such techniques are implemented at the receiver and they attempt to achieve (in the best case) what orthogonal codes provide at the transmitter in an orthogonal multiple access system, i.e. to eliminate the other user interference.

Each of these two approaches is more efficient if it is used in the appropriate application. For example, Orthogonal CDMA (O-CDMA) can be used more efficiently in fixed service or low mobility wireless applications where synchronization is easier to achieve. Also, the O-CDMA is preferable in the forward wireless link (base-to-mobile), since no synchronization is required in this case. Asynchronous CDMA, on the other hand, is more appropriate in the reverse link (mobile-to-base) high mobility environment.

The use of different access methods, however, led to the development of incompatible technologies and communication standards. In this chapter we attempt to provide an approach for unifying the multiple access communications. This approach is based on a *user encoding process* which is applied in order to integrate different access methods. Based on the proposed point of view, we represent a transmitter by a *symbol encoder*, and a *user encoder*, as illustrated in Figure 1.1. The symbol encoding provides channel encoding and symbol keying, while the user encoding provides the system and the user access into the communication link.

The *user encoding*, in particular, is defined as *the process in which a code sequence is used for both (1) to 'spread' the operating domain (i.e. time or spectrum), and (2) to identify each particular user in that domain.* In this process the operation of *spreading* is required in order to create a 'space' in the channel which will contain all accessing or multiplexed users.

The encoded signal will then depend upon:

(1) The type of code sequence used. That is, the code sequences may be mutually orthogonal or pseudo-orthogonal, real or complex.
(2) The type of spreading. Spreading may take place either in the frequency domain, called *spread-spectrum*, or in the time domain, called *spread-time.*
(3) The pulse-shape of the data symbol. The pulse-shape, for example, may be time-limited or bandwidth-limited.

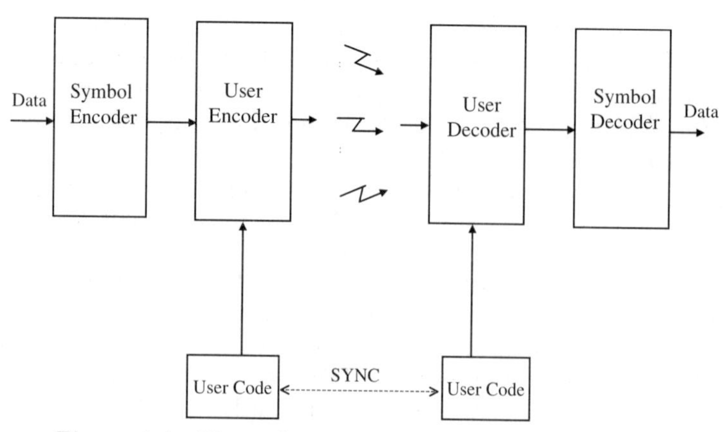

Figure 1.1 The multiuser data communications process.

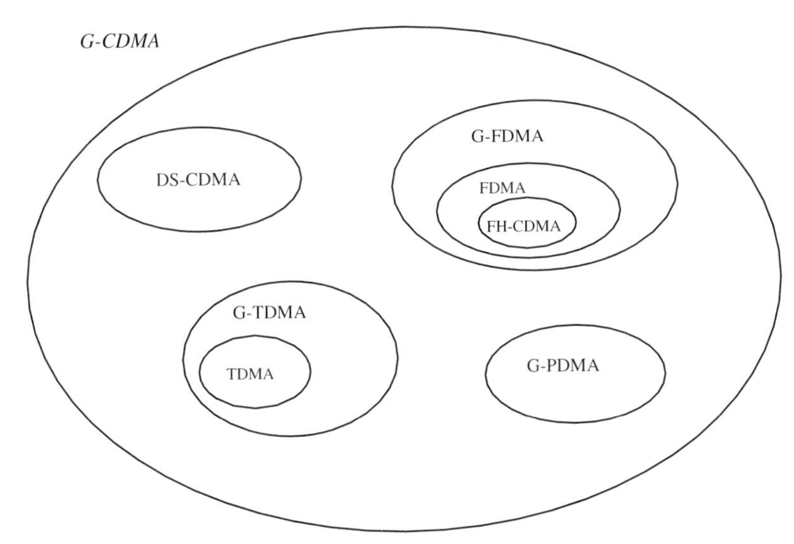

Figure 1.2 The G-CDMA as the super-set of the multiple access methods.

Each set of parameters (1), (2) and (3) defines a multiple access method or a type of user encoder. The combination of these parameters, (1), (2) and (3), will then create a large set of multiple accesses in which the conventional methods are only special cases, as illustrated in Figure 1.2. This super-set multiple access method is called Generalized CDMA (G-CDMA). Using this approach, in addition to the conventional methods, new multiple access methods have been created, such as the Generalized-TDMA and the Generalized-FDMA. Our purpose in this chapter, however, is not to examine and compare the performance of the new access methods, but to use them for demonstrating the continuum of the user encoding process.

In the next section we present user encoding by real sequences, with spread-spectrum or spread-time, having synchronous or asynchronous access. We have reviewed the conventional asynchronous CDMA and have derived the traditional time division multiple access from the orthogonal spread-time CDMA. In Section 1.3 we present user encoding by complex sequences, with spread-spectrum or spread-phase, having synchronous or asynchronous access. In this case we have defined the generalized Frequency Division Multiple Access (FDMA) as a complex CDMA scheme, and from it we have derived the traditional FDMA and the frequency hopping CDMA. We have also presented a spread-phase CDMA and a Phase Division Multiple Access (PDMA) scheme. In Section 1.4 we present composite multiple access methods such as the spread-spectrum and spread-time multiple access using the method of extended orthogonal sequences presented in Chapter 2.

This work was originally presented in reference [1].

1.2 User Encoding by Real Sequences

Let us now consider user encoding by sequences which are real numbers. First we assume the case of square pulse (time-limited) waveforms and binary (± 1) sequences.

In particular, let a signal $d_i(t)$ of a data sequence of K symbols of user i,

$$d_i(t) = \sum_{k=0}^{K-1} d_{i,k}\; p_{T_d}(t - kT_d) \quad \text{where} \quad p_T(t) = \begin{cases} 1 & \text{for } 0 \le t < T \\ 0 & \text{otherwise} \end{cases}$$

Also, let the code-sequence $c_i(t)$ assigned to user i be given by

$$c_i(t) = \sum_{l=0}^{L-1} c_{i,l}\; p_{T_c}(t - lT_c) \qquad 1 \le i \le M$$

where L is the length of the sequence, M is the number of sequences, T_d is the duration of the data symbol and T_c is the duration of the code symbol, and $R_d = 1/T_d$ is the data rate and $R_c = 1/T_c$ is the code rate.

The encoded signal of user i is then $s_i(t) = d_i(t) \star c_i(t)$. The symbol $\star$ indicates the operation of user encoding, and is specified in each case we examine. As a result of encoding, $s_i(t)$ may be a spread-spectrum or a spread-time signal. Hence, we may distinguish the cases of spread-spectrum and spread-time described in the following subsections.

1.2.1 Spread-Spectrum

In the case of spread-spectrum, the length of the data symbol is N times longer than the length of the encoding symbol T_c. Hence, we define the ratio

$$N_{SS} \equiv \frac{T_d}{T_c} = \frac{R_c}{R_d} = N$$

to be the *spreading factor*, where N is an integer $N > 1$, and $T_d = NT_c$. The rate of $s_i(t)$ is then $R_c > R_d$, which means that the required bandwidth has to be spread to accommodate the rate $R_c = NR_d$. The encoded symbol or the spread time-pulse is called a *chip*.

Considering a spread-spectrum process, we may again distiguish two cases. In the first case, spreading is achieved with orthogonal squences, and such a system is called

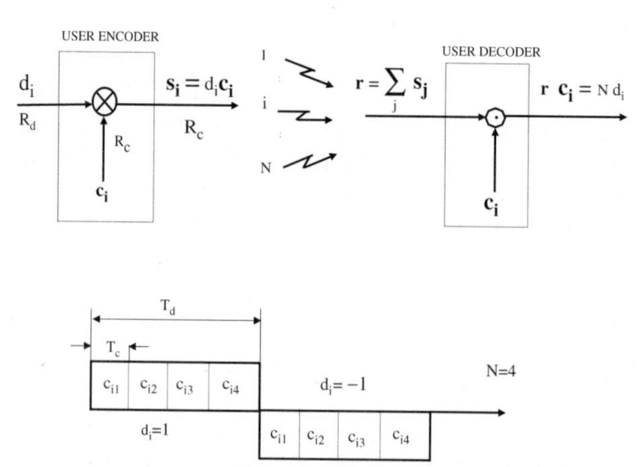

Figure 1.3 The Spreading Process in Orthogonal CDMA.

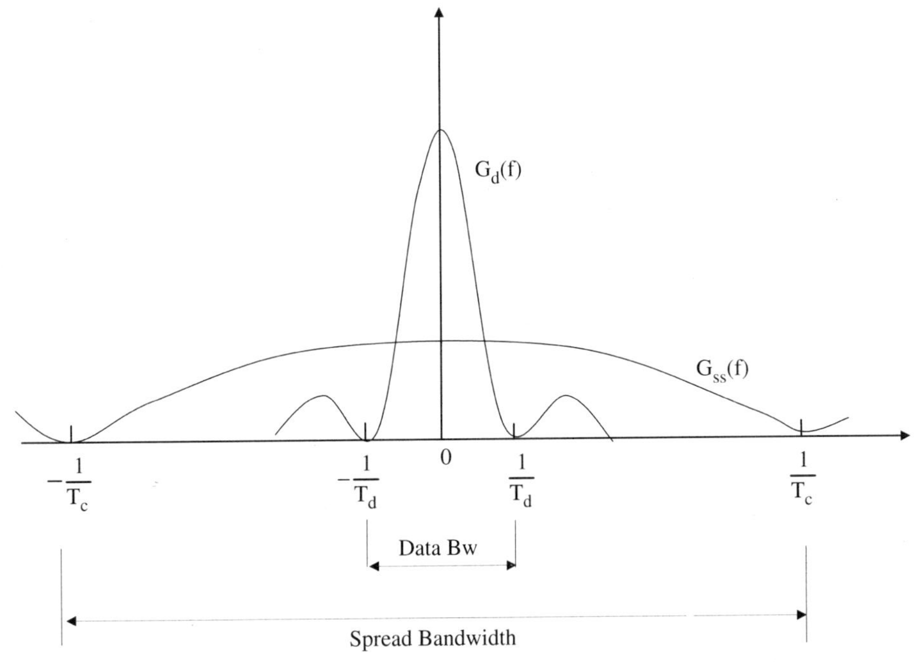

Figure 1.4 The power spectrum of data and spread signal.

orthogonal or synchronous CDMA. In the second case, spreading is achieved with Pseudo-random Noise (PN) sequences. Then we have the conventional asynchronous CDMA, also called direct sequence CDMA (DS/CDMA).

The Orthogonal CDMA

Orthogonal CDMA (O-CDMA) is based on binary orthogonal sequences of length N. That is, the spreading factor is equal to the sequence length, which is also equal to the number of sequences. Hence, $M = N = L$. Let d_i be a data symbol of user i, and $\mathbf{c_i} \equiv [c_{1i}, c_{2i}, .., c_{Ni}]$ be the ith orthogonal code vector (sequence), $i = 1, .., N$; $d_{i,j}, c_{ij} \in \{-1, +1\}$. The encoded data vector of user i, $\mathbf{s_i}$ is defined as follows:

$$\mathbf{s_i} \equiv d_i\mathbf{c_i} \equiv [d_i c_{0,i}, d_i c_{1,i}, .., d_i c_{N-1,i}].$$

Assuming K consecutive data symbols, the transmitted signal of the O-CDMA is described by the equation

$$s_i(t) = \sum_{k=0}^{K-1} d_{k,i}\; c_i(t - kT_d) \quad \text{where} \quad c_i(t) = \sum_{l=0}^{N-1} c_{l,i}\; p_{T_c}(t - lT_c) \qquad 1 \le i \le N$$

The transmitted signal $s_i(t)$ has a rate $R_c = 1/T_c = N/T_d = NR_d$, since $T_d = NT_c$. This means that the required bandwidth of the transmitted signal is N times wider than the bandwidth of the data $d_i(t)$, (spread-spectrum). Hence, the spreading factor is $N_{ss} = \frac{R_c}{R_d} = \frac{T_d}{T_c} = N > 1$. The spreading process is illustrated in Figure 1.3. Assuming that each chip is a square time pulse with duration T_c, the spectrum of the spreaded signal is (see Figure 1.4)

$$G_{ss}(f) = T_c \left(\frac{\sin\ \pi f T_c}{\pi f T_c} \right)^2$$

That is, the chip pulse is time-limited but spectrally unlimited. Therefore, a band-limiting filter (LPF) has to be used to limit the bandwidth in this case. Now, we assume that all N users accessing the system are synchronized to a reference time so that chips and symbols from all users are aligned at the receiver. Also, omitting the thermal noise and the impact of the band-limiting filter, the received signal at the input of the decoder is given by

$$r(t) = \sum_{j=1}^{K} s_j(t) = \sum_{j=1}^{K} \sum_{l=0}^{N-1} d_{k,j}\ c_{l,j} p_{T_c}(t - lT_c)$$

After the A/D converter the received signal may be represented by

$$\mathbf{r} = \sum_{j=1}^{N} \mathbf{s_j} = \sum_{j=1}^{N} d_j \mathbf{c_j}$$

The decoding process consists of taking the inner product between vectors $\mathbf{r}$ and $\mathbf{c_i}$. That is,

$$\mathbf{r} \cdot \mathbf{c_i} = \sum_{i=1}^{N} \sum_{j=1}^{N} d_j \mathbf{c_j} \cdot \mathbf{c_i} = L d_i \qquad \text{since} \quad \mathbf{c_j} \cdot \mathbf{c_i} = \sum_{k=1}^{N} c_{kj} c_{ik} = \begin{cases} L & if\ \ i = j \\ 0 & if\ \ i \neq j \end{cases}$$

The Asynchronous DS/CDMA

In the asynchronous DS/CDMA we use Pseudo-random Noise (PN) sequences with length L, where $L \geq N$ ($T_d = NT_c$). PN-sequences are defined in Chapter 2 and are represented here by a continuous time function $c_i(t) = \sum_{l=0}^{L-1} c_{l,i}\ p_{T_c}(t - lT_c)$, where $p_{T_c}(t)$ is a square time-pulse with length T_c, and $c_{l,i} \in \{-1, +1\}$. The continuous time autocorrelation function $R_i(\tau)$ of $c_i(t)$ is then defined by

$$R_i(\tau) = \frac{1}{L} \int_0^L c_i(t) c_i(t + \tau) dt$$

$R_i(\tau)$ has been evaluated and is equal to

$$R_i(\tau) = \sum_l q(\tau - lLT_c)$$

where

$$q(\tau) = \begin{cases} 1 - \frac{|\tau|}{LT_c}(1 + \frac{1}{L}) & \text{for } |\tau| \leq T_c \\ -\frac{1}{L} & \text{for } T_c \leq |\tau| \leq LT_c/2 \end{cases}$$

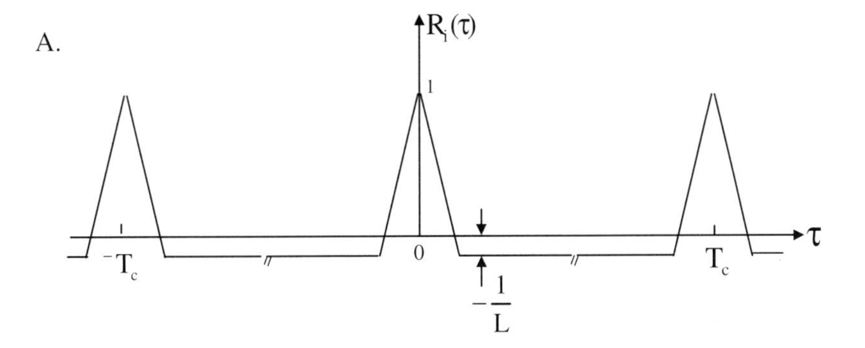

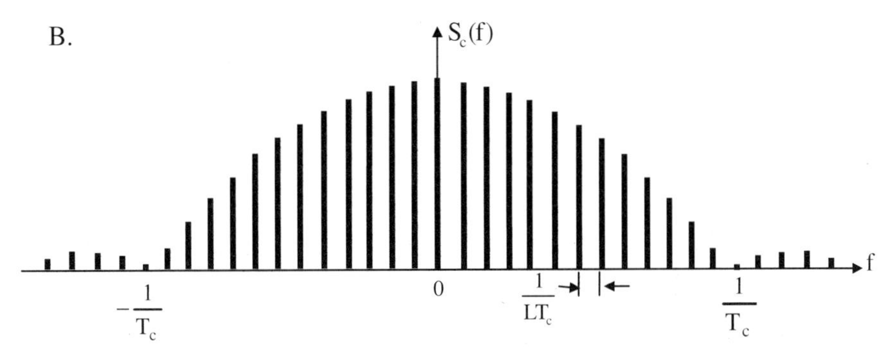

Figure 1.5 The power spectrum of the data and the spread signal.

$R_i(\tau)$ is shown Figure 1.5-A. The power spectral density $S_c(f)$ of $c_i(t)$ is then the Fourier transform of $R_i(\tau)$, and is given by

$$S_c(f) = \frac{L+1}{L^2}\left(\frac{\sin\, \pi f T_c}{\pi f T_c}\right)\left[\sum_{n=-\infty, n\neq 0}^{\infty} \delta(f - n/LT_c)\right] + \frac{1}{L^2}\delta(f)$$

Since $R_i(\tau)$ is a periodic function with period L, $S_c(f)$ is a line spectrum. As L increases the spectral lines get closer together. $S_c(f)$ is shown in Figure 1.5-B.

Now let $c_i(t)$ be assigned to the ith user. Also, let a sequence of K data symbols

$$d_i(t) = \sum_{k=0}^{K-1} d_{k,i}\; p_{T_d}(t - kT_d)$$

where $d_{k,i} \in \{-1, +1\}$. The encoded signal of user i is then

$$s_i(t) = d_i(t)c_i(t) = \sum_{k=0}^{K-1} d_{k,i}\; c_i(t - kT_d) = \sum_{k=0}^{K-1}\sum_{l=0}^{N-1} d_{k,i}c_{l,i}\; p_{T_c}(t - lT_c)$$

The signal $s_i(t)$ is transmitted at a carrier frequency f_o ($f_o \gg 1/T_c$), which is $s_i(t)\sqrt{2P_i}\cos(2\pi f_o t + \theta_i)$, where P_i is the power of the transmitted signal of user i.

Assuming M transmitting users, and omitting the thermal noise component, the received signal is given by

$$r(t) = \sum_{j=1}^{M} \sqrt{2P} d_j(t-\tau_j) c_j(t-\tau_j) \cos(2\pi f_o t + \phi_j)$$

Since all users are transmitting asynchronously, the time delays (τ_j, for $j = 1, 2, .., M$) are different from each other. Also, $\phi_j = \theta_j - 2\pi\tau_j$. Without loss of generality, we may assume $\theta_i = 0$ and $\tau_i = 0$, since we are only concerned with the relative phase shifts modulo 2π and time delays modulo T_d. Then, $0 \leq \tau_j < T_d$ and $0 \leq \theta_j < 2\pi$ for $j \neq i$. We have also assumed that each signal presents the same power P to the receiver. This assumption is satisfied with a power control mechanism.

The transmitted signal $s_i(t)$, is recovered by correlating the received signal $r(t)$ with the locally generating signal $c_i(t) \cos 2\pi f_o t$ of user i, over the period of the symbol $k = 0$:

$$\begin{aligned} Z_i &= \int_0^{T_d} r(t) c_i(t) \cos 2\pi f_o t dt \\ &= \sqrt{P/2} \left\{ d_{i,0} T_d + \sum_{j=1 (j \neq i)}^{M} [d_{j,-1} R_{j,i}(\tau_k) + d_{j,0} R'_{j,i}(\tau_k)] \cos \phi_j \right\} \end{aligned}$$

The first term in the above expression $d_{i,0}T_d$ is the desired signal of user i, while the summation term represents the interference from all other users j, to user i. The interference is expressed in terms of the continuous-time partial cross-correlation functions $R_{j,i}$ and $R'_{j,i}$, defined by

$$R_{j,i}(\tau) = \int_0^{\tau} c_j(t-\tau) c_i(t) dt \quad \text{and} \quad R'_{j,i}(\tau) = \int_{\tau}^{T_d} c_j(t-\tau) c_i(t) dt$$

In order to evaluate the interference term we consider the phase shifts, time delays and data symbols as mutually independent random variables. The interference term in the above equation of Z_i is random and may be treated as noise. Now, to evaluate the variance of Z_i, we assume, without loss of generality, that $\phi_i = 0$, $\tau_i = 0$ and $d_{i,0} = 1$. Then,

$$\begin{aligned} Var\{Z_i\} &= \frac{P}{4T_d} \sum_{j=1}^{M} \int_0^{\tau} [R_{j,i}^2(\tau) + {R'_{j,i}}^2(\tau)] d\tau \\ &= \frac{P}{4T_d} \sum_{j=1}^{M} \sum_{l=0}^{N-1} \int_{lT_c}^{(l+1)T_c} [R_{j,i}^2(\tau) + {R'_{j,i}}^2(\tau)] d\tau \end{aligned}$$

for $0 \leq lT_c \leq \tau \leq (l+1)T_c \leq T_d$. The expected values have been computed with respect to the mutually independent random variables ϕ_j, τ_j, $d_{j,-1}$ and $d_{j,0}$ for $1 \leq j \leq M$ and $j \neq i$. We have assumed that ϕ_j is uniformly distributed on the interval $[0, \pi]$ and τ_j is uniformly distributed on the interval $[0, T_d]$ for $j \neq i$. Also, the data symbols $d_{j,k}$ are assumed to take values $+1$ and -1 with equal probability.

The $Var\{Z_i\}$ has been evaluated approximately in [2], and is found to be

$$Var\{Z_i\} \approx PT_d^2(M-1)/6N$$

The Signal-to-Interference Ratio (SIR) is defined as the ratio of the desired signal $\sqrt{P/2}\,T_d$ divided by the rms value of the interference, $\sqrt{Var\{Z_i\}}$. Then we have,

$$SIR_i \equiv \frac{\sqrt{P/2}\,T_d}{\sqrt{Var(Z_i)}} = \frac{\sqrt{P/2}\,T_d}{\sqrt{PT_d^2(M-1)/6N}} \approx \sqrt{\frac{3N}{M-1}}$$

where N is the spreading factor and M is the number of accessing users.

1.2.2 Spread-Time

As in the case of spread-spectrum, spreading in time creates the 'space' in which multiple users may access the communication medium. In Spread-Time (ST) each encoding symbol may span one or more data symbols and each data symbol is repeated on every encoding symbol for the length of the sequence.

Orthogonal Spread-Time CDMA

Let d_i be the kth symbol of user i and $\mathbf{c_i}$ an orthogonal code sequence given by the vector

$$\mathbf{c_i} \equiv [c_{1i}, c_{2i}, .., c_{Ni}] \qquad \text{for} \quad i = 1, .., N$$

where $d_i, c_{ji} \in \{-1, +1\}$. The encoded time-spread symbol is then given by the vector

$$\mathbf{s_i} = d_i \mathbf{c_i} = [d_i c_{1i}, d_i c_{2i}, .., d_i c_{Ni}]$$

(Since this is an orthogonal system $N = M = L$, L is the sequence length.) The transmitted signal $s_i(t)$ is then given by

$$s_i(t) = d_i \sum_{n=0}^{N-1} c_{ni} P_{T_d}(t - nT_d) \quad \text{for} \quad 0 \le t \le NT_d$$

$s_i(t)$ has the same rate $R_d = 1/T_d$ as the data signal $d_i(t)$, while the rate of the code sequence is $R_c = R_d/N$. This means that the required bandwidth of the transmitted signal is the same as $d_i(t)$, while the required time for the transmission of its data symbols is N times longer (spread-time). Hence, given the length of the encoding symbol T_c, and the length of the data symbol, T_d, we define the *ST-Spreading Factor* to be the ratio

$$N_{ST} = \frac{T_c}{T_d} = \frac{R_d}{R_c} = N > 1$$

At the receiving end the signal is given by

$$r(t) = \sum_{j=1}^{N} d_j \sum_{n=0}^{N-1} c_{ni} P_{T_d}(t - nT_d)$$

In the above equation we have assumed that the symbols from all transmitting users are synchronized at the input of the receiver. We have also assumed that all arriving signals present equal power to the receiver. Also, the thermal noise component has been omitted and the impact of band-limiting filter has been ignored. After the A/D converter the received signal can be represented by the vector

$$\mathbf{r} = \sum_{j=1}^{N} \mathbf{s_j} = \sum_{j=1}^{N} [d_j c_{1j}, d_j c_{2j}, .., d_j c_{Nj}]$$

The transmitted symbol d_i will then be recovered by taking the inner product of the vector $\mathbf{r}$ with the corresponding code vector $\mathbf{c_i}$ of user i

$$\mathbf{r} \cdot \mathbf{c_i} = \sum_{j=1}^{N} \mathbf{s_j} \cdot \mathbf{c_i} = \sum_{j=1}^{N} d_j \mathbf{c_j} \cdot \mathbf{c_i} = N d_i \quad \text{since} \quad \mathbf{c_j} \cdot \mathbf{c_i} = \sum_{k=1}^{N} c_{kj} c_{ki} = \begin{cases} N & \text{if } i = j \\ 0 & \text{if } i \neq j \end{cases}$$

Now, let us consider having a sequence of K data symbols of user i represented by the vector $\mathbf{d_i} \equiv [d_{1i}, d_{2i}, .., d_{Ki}]$. The encoded data vector of user i, $\mathbf{s_i}$, is then the

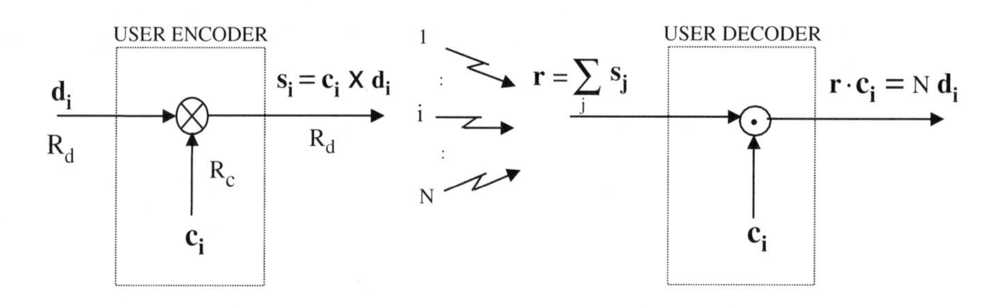

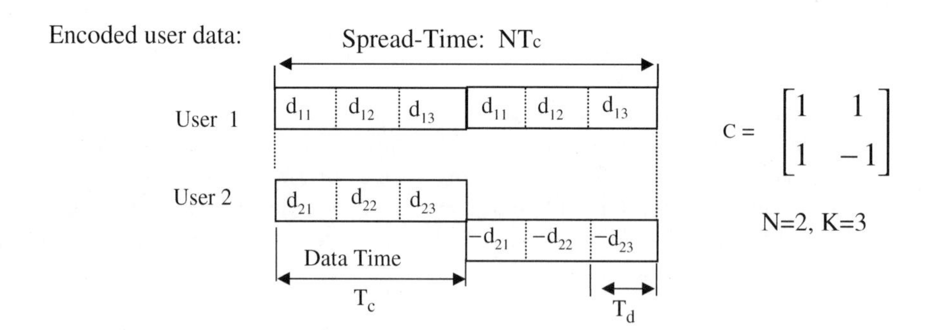

$\mathbf{c_i}$: User Code Vector, size N
$\mathbf{d_i}$: User Data Vector, size K
R_c : Code Rate
T_c : Code Symbol Length
R_s : Symbol Rate
T_d : Data Symbol Length
C: Orthogonal Code Matrix

Figure 1.6 The Generalized Time Division Multiple Access (G-TDMA).

Kronecker product of vectors $\mathbf{d_i}$ and $\mathbf{c_i}$, defined as

$$\mathbf{s_i} \equiv \mathbf{c_i} \times \mathbf{d_i} \equiv [c_{1i}\mathbf{d_i}, c_{2i}\mathbf{d_i}, .., c_{Ni}\mathbf{d_i}]$$

The time period of the K code symbols over which the user data are spread, is called the *frame* or the *time-width*, while the time interval of the K symbols is called a *time slot*. The spread-time access of this type is also called *Generalized Time Division Multiple Access* (G-TDMA). The transmitted signal of the G-TDMA is illustrated in Figure 1.6, and is described by the equation

$$s_i(t) = \sum_{l=0}^{N-1} c_{l,i}\; d_i(t - lT_c) \quad \text{where} \quad d_i(t) = \sum_{k=0}^{K-1} d_{k,i}\; p_{T_d}(t - kT_d) \qquad 1 \leq i \leq N$$

The ST-Spreading Factor in this case is

$$N_{ST} = \frac{NT_c}{KT_d} = \frac{NKT_d}{KT_d} = N$$

Assuming perfect synchronization and power control, the signal at the input of the receiver is

$$r(t) = \sum_{j=1}^{N} s_j(t) = \sum_{j=1}^{N} \sum_{l=0}^{N-1} c_{l,j}\; d_j(t - lT_c)$$

This signal after the A/D converter may be written as $\mathbf{r} = \sum_{j=1}^{N} \mathbf{s_j}$. The decoding process consists of taking the inner product of vector $\mathbf{r}$, with the corresponding code vector, of user i, $\mathbf{c_i}$ (see Figure 1.6). Then, as shown below, at the output of the decoder we receive the data symbols of user i:

$$\begin{aligned}
\mathbf{r} \cdot \mathbf{c_i} = \sum_{j=1}^{N} \mathbf{s_j} \cdot \mathbf{c_i} &= \sum_{j=1}^{N} [c_{1j}\mathbf{d_j}, c_{2j}\mathbf{d_j}, .., c_{Nj}\mathbf{d_j}] \cdot \mathbf{c_i} \\
&= \sum_{j=1}^{N} c_{1j}\mathbf{d_j}c_{1i} + \sum_{j=1}^{N} c_{2j}\mathbf{d_j}c_{2i} + \cdots + \sum_{j=1}^{N} c_{Lj}\mathbf{d_j}c_{Li} \\
&= \sum_{k=1}^{N}\sum_{j=1}^{N} c_{kj}\mathbf{d_j}c_{ki} = \sum_{j=1}^{N}\left[\sum_{k=1}^{N} c_{kj}c_{ki}\right]\mathbf{d_j} = \sum_{j=1}^{N} [\mathbf{c_j} \cdot \mathbf{c_i}]\, \mathbf{d_j} = N\mathbf{d_i}
\end{aligned}$$

This is because vectors $\mathbf{c_i}$, $i = 1, .., N$, are mutually orthogonal.

As we discussed above, the spread-time method presented here is an orthogonal division multiple access, and therefore requires time synchronization between all transmiting users. However, the synchronization requirement in this case, unlike the spread-spectrum orthogonal CDMA, can be easily achieved since the length of the code symbol (or time slot) is N times longer than the data symbol. Also, the ST Orthogonal CDMA, like the spread-spectrum DS/CDMA, requires power control.

The use of pseudo-random (PN) sequences with this type of spread-time accesses is also possible. Such PN spread-time systems can be asynchronous (i.e. no synchronization required between accessing users). It is, however, less efficient than the orthogonal spread-time method in which synchronization can be easily provided.

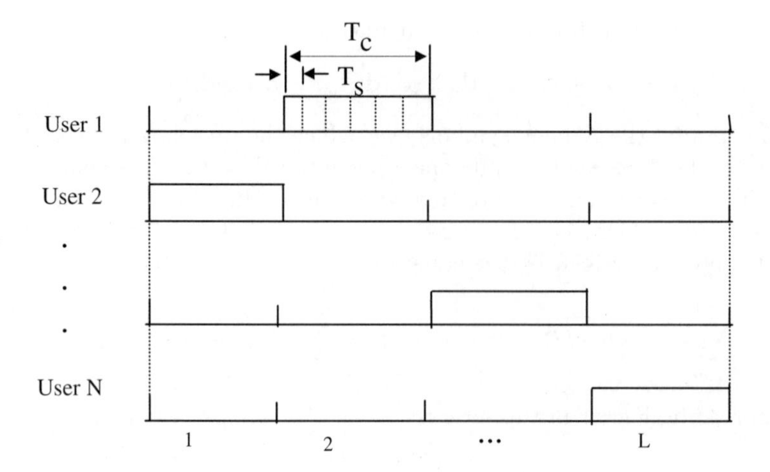

$$C = \begin{bmatrix} 0 & 1 & 0 & 0 \\ 1 & 0 & 0 & 0 \\ 0 & 0 & 1 & 0 \\ 0 & 0 & 0 & 1 \end{bmatrix}$$

Orthogonal Matrix **C** is not a Hadamard
but is a square matri (L=N)

Figure 1.7 Conventional TDMA and the corresponding encoding matrix.

Time Division Multiple Access (TDMA)

As we describe in Chapter 2, the set of code sequences $\mathbf{c_i}$, $i = 1, .., N$, is represented by a matrix $\mathbf{C} = [\mathbf{c_1}, \mathbf{c_2}, ..., \mathbf{c_N}]$, where $\mathbf{c_i} = [c_{1i}, c_{2i}, .., c_{Ni}]^T$ and $c_{ij} \in \{-1, +1\}$; matrix $\mathbf{C}$ is then orthogonal if $\mathbf{C}\ \mathbf{C^T} = N\mathbf{I}$ (where $\mathbf{I}$ is the identity matrix of size N). If we also have the property $|det\mathbf{C}| = N^{N/2}$, then $\mathbf{C}$ is a *Hadamard* matrix. Hadamard matrices exist for $N = 1, 2, 8, .., 4n, ..$ $(n = 1, 2, 3, ..)$ and have the property that every row (except one) has $N/2$ $1s$ and $N/2$ $-1s$.

In the G-TDMA described above, the matrix $\mathbf{C}$ may or may not be Hadamard. Let us now consider the special case in which $\mathbf{C}$ is a non-Hadamard orthogonal matrix of the following type:
$\mathbf{C} = [c_{ij}]$, where $c_{ij} \in \{0, 1\}$ in which each row and column has exactly one non-zero entry. Such matrices exist for any size N.

For example, let the code sequence $\mathbf{c_i} = [1, 0, .., 0]$; Then, $\mathbf{s_i} = [\mathbf{d_i}, 0, .., 0]$. The transmited signal of user i then is,

$$s_i(t) = \begin{cases} d_i(t) & \text{if } i = 1 \\ 0 & \text{if } i \neq 1 \end{cases} \qquad \text{where,} \quad d_i(t) = \sum_{k=0}^{K-1} d_{k,i}\ p_{T_d}(t - kT_d)$$

This means that user i transmits only during time slot 1. Hence, based on the above definition of matrix $\mathbf{C}$, it is equivalent to saying that each user transmits on a time slot assigned for that user only. This special case of G-TDMA is the conventional *Time Division Multiple Access* (TDMA), and is illustrated in Figure 1.7.

In TDMA the total received power during a time slot comes from a single user which has been assigned to transmit in that slot. This means that a TDMA transmitter bursts its power during its assigned slot while remaining idle during the non-assigned slots. On the other hand, in the G-TDMA using Hadamard matrices (called H-TDMA), the transmitted energy from each user is spread along the time frame. The H-TDMA may then achieve time diversity in wireless access systems, and thus avoid the channel fading. The conventional TDMA, however, does not need power control and has been used extensively because of the simplicity of its implementation.

1.3 User Encoding by Complex Sequences

Let us now consider user encoding with complex sequences. In general, a sequence $\mathbf{a_i} = \{a_i^{(\ell)}\}$ with length L, $\ell = 0, 1, 2, .., L-1$, is defined as a complex sequence if each entry $a_i^{(\ell)}$ takes any value in the set $\{e^{j(\theta+2\pi k_\ell/N)}\}$, where $k_\ell \in \{0, 1, 2, .., N-1\}$ and $j^2 + 1 = 0$. θ is a constant angle in $[0, 2\pi/N)$, N is an even number and $N \leq L$. This means that $a_i^{(\ell)}$ takes any value among the N equally spaced values on the unit circle. The minimum value of N is $N = 4$, i.e. $a_\ell \in \{\pm 1, \pm j\}$. In this case the sequence is called *quarterphase*, while for $N > 4$ it is called *polyphase*. The encoding process in this section may utilize orthogonal or pseudo-orthogonal complex sequences.

A set of orthogonal complex sequences of size N has a matrix format as $\mathbf{A} = [\mathbf{a_0}, \mathbf{a_k}, .., \mathbf{a_{N-1}}]$. $\mathbf{A}$ is a complex orthogonal matrix if $\mathbf{AA}^* = N\mathbf{I_N}$, where $\mathbf{A}^*$ denotes the Hermitian conjugate (transpose, complex conjugate) and $\mathbf{I_N}$ is the unit matrix. There are several types of such complex orthogonal matrices. Some of them are the following:

1. *Complex Hadamard matrices* are quarterphase orthogonal matrices with sizes $2n$. These matrices have elements ± 1 and $\pm j$, and can be constructed for even sizes (see Chapter 2).
2. *Polyphase Orthogonal Matrices* (POM) have N phases ($N \geq 4$), and size L, where $L \geq N$ (N and L are even numbers). A particular type of POM is constructed using a real binary Hadamard matrix $\mathbf{H} = [h_{nm}]$ and the vector $\mathbf{a} = [a_n] = [1, e^{j2\pi/N}, .., e^{j2\pi(KN-1)/N}]$, where $KN = L$. Then, the matrix $\mathbf{W} = [w_{nm}]$, where $w_{nm} = h_{nm}a_n$ is a POM with $N \leq L$. See Chapter 2 for details.
3. *Fourier Orthogonal Transformation* (FOT) is a particular type of POM based on the Discrete Fourier Transform (DFT), in which $N = L$. The FOT matrix is given by $\mathbf{W} = [w_{nm}]$, where $w_{nm} = e^{j2\pi nm/N}$ and $n, m = 0, 1, .., N-1$.

A pseudo-orthogonal complex sequence is any sequence $\mathbf{a_i} = \{a_i^{(\ell)}\}$ with length L, in which each element $a_i^{(\ell)}$ ($\ell = 0, 1, 2, .., L-1$) takes any values in the set $\{e^{j(\theta+2\pi k_\ell/N)}$ for $k_\ell = 0, 1, 2, .., N-1\}$ with equal probability. A particular type of pseudo-orthogonal complex sequence is constructed by taking $a_i^{(\ell)} = w_i^{(\ell)} e^{j(\ell 2\pi/N)}$, where $w_i^{(\ell)}$ ($\ell = 0, 1, 2, .., L-1$) is a real binary PN-sequence ($w_i^{(\ell)} \in \{+1, -1\}$), with $L \gg N$.

1.3.1 Spread-Spectrum

In this section, as in that for the Spread-Spectrum (SS) CDMA with real encoding sequences, we examine the orthogonal and pseudo-orthogonal SS-CDMA, but with complex sequences. Here, we also derive the conventional Frequency Division Multiple Access (FDMA) and the frequency hopping CDMA as special cases of a more general approach called generalized FDMA.

The Orthogonal Complex CDMA

Let $x_{n,k}$ represent the kth symbol of user n. $x_{n,k}$ is assumed to have the format $x_{n,k} = a_{n,k} + jb_{n,k}$.

Now we form the vector $\mathbf{x_{n,k}} = [x_{n,k}^{(\ell)}]$, where

$$x_{n,k}^{(\ell)} = \begin{cases} a_{n,k} + jb_{n,k} & \text{for } \ 0 \le \ell \le N/2 - 1 \\ a_{n,k} - jb_{n,k} & \text{for } \ -N/2 \le \ell \le -1 \end{cases}$$

N is an even number ($N = 2N'$). Since this process is repeated in every successive symbol, we may drop the index k. Then we may write

$$\mathbf{x_n} = [x_n^{(-N/2)}, ..., x_n^{(-1)}, x_n^{(0)}, x_n^{(1)}, .., x_n^{(N/2-1)}]$$

where $x_n^{(\ell)} = \{x_n^{(-\ell)}\}^* = \alpha_n e^{j\theta_n}$ and (*indicates complex conjugate). This will ensure that the encoded data signal, given below, is a real function.

Let us now consider a particular type of complex encoding sequence (encoding with other types of complex sequences is also possible), which is given by the vector

$$\mathbf{h_n^+} = [w_{\ell,n}^+ \ e^{j2\pi\ell/N}] \quad \text{for } \ 0 \le \ell \le N/2 - 1 \quad \text{and } \ 0 \le n \le N/2 - 1$$

where each entry $w_{\ell,n}$ is an entry of a real binary Hadamard matrix $\mathbf{W} = [w_{\ell,n}]$ of size $N/2$. The code matrix $\mathbf{H}^+ = [\mathbf{h_0^+}, .., \mathbf{h_{N/2-1}^+}]^T$, as shown in Chapter 2, is a polyphase orthogonal matrix. Also, let

$$\mathbf{h_n^-} = [w_{\ell,n}^- \ e^{j2\pi\ell/N}] \quad \text{for} \quad -N/2 \le \ell \le 0$$
$$\text{where} \quad w_{\ell,n}^+ = w_{-\ell-1,n}^- \quad \text{and} \quad w_{\ell,n}^+ = w_{\ell,-n-1}^-$$

($\mathbf{w_n^+} = [w_{\ell,n}^+]$ and $\mathbf{w_n^-} = [w_{\ell,-n-1}^-]$ are *mirror image sequences*). Then we form the vector

$$\mathbf{h_n} = [\mathbf{h_n^+}, \ \mathbf{h_n^-}] = [w_{-N/2,n} e^{j2\pi(-N/2)/N}, .., w_{0,n}, .., w_{N/2-1,n} \ e^{j2\pi(N/2-1)/N}]$$

Let us now assume that the vector $\mathbf{h_n}$ is assigned to the nth user. Then, user encoding is achieved by taking the inner product between vectors $\mathbf{x_n}$ and $\mathbf{h_n} e^{j\pi/N}$. The encoded symbol k of user n is then,

$$s_n = \mathbf{x_n} \cdot \mathbf{h_n} e^{j\pi/N} = \sum_{\ell=-N/2}^{(N/2)-1} x_n^{(\ell)} w_{\ell,n} e^{j2\pi(\ell+\frac{1}{2})/N}$$

where s_n is a real function. This discrete time signal is then converted into analog format shown by the equation

$$s_n(t) = g(t) \sum_{\ell=-N/2}^{(N/2)-1} x_n^{(\ell)} w_{\ell,n} \, e^{j2\pi f_c(\ell+1/2)t}$$
$$= 2g(t) \sum_{\ell=0}^{(N/2)-1} \alpha_n w_{\ell,n} \, \cos[2\pi f_c(\ell+1/2)t + \theta_n]$$

where $g(t)$ is the pulse-shape waveform of the data signal and $f_c = 1/N$. α_n is the magnitude and θ_n is the phase of $x_n^{(\ell)}$. Taking the Fourier transform of the above expression, we have

$$S_n(f) = \sum_{\ell=-N/2}^{(N/2)-1} \alpha_n w_{\ell,n} \, G(f - (\ell+1/2)f_c)$$

where $S_n(f)$ and $G(f)$ are the Fourier transforms of $s_n(t)$ and $g(t)$, respectively. We assume that $G(f)$ is band-limited to the frequency f_w ($f_w \leq f_c/2$). The frequency spectrum of $S_n(f)$ in this case is shown in Figure 1.8. If we assume that $g(t)$ is a raised cosine type function,

$$g(t) = 2f_w \sin c(2\pi f_w t) \frac{\cos 2\pi \rho f_w t}{1 - 4\rho^2(2f_w t)^2}$$

then $g(t)$ is time limitless. Therefore we place a time-limiting filter before the transmission of the signal. If the roll-off factor is $\rho = 0$, then the frequency spectrum $S_n(f)$ is flat over all frequencies $|f| \leq (N-1)f_c/2$. This is analogous to the case of user encoding with real sequences in which the pulse-shape is time-limited (square-pulse) and the filter is bandwidth-limiting.

The encoding process shown above results in spreading the bandwidth of the user data. If the data-signal before encoding has a bandwidth $B_d = f_c$ (at baseband), then the encoded signal has a bandwidth $B_c = (N-1)f_c$. Hence the spreading factor is $B_c/B_d = N-1$. This value of the spreading factor is verified by the fact that the code rate is $N-1$ times higher than the symbol rate, (each symbol is encoded by a code vector of size $N-1$). A similar type of spreading system has been presented in reference [3], which is called spread-time CDMA. The analysis given above, however, concludes that such system is actually spread-spectrum. 'Spread-time' results from the band-limited shape of the data-pulse and not from the encoding process. The spread baseband signal may then be translated to a desirable carrier frequency f_o for transmission. Hence, the transmitted signal of user n will be $s_n(t)\cos(2\pi f_o t + \phi)$.

The spread signal has the spectral form of the data-pulse (before encoding) translated to $(N-1)$ frequency bins, i.e. $-(N/2)f_c, ..., -f_c, 0, f_c, .., (N/2-1)f_c$. For this reason, we call this type of system *Generalized Frequency Division Multiple Access* (G-FDMA). The energy of each transmited symbol is distributed over all these frequency bins if the binary orthogonal matrix $\mathbf{W} = [w_{\ell,n}]$ is a Hadamard matrix. In this case, all users transmit simultaneously in every frequency bin.

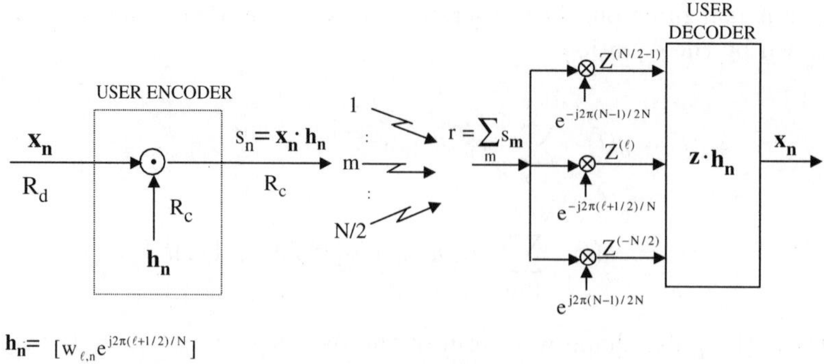

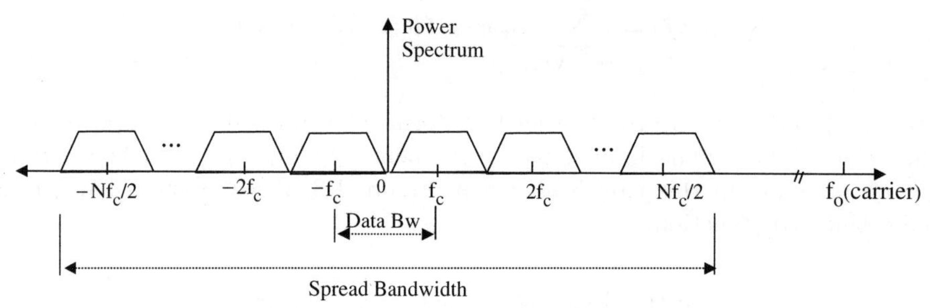

Figure 1.8 The Generalized Frequency Division Multiple Access (G-FDMA).

The signal at the receiving end is the sum of the transmitted signals

$$r(t) = \sum_{m=0}^{N/2-1} s_m(t) = 2 \sum_{m=0}^{N/2-1} g(t-\tau_m)\alpha_m \sum_{\ell=0}^{N/2-1} w_{m,\ell} \cos[2\pi f_c(\ell+1/2)t + \theta_m - \phi_m]$$

(τ_m and ϕ_m are randon variables). Assuming ideal synchronization of time and phase of all transmitting users, we may set $\tau_m = 0$ and $\phi_m = 0$. We also assume ideal power control. That is, all signals have equal power at the receiver. Then

$$r(t) = 2g(t) \sum_{m=0}^{N/2-1} \sum_{\ell=0}^{N/2-1} \alpha_m w_{m,\ell} \cos[2\pi f_c(\ell+1/2)t + \theta_m]$$

After the A/D conversion we write the signal in discrete form as follows:

$$r = \sum_{m=0}^{N/2-1} \sum_{\ell=-N/2}^{N/2-1} x_m^{(\ell)} w_{\ell,m} e^{j2\pi(\ell+\frac{1}{2})/N}$$

The desired signal of the nth user is recovered by the user decoder in two steps.

Step 1: We multiply r with $e^{-j2\pi(\ell_1+\frac{1}{2})/N}$,

$$re^{-j2\pi(\ell_1+\frac{1}{2})/N} = \sum_{m=-N/2}^{N/2-1} \sum_{\ell=-N/2}^{N/2-1} x_m^{(\ell)} w_{m,\ell} e^{j2\pi(\ell-\ell_1)/N}$$

In the above summation, let us consider the term for which $\ell = \ell_1$. Then we obtain

$$z^{(\ell_1)} = \sum_{m=0}^{N/2-1} x_m^{(\ell_1)} w_{m,\ell_1}$$

The terms for which $\ell - \ell_1 \neq 0$ will be rejected as high frequency (rate) terms. This process is repeated for each value of $\ell_1 = -N/2, .., 0, .., N/2 - 1$. Now, let the vectors $\mathbf{z}^+ = [z^{(0)}, .., z^{(N/2-1)}]$ and the vector $\mathbf{z}^- = [z^{(-N/2)}, .., z^{(-1)}]$, the components of which are obtained above.

Step 2: We decode the data symbol of user n by taking the inner product of vector $\mathbf{z}^+$ with the vector $\mathbf{w_n^+} = [w_{0,n}, ..., w_{(N/2-1),n}]$ or the inner product of vector $\mathbf{z}^-$ with the vector $\mathbf{w_n^-} = [w_{-n/2,n}, ..., w_{(-1),n}]$:

$$\mathbf{z}^+ \cdot \mathbf{w_n^+} = \sum_{\ell=0}^{N/2-1} \sum_{m=0}^{N/2-1} x_m^{(\ell)} w_{m,\ell} w_{\ell,n} = \sum_{\ell=0}^{N/2-1} x_n^{(\ell)} = \frac{N}{2} x_n = \frac{N}{2}(a_n + jb_n)$$

Similarly $\mathbf{z}^- \cdot \mathbf{w_n^-} = \frac{N}{2}(a_n - jb_n)$. The total energy of the received data is the sum of energies of each ℓ_1 (i.e., each frequency bin). The total number of users accessing the system will be $N' = N/2$.

The above process is illustrated in Figure 1.8. Due to the orthogonal nature of the G-FDMA system, it is necessary to have synchronization at the receiving end between all transmitting users. That is, all received signals must be synchronized with both a reference time and a reference phase. Time-synchronization may be easier to achieve than phase-synchronization.

Frequency Division Multiple Access (FDMA)

Now let us consider a special case in which $\mathbf{W} = [w_{\ell,n}]$ is a non-Hadamard orthogonal matrix with the following form: each entry $w_{\ell,n} \in \{0, 1\}$ and each row and column in $\mathbf{W}$ has exactly one non-zero entry. Then the above spread-spectrum system reduces to the *Frequency Division Multiple Access* (FDMA). In FDMA the transmitted signal is given by

$$s_n(t) = \sum_{\ell=\ell_1,\ \ell_1} x_n^{(\ell)} e^{j2\pi(\ell+\frac{1}{2})/N} = 2g(t)\alpha_n \cos\left[2\pi f_c\left(\ell + \frac{1}{2}\right)t + \theta_n\right]$$

As shown above, the FDMA user n transmits only the assigned frequency bin ℓ, indicated by its code vector. That is, the FDMA has all its transmitted power in the ℓth frequency bin, while the G-FDMA with Hadamard matrices has its power distributed over all bins, thus providing diversity in frequency selective fading channels. On the other hand, FDMA does not require power control. We also assume that all received signals are synchronized in both time and phase. Therefore, this type of FDMA is a *synchronous* one and it does not require a 'guard-band' between the frequency bins. If we assume asynchronous complex CDMA (described below), then the resulting FDMA is the conventional one which will require a guard-band between the frequency bins.

Let us now consider the FDMA transmitted signal $s_n^{(k)}(t)$ in which the kth symbol, of the nth user is assigned to the ℓth frequency bin, as described by the above equation. If the transmitted signal $s_n^{(k+1)}(t)$ of the $(k+1)^{th}$ symbol is given by the equation

$$\begin{aligned} s_n^{(k+1)}(t) &= \sum_{\ell=\ell_1,-\ell_1} x_{n,k} e^{j2\pi(\ell+i_{n,k}+\frac{1}{2})/N} \\ &= 2g(t)\alpha_{n,k+1} \cos\left[2\pi f_c\left(\ell+i_{n,k}+\frac{1}{2}\right)t+\theta_n^{(k+1)}\right] \end{aligned}$$

then the frequency bin for that symbol will be on the $(\ell+i_{n,k})^{th}$, (mod$(N/2)$). This type of system is called *Frequency-Hopping CDMA*, where $i_{n,k}$ indicates the frequency bin of the next hop for the $(k+1)^{th}$ symbol. The values of $i_{n,k}$ are determined by an orthogonal or PN-code which is assigned to the nth user. If the codes are orthogonal, consecutive symbols of different users may 'hop' simultaneously without 'hitting' the same bin. A simple approach is to set $i_{n,k}=1$.

Asynchronous Complex CDMA

In asynchronous complex CDMA we assume that the encoding sequence is a pseudo-orthogonal polyphase sequence $\mathbf{c_n}$ defined by the vector

$$\mathbf{c_n} = [c_{n\ell}\, e^{j2\pi\ell/N}] \quad \text{for} \quad -N/2 \le \ell \le N/2-1$$

where $\{c_{n\ell}\}$ is a binary PN-sequence for $\ell = 0, 1, .., L-1$ and $L = kN$, k is an integer. Also, we let $c_{n,\ell} = c_{n,-\ell-1}$. (The sequences $\mathbf{c_n^+} = [c_{\ell,n}]$ and $\mathbf{c_n^-} = [c_{\ell,-n-1}]$ are mirror image sequences.) Now we consider any data symbol $a_n + jb_n$ of the nth user. Then we define the vector $\mathbf{x_n} = [x_n^{(\ell)}]$ for $-N/2 \le \ell \le N/2-1$, where $x_n^{(\ell)} = x_n^{(-\ell-1)^*} = a_n + jb_n$ (* indicates complex conjugate), so that the encoded signal is real. The encoded symbol s_n is then defined by the inner product of vectors $\mathbf{x_n}$ and $\mathbf{c_n}e^{j\pi/N}$ as follows:

$$s_n = \mathbf{x_n} \cdot \mathbf{c_n} e^{j\pi/N} = \sum_{\ell=-N/2}^{N/2-1} c_{n,\ell}\, x_n^{(\ell)}\, e^{j2\pi(\ell+\frac{1}{2})/N}$$

where s_n is a real function. Then, since $x_n^{(\ell)} = a_n + jb_n$ for $0 \le \ell \le N/2-1$, the continuous time waveform is given by

$$\begin{aligned} s_n(t) &= g(t) \sum_{\ell=-N/2}^{(N/2)-1} c_{n,\ell}\, x_n^{(\ell)}\, e^{j2\pi f_c(\ell+\frac{1}{2})t} \\ &= 2g(t) \sum_{\ell=0}^{(N/2)-1} c_{n,\ell}[a_n \cos 2\pi f_c(\ell+1/2)t - b_n \sin 2\pi f_c(\ell+1/2)t] \end{aligned}$$

$g(t)$ is the time pulse which should be bandwidth-limited to a frequency $f_w \leq f_c/2$. (A similar system has been presented in [4] using complementary orthogonal sequences.)

Assuming asynchronuous transmission of $M-1$ users ($M-1 \leq N/2$), but perfect power control, the in-phase component (cos-terms) of the received signal is

$$r_I(t) = \sum_{m=0}^{M-1} s_m(t-\tau_m) = \sum_{m=0}^{M} \sum_{\ell=0}^{(N/2)-1} g(t-\tau_m) a_m c_{m,\ell} \cos[2\pi f_c(\ell+1/2)t - \phi_m^{(\ell)}]$$

Assuming that the user of interest $m=0$ has its receiver and transmitter synchronized at $\tau_0 = 0$ and $\phi_0^{(\ell)} = 0$ for all ℓ, we may recover its data as shown below:

$$r_I(t)\cos[2\pi f_c(\ell+1/2)t] = \mathbf{c_0}/2 + \sum_{m=1}^{M-1} \mathbf{c'_m}$$

where the vectors $\mathbf{c_0}$ and $\mathbf{c'_m}$ are defined as

$$\mathbf{c_0} = [a_0 c_{0,\ell}(t)] \quad \text{for} \quad 0 \leq \ell \leq N/2-1 \quad \text{and}$$
$$\mathbf{c'_m} = [a_m \, cos\phi_m^{(\ell)} \, c_{m,\ell}(t-\tau_m)] \quad \text{for} \quad 0 \leq \ell \leq N/2-1 \quad \text{and} \quad m = 1, 2, .., M-1$$

$c_{m\ell}(t) = c_{m\ell} g(t)$, ($c_{m\ell} = \pm 1$). The data of user $m=0$ may then be recovered by taking the inner product of the above vectors with the vector $\mathbf{c_0}$. The result will then be the user data plus the interference, which is

$$I_0 = \sum_{m=1}^{M-1} \sum_{\ell=1}^{N/2-1} a_m \, \cos\phi_m^{(\ell)} \, c_{m,\ell}(t-\tau_m) \, c_{0,\ell}(t)$$

1.3.2 Spread-Phase

The access method of spread-time when the encoding sequences are real is translated into Spread-Phase (SP) when the spreading sequences are complex. As a result of spread-phase CDMA (following an equivalent approach as in the previous section), we derive the Phase Division Multiple Access (PDMA) methods.

Orthogonal Spread-Phase CDMA

As in the case of spread-time with real sequences, the encoded signal in this case has the following form:

$$s_n(t) = x_n^\star \sum_{\ell=-N/2}^{-1} w_{n,-\ell-1} \, q_{f_c}(-t+(\ell+1/2)T_d) + x_n \sum_{\ell=0}^{N/2-1} w_{n,\ell} \, q_{f_c}(t-(\ell+1/2)T_d)$$

where $x_n = \alpha_n e^{j\theta_n}$, $x_n^\star = \alpha_n e^{-j\theta_n}$, $w_{n,\ell}$ are the elements of an $N/2 \times N/2$ binary (±1) orthogonal matrix, and $w_{n,\ell} = w_{n,-\ell-1}$, $q_{f_c}(t) = e^{j2\pi f_c t}$, and T_d is the time duration of the data pulse. Then

$$q_{f_c}(t-(\ell+1/2)T_d) = e^{j2\pi f_c(t-(\ell+1/2)T_d)} = e^{j2\pi f_c t} e^{j\phi_\ell}$$

where $\phi_\ell = 2\pi(\ell+1/2)f_c T_d$, $\phi_\ell = -\phi_{-\ell-1}$.

We may then rewrite the above equation as

$$s_n(t) = x_n^\star \; e^{-j2\pi f_c t} \sum_{\ell=-N/2}^{-1} w_{n,-\ell-1} \; e^{j\phi_{-\ell-1}} + x_n \; e^{j2\pi f_c t} \sum_{\ell=0}^{N/2-1} w_{n,\ell} \; e^{-j\phi_\ell}$$

$s_n(t)$ is then a real function, and is given by

$$s_n(t) = 2x_n g(t) \sum_{\ell=0}^{N/2-1} w_{n,\ell} \; \cos(2\pi f_c t - \phi_\ell + \theta_n) \quad \text{for} \;\; 0 \le t \le (N/2)T_d$$

$g(t)$ is the pulse waveform of the data signal bandwidth $f_w \le 2f_c$. $s_n(t)$ can also be written in discrete time format as

$$s_n = \sum_{\ell=-N/2}^{-1} x_n^{(-\ell-1)} w_{n,-\ell-1} \; e^{-j(2\pi/N-\phi_{-\ell-1})} + \sum_{\ell=0}^{N/2-1} x_n^{(\ell)} w_{n,\ell} \; e^{j(2\pi/N-\phi_\ell)} = \mathbf{x_n} \cdot \mathbf{h_n}$$

s_n represents the encoded symbol of user n, $\mathbf{x_n}$ is the code vector of symbol k of user n ($\mathbf{x_n} = [x_n^{(\ell)}]$ for $-N/2 \le \ell \le N/2-1$, where $x_n^{(\ell)} = x_n^{(-\ell-1)^*}$), and the vector $\mathbf{h_n}$ indicates the encoding sequence

$$\mathbf{h_n} = [\mathbf{h_n^+}, \; \mathbf{h_n^-}] \;\; \text{where} \;\; \mathbf{h_n^+} = [w_{n,\ell} \; e^{-j\phi_\ell}] e^{j2\pi/N} \;\; \text{for} \;\; -N/2 \le \ell \le -1$$
$$\text{and} \;\; \mathbf{h_n^-} = [w_{n,-\ell-1} \; e^{j\phi_{-\ell-1}}] e^{-j2\pi/N} \;\; \text{for} \;\; 0 \le \ell \le N/2-1$$

The code matrices $\mathbf{H}^+ = [w_{n,\ell} \; e^{-j\phi_\ell}]$ and $\mathbf{H}^- = [w_{n,\ell} \; e^{j\phi_\ell}]$ must be orthogonal matrices.

The above equations indicate that a data symbol of user n is repeated $N-1$ times, and each is placed on one of the $N-1$ different phases. Therefore, the data rate is the same as the rate of phase change, which means that no spread-spectrum occurs. The bandwidth of the encoded signal then is $2f_w$. Hence, since the same symbol appears with different phases, this method is called *Spread-Phase* (SP). The spreading factor is then $N_{SP} = (N-1)T_d/T_d = N-1$. Spread-phase is actually another form of spread-time. Since each user is identifined with an orthogonal sequence, this access method is called *Orthogonal Spread-Phase CDMA*.

The Orthogonal SP-CDMA may be extended by considering K consecutive symbols of user n as defined by the vector $\mathbf{x_n} = [x_{n,k}]$ for $0 \le k \le K$, where $x_{n,k} = a_{n,k} + jb_{n,k}$. Now we define the vector $\mathbf{x'_n} = [\mathbf{x_n^{(\ell)}}]$, where each vector $\mathbf{x_n^{(\ell)}}$ is defined as

$$\mathbf{x_n^{(\ell)}} = \begin{cases} \mathbf{x_n} & \text{for} \;\; 0 \le \ell \le N/2-1 \\ \mathbf{x_n^\star} & \text{for} \;\; -N/2 \le \ell \le -1 \end{cases}$$

The encoded symbol vector then is $\mathbf{s_n}$, given below:

$$\mathbf{s_n} = \mathbf{x'_n} \cdot \mathbf{h_n} = \sum_{\ell=-N/2}^{-1} \mathbf{x_n^{(-\ell-1)}} w_{n,-\ell-1} \; e^{-j(2\pi/N-\phi_{-\ell-1})} + \sum_{\ell=0}^{N/2-1} \mathbf{x_n^{(\ell)}} w_{n,\ell} \; e^{j(2\pi/N-\phi_\ell)}$$

In this case, the phase change will occur after K consecutive data symbols. The spreading factor, however, remains the same ($N_{SP} = (N-1)KT_d/KT_d = N-1$).

This type of access method is called *Generalized Phase Division Multiple Access* (G-PDMA). As in the case of G-FDMA, the G-PDMA also requires synchronization of all users in time and phase, and perfect power control.

The received signal then is $\mathbf{r} = \sum_{m=0}^{N/2-1} \mathbf{s_m}$. The signal of desired user $\mathbf{s_n}$ can be recovered from $\mathbf{r}$ in a similar manner as in the case of G-FDMA. A particular case of G-PDMA occurs when the orthogonal matrix $[w_{n,\ell}]$ (for $0 \le n, \ell \le N/2-1$) is a $\{0,1\}$ matrix with exactly one non-zero entry in each row and column. Then, the resulting encoded signal is given by

$$\mathbf{s_n}(t) = 2\mathbf{x_n} g(t) w_{n,\ell} \cos(2\pi f_c t - \phi_\ell + \theta_n) \quad \text{for} \quad 0 \le t \le (N/2)KT_d$$

and is called *Phase Division Multiple Access* (PDMA).

1.4 Composite Multiple Access Systems

Composite access methods can be defined from basic ones, such as those presented in previous sections. The method of constructing a composite access is based on the idea of *overspreading*, in which each spread symbol is taken as an input symbol of another spreader (encoder). This means that a number of encoding sequences (two or more), each providing a certain type of access, can be combined into one, using the method of the *Kronecker product* (described in Chapter 2), which defines the composite multiple access method.

Similarly, a composite orthogonal access can be defined from two or more basic orthogonal accesses by using an *extended orthogonal* code. As shown in Chapter 2, the Kronecker product $E_z = G_x \times H_y$ between orthogonal matrices G_x and H_y, is also an orthogonal matrix E_z with size $z = xy$, which is called an extended orthogonal matrix. Therefore, a composite multiple access defined by two or more orthogonal accesses using the method of extended codes is also an orthogonal access.

As examples of composite multiple access we may consider the following: (1) spectrum overspreading; (2) time overspreading; and (3) spectrum and time spreading using real or complex sequences in each step of spreading.

Spectrum overspreading is formed by spreading and overspreading in the spectrum for the purpose of providing access to individual users as well as groups of users. A group may be formed on the basis of common location or common services of the users. Such a double access scheme will then separate and provide access to users both as individuals and as groups. Spectrum overspreading applications are presented in Section 1.4.2. The method of spectrum-overspreading is also applied in a multibeam satellite network, presented in reference [5], in order to isolate the accessing users in different satellite beams. A similar process can be employed for the case of time overspreading. The concept of spreading in spectrum and time (called spread-spectrum and time multiple access) is examined in the section that follows.

1.4.1 Spread-Spectrum and Time Multiple Access

Let us consider the real (binary) orthogonal code matrices $\mathbf{a}$ and $\mathbf{b}$ with sizes N_a and N_b, respectively. Also, let the code vectors (sequences) $\mathbf{a_i} \in \mathbf{a}$ and $\mathbf{b_k} \in \mathbf{b}$. Spread-

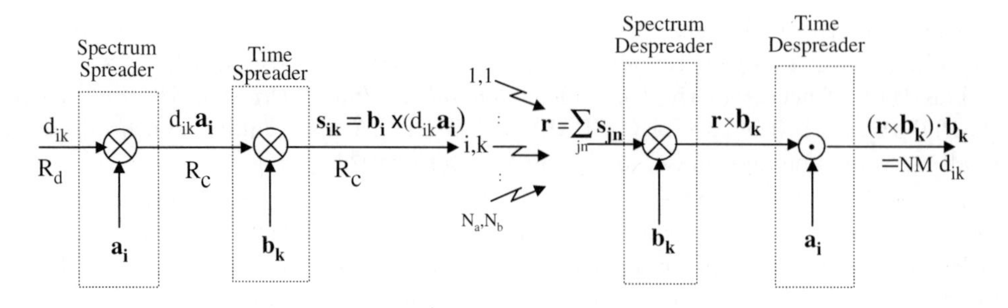

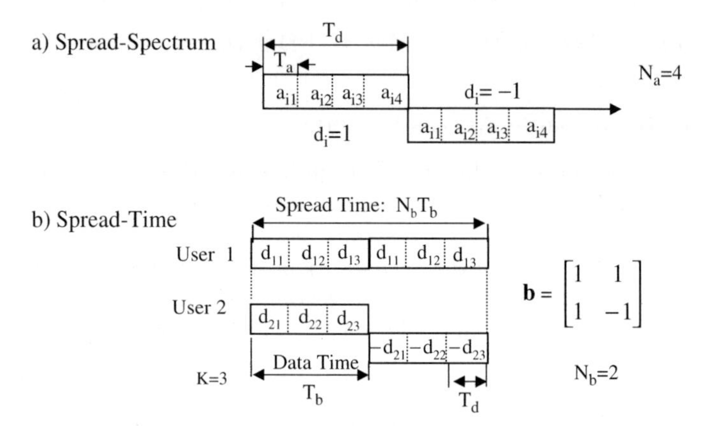

Figure 1.9 The Spread-Spectrum and Time Multiple Access (SS&TMA).

spectrum and time multiple access (SS&TMA) is then defined by assigning to a user (i, k) the sequence $\mathbf{a_i}$ for spectrum spreading and the sequence $\mathbf{b_k}$ for time spreading. Assuming any data symbol d_{ik}, the SS-signal (i.e. the signal at the output of the SS encoder) is given by

$$\mathbf{s'_i} = d_{ik}\mathbf{a_i} = [d_{ik}a_{0,i}, d_{ik}a_{1,i}, .., d_{ik}a_{N_a-1,i}]$$

which corresponds to the time signal

$$s'_i(t) = \sum_{l=0}^{N_a-1} d_{ik}a_{l,i}\ p_{T_a}(t - lT_a) \quad \text{for} \quad 1 \leq i \leq N_a$$

This signal has a rate $R_a = N_a R_d$, where R_d is the data rate. $T_d = N_a T_a$, where T_d, T_a are the lengths of the data symbol and SS-code symbol, respectively. The data time interval is not spread. This signal then enters the spread-time encoder, the output of which is given by

$$\mathbf{s_{ik}} = \mathbf{b_k} \times \mathbf{s'_i} = [b_{1i}\mathbf{s'_i}, b_{2i}\mathbf{s'_i}, .., b_{N_b i}\mathbf{s'_i}]$$

The continuous time signal is then given by

$$s_{ik}(t) = \sum_{l=0}^{N-1} b_{l,i}\ s'_i(t - lT_b) \quad \text{where} \quad s'_i(t) = \sum_{l=0}^{N_a-1} d_{ik}a_{l,i}\ p_{T_a}(t - lT_a)$$

for $1 \leq i \leq N_a$ and $1 \leq k \leq N_b$. The above spread-spectrum and spread-time processes and the corresponding time signals are illustrated in Figure 1.9. The output signal $s_{ik}(t)$ has the same rate as $s_i'(t)$, i.e. $R_b = R_a = N_a R_d$. The data time interval is spread by N_b, i.e. $T_b = N_b T_d$. Hence, the signal $s_{ik}(t)$ has a spectrum spreading factor $N_{SS} = N_a$ and a time spreading factor $N_{ST} = N_b$. If we assume that the SS&TMA is a synchronized orthogonal access system, then its capacity will be $N_a N_b$ users. The despreading operation is performed by despreading the last spreading first (i.e. time-despreading) and the the first last (i.e. spectrum-despreading). Each despreading is performed as described in Section 1.2.

This type of access may be used in applications where there is a bandwidth limitation. The SS&TMA can accommodate $N_a N_b$ users while satisfying the bandwidth constraint by distributing the spreading in both spectrum and time. Also, the SS&TMA, as other composite multiple access cases, can be used to provide access to each individual user as well as groups of users, as we describe in next subsection.

A similar process applies in the case of spread-spectrum and spread-time using complex encoding sequences. Let us assume that spread-spectrum is achieved with complex sequences (as those presented in Section 1.3.1, called G-FDMA), while the time is spread with real sequences. A special case of the above system is the conventional TDMA cellular network, in which spread-time is reduced to the conventional TDMA, while spread-spectrum (or G-FDMA) is reduced to the conventional FDMA. Then, TDMA is applied for user access within the cell, while FDMA is applied to separate each cell in a cluster. Frequency bins are reused in each cluster of cells (frequency-reuse). Other applications for composite accesses are presented in the following subsection.

1.4.2 Applications of Composite Access Methods

The method of composite access provides us with an approach in which access is achieved in one or more steps. For example, accessing users may form groups on the basis of common services or location. A two-step process may then be used for the access of users as individuals and as groups. Here we present two applications: the first is a two-step access for wireless networks; and the second is a method of multiplexing multiple symbol rates.

A: Two-Step Access Method

In this application a composite access method may be utilized as follow: let us consider a wireless access node with multiple antenna beams. Such a node may be a base station with multiple sectorized antennas or a multibeam satellite. In these networks, each user will experience interference from other users within the beam or cell sector, as well as interference from other beams or cell sectors. A composite access may then be used first to separate users within a beam, and secondly, to separate the beams. Each user is double indexed to indicate both the beam and the user in it. The first spreading (time or spectrum) is then applied to identify and isolate beams from each other, while the second spreading may identify and isolate each individual user within the beam.

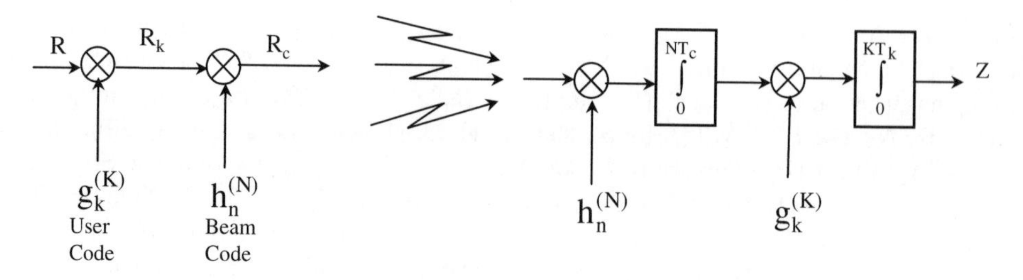

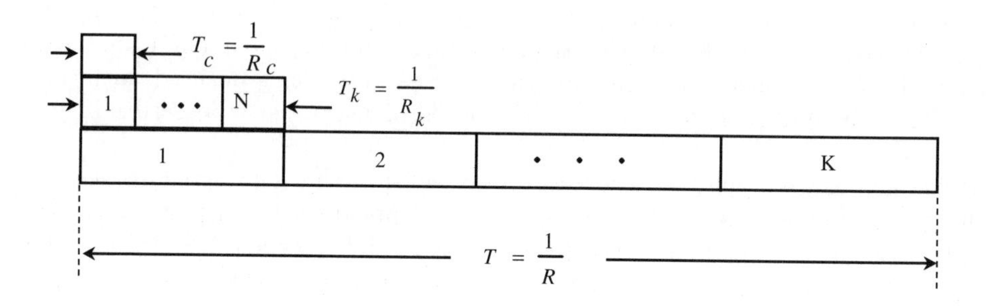

Figure 1.10 The 'overspreading' method for a two-step access system.

Let us assume that that both spreading steps are in the frequency domain (spread-spectrum) and that we apply orthogonal (binary) sequences. Then, the interference between users within the beam will be rejected by the first set of orthogonal codes, while the interference between beams will be rejected by the second set of orthogonal codes. As illustrated in Figure 1.10, a user k in cell n is identified by two sequences, the *user* sequence $g_k(t)$ and the *beam* sequence $h_n(t)$. The sequences $g_k(t)$ and $h_n(t)$ are concatenated in such a way that the length of the second one equals excactly one chip length of the first one. Hence, if K is the length of $g_k(t)$ and N is the length of $h_n(t)$, the combined code has length KN. The *over-spreading* process then corresponds to the Kronecker product $H_K \times G_N$ of sequences $g_k(t)$ and $h_n(t)$ (see Chapter 2). Let $d_k(t)$ represent the data signal of user k. $d_k(t)$ is a sequence of rectangular pulses with unit amplitude (positive or negative) and duration T,

$$d_k(t) = \sum_{l=-\infty}^{\infty} d_{k,l} p_T(t - lT) \quad \text{where} \quad p_T(t) = \begin{cases} 1 & \text{for } 0 \le t < T \\ 0 & \text{otherwise} \end{cases}$$

and $d_{k,l} \in \{1, -1\}$. The kth user is assigned to the user $g_k(t)$ and beam h_n orthogonal sequences. The transmitted signal then is

$$g_k(t) = d_k \sum_{i=0}^{K} g_{k,i} \sum_{i=0}^{N} h_{n,j} p_{T_c}(t - iT_c) \quad \text{for} \quad 0 \le t \le T$$

The received modulated signal $r(t)$ from every user and every beam in then given by

$$r(t) = \sum_{n=1}^{N} \sum_{k=1}^{K} A_{n,k} d_{n,k}(t - \tau_{n,k}) g_k(t - \tau_{n,k}) h_n(t - \tau_{n,k}) \cos[(w_o(t - \tau_{n,k}) + \theta_{n,k}]$$

If we assume that all transmissions are synchronized, then we may set $\tau_{n,k} = 0$ for all k and n.

The signal of the particular user-1 in cell-1 will then be reconverted after coherent demodulation and despreading the user and cell sequences, as illustrated in Figure 1.10. Then,

$$\begin{aligned} Z &= \int_0^T \left[\int_0^{T_k} r(\mu) h_1(\mu) \cos(w_o \mu) d\mu \right] g_1(\nu) d\nu \\ &= \int_0^T \sum_{k=1}^{K} \left[\int_0^{T_k} \sum_{n=1}^{N} A_{n,k} d_{n,k}(\mu) h_n(\mu) h_1(\mu) \cos(\theta_{n,k}) d\mu \right] g_k(\nu) g_1(\nu) d\nu \end{aligned}$$

In the above equation, $T_k = NT_c$, $T = KT_k = NKT_c$ and $\mu = \mu(\nu)$. Then expanding this expression we obtain

$$\begin{aligned} Z = A_{1,1} d_{1,1} \cos(\theta_{1,1}) & \int_0^T \left[\int_0^{T_k} h_1(\mu) h_1(\mu) d\mu \right] g_1(\nu) g_1(\nu) d\nu \\ + \sum_{k=2}^{K} A_{1,k} d_{1,k} \cos(\theta_{1,k}) & \int_0^T \left[\int_0^{T_k} h_1(\mu) h_1(\mu) d\mu \right] g_k(\nu) g_1(\nu) d\nu \\ + \sum_{n=2}^{N} A_{n,1} d_{n,1} \cos(\theta_{n,1}) & \int_0^T \left[\int_0^{T_k} h_n(\mu) h_1(\mu) d\mu \right] g_1(\nu) g_1(\nu) d\nu \\ + \sum_{k=2}^{K} \sum_{n=2}^{N} A_{n,k} d_{n,k} \cos(\theta_{n,k}) & \int_0^T \left[\int_0^{T_k} h_n(\mu) h_1(\mu) d\mu \right] g_k(\nu) g_1(\nu) d\nu \end{aligned}$$

Considering the orthogonality properties of the sequences g_k and h_n, we have

$$\int_0^{T_k} h_n(\mu) h_1(\mu) d\mu = T_c \sum_{i=1}^{N} h_{ni} h_{1i} = NT_c \delta_{n,1}$$

and $$\int_0^{T} g_k(\nu) g_1(\nu) d\nu = T_k \sum_{i=1}^{K} g_{ki} g_{1i} = KT_k \delta_{k,1}$$

where $\delta_{n,j} = \{1 \; if \; n = j, \; 0 \; if \; n \neq j\}$ and $h_{ni} \in \{1, -1\}$ $g_{ki} \in \{1, -1\}$.

We observe that the first row in the above equation is non-zero while the other rows are zero. Hence, $Z = \pm A_{1,1} KNT_c$, if we assume coherent demodulation ($\theta_{1,1} = 0$). Therefore, by applying the proposed approach we may allow the beams to overlap without causing interference to the indidual users. The penalty for this is the bandwidth expansion that results from the overspreading, while the advantage is higher frequency reuse. This means that cell sectors or beams may partially overlap without interfering with each other.

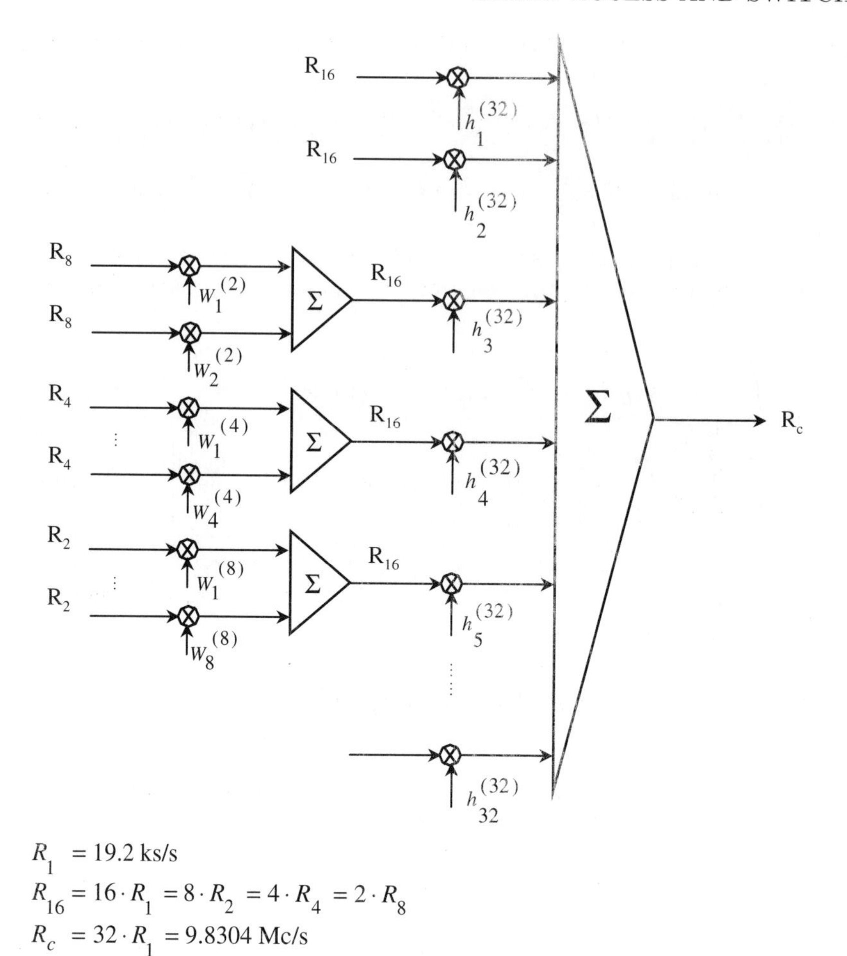

Figure 1.11 Example of multiplexing the rates, $R_2,...,R_{16}$, into a O-CDMA channel.

B: Multiplexing Multiple Symbol Rates

Another application of the above process allows multiplexing of multiple symbol rates within the orthogonal CDMA channel. Let us consider a basic symbol rate R and multiples of R, $R_k = kR$ for $k = 1, 2, 4, 8, 12, 16, 20, ...$ Also let R_c ($R_k < R_c$) be the chip rate (spreading rate) of the O-CDMA. Then we can use the method of extended sequences to accommodate multiple transmission rates in the O-CDMA network.

This method consists of concatenating two orthogonal sequences for each transmiting user with a given rate, as illustrated in the example of Figure 1.11. Sequence $w_i^{(k)}$ identifies the particular user i in the group of users transmiting at the rate R_k and the sequence $h_j^{(N)}$ identifying the group j among the other groups of transmission rates. The rate of $w_i^{(k)}$ is R_k and the rate of $h_j^{(N)}$ is the O-CDMA chip rate $R_c = KR$. The data of user (i, j) will then be recovered by despreading first the sequence $h_j^{(N)}$ and then the sequence $w_i^{(k)}$, then

$$Z = \int_0^T \left[\int_0^{T_k} r(\ell) h_j(\ell) \cos(\omega_o \ell) d\ell \right] w_i(\nu) d\nu = \pm A_{i,j} k N T_c$$

where the symbol length $T_k = 1/R_k$ and the chip length $T_c = 1/R_c$. Let N_k be the number of chips per symbol of the given symbol rate R_k, $N_k = T_k/T_c = R_c/R_k$ for $k = 1, 2, 4, \ldots$

We may then provide N_k users of a specific rate R_k in the O-CDMA channel or a mixture of users with different rates. Let n_k be the number of users having the rate R_k, then $n_k < N_k$ for each $n = 1, 2, 4, \ldots$ If $n_k = N_k$ we would only have users of rate R_k. We may have, for example, n_4 users of rate R_4, n_2 of rate R_2 and n_1 of rate R, then

$$n_4 = N_4 - 4[N_2 - 2(N_1 - n_1)] < N_4, \quad n_2 = N_2 - 2(N_1 - n_1) < N_2 \text{ and } n_1 < N_1$$

In a more general application, each user within the O-CDMA channel may be identified by three orthogonal codes w_i, h_j and g_k, defining its rate, the user with the given rate, and the cell or beam the user belongs to. The extended code resulting from these sequences can be represented by a matrix $\mathbf{Q} = W \times H \times G$ where, $W = [w_1, w_2, ..., w_I]^T$, that is each sequence w_i is a row vector in W. Similarly, $H = [h_1, h_2, ..., h_J]^T$ and $G = [g_1, g_2, ..., g_K]^T$.

1.5 Conclusion

In this chapter, we have presented an introduction to multiple access communications by presenting a unified multiple access method called Generalized CDMA. The G-CDMA represents a super-set of multiple accesses created by the process of user encoding which is performed with different types of encoding sequences. The choice of spread-spectrum with real encoding sequences provides the conventional (synchronous and asynchronous) CDMA. The choice of spread-time with real encoding sequences creates a new type of multiple access that we call Generalized TDMA (G-TDMA). The conventional TDMA is a special case of the G-TDMA in which a user may only transmit in an assigned time slot, and not in every slot as in G-TDMA. The choice of spread-spectrum with complex encoding sequences generates another type of multiple access that we call Generalized FDMA (G-FDMA). The conventional FDMA is a special case of the G-FDMA in which a user may only transmit in an assigned frequency bin, and not in every bin as in G-TDMA. Finally, the choice of spread-time with complex encoding sequences translates into spread-phase, which generates the Phase Division Multiple Access (PDMA). For each of the above multiple accesses we have verified the functional correctness by examining the despreading and recovery process, but we did not provide any performance analysis for noisy channel conditions. In addition we have presented composite multiple access methods such as spread-spectrum and time multiple access (SS&TMA) using the method of extended sequences given in Chapter 2. We have also presented applications of composite access methods for wireless two-step access networks, and for multiplexing multiple symbol rates in orthogonal CDMA channels.

References

[1] D. Gerakoulis 'G-CDMA: a Unifying Approach to Multiple Access Communications' AT&T Labs-Research Technical Memorardum HA1360000-000306-01TM.

[2] M. B. Pursley 'Performance Evaluation for Phase-Coded Spread-Spectrum Multiple Access Communications-Part I: System Analysis' *IEEE Transactions on Commun.* Vol. 25, No. 8, August 1977, pp. 795–799.

[3] P. M. Crespo, M. L. Honig and J. A. Salehi 'Spread-Time Code-Division Multiple Access' *IEEE Transactions on Commun.* Vol. 43, No. 6, June 1995, pp. 2139–2148.

[4] S.-M. Tseng and M. R. Bell 'Asynchronous Multicarrier DS-CDMA Using Mutually Orthogonal Complementary Sets of Sequences' *IEEE Transactions on Commun.* Vol. 48, No. 1, January 2000, pp. 53–59.

[5] D. Gerakoulis, E. Geraniotis, R. Miller and S. Ghassemzadeh 'A Satellite Switched CDMA System for Fixed Service Communications' *IEEE Communications Magazine*, July 1999, pp. 86–92.

2

Spreading Sequences

2.1 Overview

In this chapter we study the structures and properties of orthogonal and pseudo-orthogonal sequences. Firstly, we examine several types of pseudo-orthogonal (PN) and Quasi-Orthogonal (QO) sequences, and present their cross-correlation properties under synchronous and asynchronous conditions.

Secondly, we survey basic methods of constructing orthogonal code sets. Orthogonal binary (Hadamard) codes may exist for lengths $1, 2$ and $4k$ (for $k = 1, 2, 3, \ldots$). Methods for generating all lengths up to 256 are presented. We also present complex, polyphase and other orthogonal code designs. In particular, we focus our attention on Kronecker product of orthogonal matrices (called extended orthogonal sequences), and their applications in the design of CDMA systems.

Thirdly, we examine the properties of orthogonal and quasi-orthogonal sequences when there is timing jitter or misalignment amongst them. That is, we investigate the performance impact on a system, when the time-pulses (representing binary ± 1 code entries) are not perfectly aligned to a common time reference. Performance results indicate that the inter-user interference power is parabolically proportional to the time jitter as a percentage of the code-symbol or the chip length.

Finally, we examine the impact of band-limited pulse-phases on interference. That is, when the code-symbol or chip waveform is not an ideal square-pulse (time-limited), but is the result of a band-limit filtering. In this case we evaluate the inter-chip and inter-user interference when there is an equalizing matched filter. The interference is evaluated when we have orthogonal or PN-sequences and under synchronous and asynchronous conditions.

2.2 Orthogonal and Pseudo-Orthogonal Sequences

2.2.1 Definitions

A *binary sequence* is defined as a vector $\mathbf{x} = \{x_1, x_2, \ldots, x_L\}$ in which $x_i \in \{-1, +1\}$ and L is the *sequence-length*. A *code* (or *code-book* or *binary-array*) is a set of N vectors $\mathbf{x}$, from the L-dimensional vector space.

The *correlation* (or *normalized cross-correlation*) $\rho(x, y)$ of two $L-$dimensional sequences $\mathbf{x}, \mathbf{y}$ is defined by

$$\rho(\mathbf{x}, \mathbf{y}) = \frac{1}{L} \sum_{i=1}^{L} x_i y_i = \frac{1}{L} \mathbf{x} \cdot \mathbf{y}$$

The *autocorrelation* function $\rho_x(j)$ of sequence $\mathbf{x}$ is defined by

$$\rho_x(j) = \frac{1}{L}\sum_{i=1}^{L} x_i x_{i+j}$$

where $x_{L+k} = x_k$ by definition.

Also, a code having

$$\rho(v_i, v_j) = \begin{cases} \frac{-1}{n-1} & \text{if } n \text{ is even} \\ \frac{-1}{n} & \text{if } n \text{ is odd} \end{cases}$$

is called *simplex* code.

2.2.2 Pseudo-random Noise (PN) Sequences

PN-sequences are sequences with autocorrelation function

$$\rho_x(j) = \begin{cases} 1 & \text{for } j = 0 \\ \frac{-1}{L} & \text{for } j \neq 0 \end{cases}$$

Methods of constructing PN-sequences are given below:

(1) **Maximum Length (M) Sequences**: let $h(x) = c_0x^p + c_1x^{n-1} + \cdots + c_{p-1}x + c_p$ denote a binary polynomial of degree p, where $c_0 = c_p = 1$ and the other coefficients take values either 0 or 1. A binary sequence $\{v_k\}$ is said to be a sequence generated by $h(x)$ if for all integers k $c_0v_k \oplus c_1v_{k-1} \oplus c_2x_{k-2} \oplus \cdots \oplus c_px_{k-p} = 0$, where $\oplus$ denotes modulo 2 addition (i.e. Exclusive OR operation). Then using the fact that $c_0 = 1$, we obtain

$$v_k = -(c_1v_{k-1} \oplus c_2v_{k-2} \oplus \cdots \oplus c_pv_{k-p}) = -\sum_{n=1}^{p} c_nv_{k-n} \quad (\oplus,\ \text{mod} - 2 \text{ addition})$$

From this it follows that the sequence $\{v_k\}$ can be generated by a p-stage binary Linear Shift Register (LSR) which has a feedback tab connected to the ith cell if $c_i = 1$, $0 < i \leq p$ and $c_p = 1$, (p is the degree of the linear recursion). The linear shift register (LSR) circuit using the design described by the above recursion formula is shown in Figure 2.1-A (this LSR is said to have Fibonacci's form). An alternative logic which also generates the sequence $\{v_k\}$ is shown in Figure 2.1-B (this LSR is said to have Galois' form). A sequence generated by such a p-stage LSR has *maximal length* if its period is $L = 2^p - 1$. That is, $v_k = v_{k+L}$ (except for all-zero cases).

If $L = 2^p - 1$ is a prime number, then every LSR using an irreducible polynomial of degree p generates maximum length sequences. (A polynomial $g(x) = \sum_{n=1}^{p} c_nx_{k-n}$ is *irreducible* if it cannot be factored, that is, divided by another polynomial of degree $n < p$.) If, however, we require a maximum length sequence for every p we must restrict our polynomials to be *primitive*. (An irreducible polynomial of degree p is *primitive* if and only if it divides $x^m - 1$ for no m less than $2^p - 1$.)

The number of maximum length sequences N_m of length $L = 2^p - 1$ is given by $N_m = \frac{\phi(L)}{p}$, where $\phi(L)$ is the Euler ϕ-function and is equal to the number of numbers relatively prime to L which are less than L.

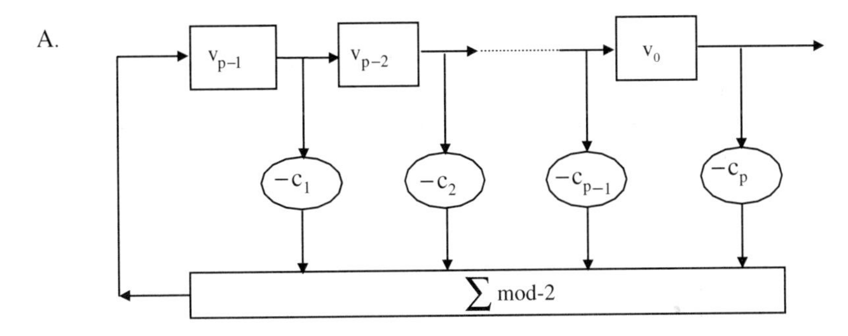

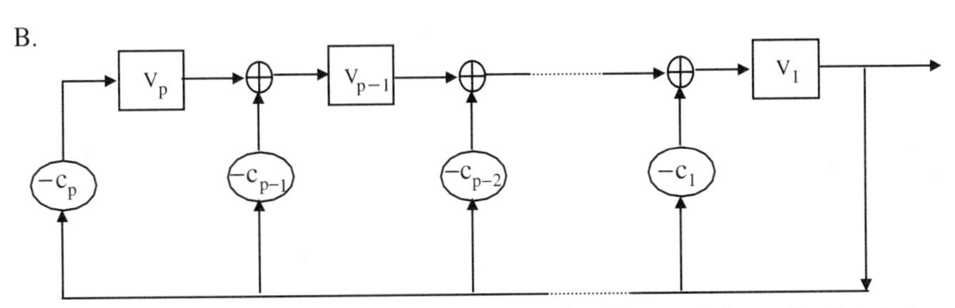

Figure 2.1 PN-Sequence generators by LSRs. A. Fibonacci form, B. Galois form.

A maximum length sequence with length $L = 2^p - 1$ has the following properties:

(a) In every sequence period the number of $+1's$ differs from the number of $-1's$ by 1.

(b) In every sequence period the number of *Runs* with length r, n_r, is given by

$$n_r = \begin{cases} 2^{p-r-1} & \text{for } \ r = 1, 2.., p-1 \\ 1 & \text{for } \ r = p \end{cases}$$

(we call *Run* the occurrence a number of $1's$ (or $-1's$) in succession). For more details on the properties of M-sequences, see reference [1].

(2) **Quadratic-Residue Sequences**

(a) The Quadratic-Residue(QR) sequences exist when the length $\ell = q = 3(\text{mod}4) = 4t - 1$ is a prime number (see [2]). The integer i is a QR modulo ℓ, if there exists an integer k such that $k^2 = i(\text{mod } \ell)$ and the greatest common denominator $GCD(i, \ell) = 1$ ((i/ℓ) is the Legendre symbol for ℓ odd prime). Thus a binary sequence $a_i \in (1, -1)$ can be constructed as follows:

$$a_i = \begin{cases} 1 & \text{If } \ i \epsilon \ QR(\ell) \\ -1 & \text{otherwise} \end{cases} \qquad \text{for } \ i = 0, 1, ..., \ell - 1$$

(b) The Quadratic-Residue 2 (QR-2), or 2nd Paley sequences, exist when the length $\ell = 2q + 1$, where $q = 1(\text{mod}4)$ is odd prime. The construction method is similar to QR and is described in [2].

(3) **Hall sequences** exist when $\ell = 4t - 1 = 4x^2 + 27$ is a prime number. Therefore its size is a subset of the QR sizes. The construction method is given in [3].
(4) **Twin-Prime sequences** exist when the length $\ell = p(p+2)$, where both p, $p+2$ are prime numbers. The construction is similar to the QR, but is based on the *Jacobi* symbol $[\frac{i}{\ell}]$ instead of the Legendre symbol.

2.2.3 *Quasi-Orthogonal (QO) Sequences*

Quasi-Orthogonal (QO) is a class of PN-Sequences that have very small cross-correlation values. The class of QO-sequences includes the *Gold-Codes* [4], and particularly a type of them called *Preferentially-Phased Gold Codes* (PPGC) [5].

Gold-Codes have the property that the cross-correlation $R_{yz}(k)$ is bounded by

$$|R_{yz}| \leq \begin{cases} 2^{(n+1)/2} + 1 & n \text{ odd} \\ 2^{(n+2)/2} + 1 & n \text{ even}, \ n \neq 0 \text{ mod} 4 \end{cases}$$

where $R_{yz}(k) \triangleq \sum_{i=0}^{L-1} y(i)z(i-k)$. Gold codes can be generated by a shift register corresponding to the product polynomial $g_1(x)g_2(x)$, where $g_1(x)$ and $g_2(x)$ is a preferred pair of primitive polynomials of degree n. (Preferred pairs of PN-sequences have the property that they have the minimum cross correlation value [4].) The shift register corresponding to the product polynomial $g_1(x)g_2(x)$, will generate $2^n + 1$ different sequences each with period $2^n - 1$. The $2^n + 1$ distinct members will then form a family of Gold codes. The $2^n + 1$ members include the $2^n - 1$ phase shifts of one code of the product polynomial with respect to the other, plus each code itself. An example of a Gold code generator is shown in Figure 2.2. The Gold code generator may also be realized with a single shift register of length $2n$.

Gold codes have three-level cross-correlation values which have different frequencies of occurence. These values and the corresponding frequency of occurence are shown in Table 2.1.

Table 2.1 Three-level cross-correlation properties of Gold codes.

n	$R(k)$	$Prob\{R(k)\}$
even*	-1	0.75
even*	$-2^{(n+2)/2} - 1$	0.125
even*	$2^{(n+2)/2} - 1$	0.125
odd	-1	0.5
odd	$-2^{(n+1)/2} - 1$	0.25
odd	$2^{(n+1)/2} - 1$	0.25

* The even values divisible by 4 not included.

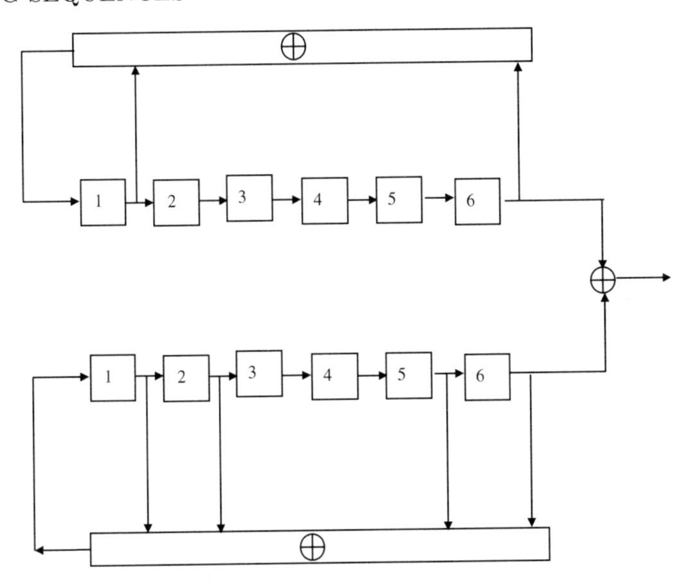

Figure 2.2 Gold code generator of length 63 by a double LSR realization.

In Table 2.1, n corresponds to a Gold code of length $L = 2^n - 1$, and the cross-correlation between sequences y, z is defined by $R(k) \triangleq \sum_{i=0}^{L-1} y(i)z(i-k)$. The $Prob\{R(k)\}$ indicates the frequency of occurence of these cross-correlation values and k is the phase offset between sequences y, z, in a number of code symbols. In Table 2.1 we assume $k \neq 0$.

Now we examine the cross-correlation properties of QO-sequencies in synchronized systems (i.e. at $k = 0$).

The criteria we use are (1) the maximum cross-correlation value $R_{\max}(0)$ (at $k = 0$), and (2) the variance of the worst-user worst-case inter-user interference $\sigma_w^2(0)$ (at $k = 0$) [6]. $\sigma_w^2(0)$ is lower bounded by

$$\sigma_w^2(0) \geq L(N - L)$$

where L is the length and N is the number of sequences. The above bound is known as the *Welch Bound*, and is presented in [6]. Given a set of code sequences $x_i^{(m)}$, for $1 \leq m \leq N$, ($\mathbf{x^{(m)}} = \{x_1^{(m)}, x_2^{(m)}, ..., x_L^{(m)}\}$, $x_i^{(m)} \in \{-1, +1\}$) the Welch bound holds with equality if and only if

$$\sum_{m=1}^{N} x_i^{(m)} x_j^{(m)} = 0 \qquad \text{for all } i, j, \quad i \neq j$$

That is, for the array of N sequences

$$\begin{matrix} x_0^{(1)} & x_1^{(1)} & \cdots & x_{L-1}^{(1)} \\ x_0^{(2)} & x_1^{(2)} & \cdots & x_{L-1}^{(2)} \\ \vdots & \vdots & & \vdots \\ x_0^{(N)} & x_1^{(N)} & \cdots & x_{L-1}^{(N)} \end{matrix}$$

Table 2.2 Comparisons of different QO-Sequences.

N	L	Code Sequences	$\sigma_w^2(0)$	$R_{max}(0)$
L+1	$2^m - 1$	Preferentially-Phased Gold Codes	1	1
$\approx L^2/2$	2^{2n}	Half of Kerdock code	$\sqrt{L-2}$	$\sqrt{L}$
L(L+2)	$2^m - 1$	All phases of Gold code sequences	$\sqrt{L-1}$	$1 + 2^{\lfloor (m+2)/2 \rfloor}$
$L\sqrt{L+1}$	$2^{2n} - 1$	All phases of Kasami sequences	$\sqrt{L+1}$	$1 + \sqrt{L+1}$
L	2^m	Hadamard (Orthogonal) codes	0	0

the Welch bound holds with equality if and only if all *columns* are orthogonal to each other (this doesn't mean that the sequences $\mathbf{x}^{(\mathbf{m})}$ are orthogonal). As shown in [5], the *Preferentially-Phased Gold Codes* (PPGC) achieve the Welch bound.

In general, a code sequence is considered 'good' in synchronous CDMA systems if the Welsh bound on $\sigma_w^2(0)$ is tight. In Table 2.2 we compare five different types of code sequences.

The *Preferentially-Phased Gold Codes* are presented in [5] (also see [7]).

The *Kerdock code* is a nonlinear, noncyclic subcode of the 2nd order Reed-Muller code: see Figure 15.7 in [8] and Appendix A in [5].

All phases of Gold code sequences are obtained by taking all L phases of the sequences in the Gold code, which results in an enlarged set with $N = L(L+2)$ sequences: see [1] and [5]. (The set of Gold codes contains $L+2$ sequences.)

All phases of the small set of Kasami sequences give a set of $L\sqrt{L+1}$ code sequences: see [1].

Finally, *Hadamard* orthogonal codes are presented in the next subsection.

QO-sequences are also more tolerant to timing jitter (or misalignment between them) in comparison with orthogonal sequences. The timing jitter properties of sequences have been examined in Section 2.3. As we have shown, orthogonal sequences are very sensitive to timing jitter ($0.1T_c$), QO-sequences are less sensitive ($0.5T_c$), and PN-sequences are insensitive to timing jitter (T_c is the chip length). However, both orthogonal and QO-sequences require synchronization.

2.2.4 Orthogonal Code Sequences

A code or binary array is *orthogonal* when it satisfies the requirement $\rho(v_i, v_j) = 0$ for any pair of sequences ($i \neq j$). For *orthogonal codes* it is usally assumed that the total number n equals the length L ($n = L$), and for that it is necessary that $n = 1, 2$ or $4t$ (see below).

An *orthogonal code* may then be represented by a $n \times n$ matrix $\mathbf{H}$ for which $\mathbf{HH}^T = n\mathbf{I}$, where $\mathbf{H}^T$ is the transpose of $\mathbf{H}$ and $\mathbf{I}$ is the identity matrix. A matrix $\mathbf{H}$ is also known as a *Hadamard* matrix. It has been shown that for any $n \times n$, ± 1 matrix $\mathbf{A} = [a_{ij}]$ with $|a_{ij}| \leq 1$, $|\det\mathbf{A}| \leq n^{n/2}$, where equality applies if and only if A is a *Hadamard* matrix [9].

In each Hadamard matrix one may interchange rows, interchange columns, change the sign of every element in a row, or change the sign of every element in a column, without disturbing the orthogonality property. If two Hadamard matrices can be transformed into each other by operations of this type, they are called *equivalent.*

A Hadamard matrix has a *normal* form if the first row and first column contain only 1s. The normal form is not unique within an equivalence class (this can be shown by example). In general, there is more than one equivalence class of Hadamard matrices for a given dimension m, $m \geq 16$.

If $m \geq 1$ is the dimension (or size) of a Hadamard matrix, then $m = 1, 2,$ *or* $4t$, (see [9]).

It has been conjectured that Hadamard matrices exist for all $m = 4t$ (it is almost certain that if m is a multiple of 4, a Hadamard matrix exists, athough this has not been proved).

If a Hadamard matrix exists for $m = 4t$, then *simplex* codes exist for $m = 4t$, $4t-1$, $2t$ and $2t-1$. If $\mathbf{H}$ is a Hadamard matrix or binary orthogonal code of size n, then its properties may be summarized as follows:

(1) $HH^T = nI_n$
(2) $|\det\mathbf{H}| = n^{n/2}$
(3) $\mathbf{H}\mathbf{H}^T = \mathbf{H}^T\mathbf{H}$
(4) Every Hadamard matrix is equivalent to a Hadamard matrix which has a *normal* form.
(5) $n = 1, 2$, or $4t$, t is an integer.
(6) If $\mathbf{H}$ has *normal* form and size $4n$, then every row (column) except the first has $2n$, $-1s$ and $2n$, $+1s$; further, n, $-1s$ in any row (column) overlap with n, $-1s$ in each other row (column).

Orthogonal Codes Based on PN-sequences

The basic types of orthogonal codes are generated from PN-sequences. Here we present four basic methods of generating PN-sequences. These methods provide the following sequence length ℓ:

(1) $\ell = 2^k - 1$: maximum length linear sequences (or m-sequences).
(2) (a) $\ell = q = 3(\text{mod}4)$ is odd prime: Quadratic Residue.
(b) $\ell = 2q + 1$, where $q = 1(\text{mod}4)$ is odd prime: Quadratic Residue $-$ 2 ($QR2$).
(3) $\ell = 4t - 1 = 4x^2 + 27$ is prime: *Hall* sequences.
(4) $\ell = p\ (p+2)$ where both p, $p+2$ are prime: *Twin-Prime* sequences.

These four types of sequences have lengths which overlap to some extent: If ℓ is a Mersenne prime then (1) and (2a) overlap. If $\ell = 31,\ 127$ then (1) and (3) overlap, and if $\ell = 15$ then (1) and (4) overlap. Also (3) is a subset of (2).

Maximum length sequences (m-sequences) are constructed by maximum length recursion using a maximum length linear feedback shift register. The Quadratic Residue sequences (QR and QR2) are known as the *first* and *second Paley* construction (see [2]). The Hall sequences are presented in [3] and the Twin-prime sequences in [10]. Given any of the above PN-sequences, a_i, we can generate orthogonal codes of length

$w = \ell + 1$, by cyclic shifting the sequence a_i and placing a leading row and column of $x = 1$, or -1, so that the number of 1s equals the number of -1s (0s) in the sequence shown below.

$$\begin{bmatrix} x & x & x & \cdots & x \\ x & c_1 & c_2 & \cdots & c_\ell \\ x & c_\ell & c_1 & \cdots & c_{\ell-1} \\ \vdots & \vdots & \vdots & \vdots & \vdots \\ x & c_2 & c_3 & \cdots & c_1 \end{bmatrix}$$

Walsh–Hadamard Sequences

The *Walsh–Hadamard* sequences are *noncyclic* orthogonal sequences having length $L = 2^k$. A *Walsh–Hadamard* (W-H) code is a square matrix with 1, 0 (-1), elements which has the design format:

$$\mathbf{H}_1 = \begin{bmatrix} 1 \end{bmatrix}, \qquad \mathbf{H}_2 = \begin{bmatrix} 1 & 1 \\ 1 & -1 \end{bmatrix} \qquad \text{and} \qquad \mathbf{H}_{2N} = \begin{bmatrix} \mathbf{H}_N & \mathbf{H}_N \\ \mathbf{H}_N & -\mathbf{H}_N \end{bmatrix}$$

The above construction was first proposed by Sylvester in [12]. Also, these matrices are associated with the discrete orthogonal functions called *Walsh functions* (see [11]).

Quaternion Type Codes

The quaternion type orthogonal codes are presented by Williamson [13]. If **A, B, C, D** are $n \times n$ (1,-1) matrices such that $\mathbf{AA}^T + \mathbf{BB}^T + \mathbf{CC}^T + \mathbf{DD}^T = 4n\mathbf{I}$ and $\mathbf{XY}^T = \mathbf{YX}^T$, for $\mathbf{X}, \mathbf{Y} \in \{\mathbf{A}, \mathbf{B}, \mathbf{C}, \mathbf{D}\}$, then

$$\mathbf{W}_{4n} = \begin{bmatrix} \mathbf{A} & -\mathbf{B} & -\mathbf{C} & -\mathbf{D} \\ \mathbf{B} & \mathbf{A} & -\mathbf{D} & \mathbf{C} \\ \mathbf{C} & \mathbf{D} & \mathbf{A} & -\mathbf{B} \\ \mathbf{D} & -\mathbf{C} & \mathbf{B} & \mathbf{A} \end{bmatrix}$$

is an orthogonal binary code of size $4n$.

Examples of Williamson matrices of sizes 5 and 7 are given below (we only show the first row of the cyclic matrices **A, B, C** and **D**):

$$\begin{matrix} \mathbf{A} \\ \mathbf{B} \\ \mathbf{C} \\ \mathbf{D} \end{matrix} \begin{bmatrix} 1 & -1 & -1 & -1 & -1 \\ 1 & -1 & -1 & -1 & -1 \\ 1 & 1 & -1 & -1 & 1 \\ 1 & -1 & 1 & 1 & -1 \end{bmatrix} \begin{bmatrix} 1 & -1 & -1 & -1 & -1 & -1 & -1 \\ 1 & 1 & -1 & -1 & -1 & -1 & 1 \\ 1 & -1 & 1 & -1 & -1 & 1 & -1 \\ 1 & -1 & -1 & 1 & 1 & -1 & -1 \end{bmatrix}$$

Additional **A, B, C** and **D** matrices exist for sizes 9, 11, 13, 15, ... We call such orthogonal codes *Quaternion Type-2 Codes* (Q2).

An extended type of quaternion codes can be constructed using the orthogonal design $OD(4t; t, t, t, t)$ called *Baumert-Hall* (B-H) arrays. For example, with $t = 3$ we provide a B-H array of size 12. Such codes are shown in Table 2.3 with the notation

Table 2.3 Hadamard matrices of all sizes up to 256 and the corresponding construction methods.

Code Length	TYPE	Code Length	TYPE	Code Length	TYPE	Code Length	TYPE
2	W	68	QR, Q	136	Q	204	QR2
4	QR, M, W	72	QR,E	140	QR	208	E
8	QR, M, W	76	Q	144	E	212	QR
12	QR	80	QR, E	148	Q	216	E
16	E, W	84	QR, Q	152	QR, E	220	QR2
20	QR, W	88	E	156	Q	224	QR, E
24	QR, E	92	Q	160	E	228	QR
28	Q	96	E	164	QR	232	E
32	QR, M, W, E	100	Q	168	QR, E	236	OD
36	TP, Q	104	QR, E	172	Q	240	QR, E
40	E	108	QR, Q	176	Q, E	244	Q
44	QR, Q	112	E	180	QR	248	E
48	QR, E	116	Q	184	Q, E	252	QR
52	Q	120	Q, E	188	Q2	256	W, M, E
56	E	124	Q	192	Q, E		
60	QR, Q	128	M, W, QR, E	196	QR2		
64	M, W, E	132	QR, Q	200	QR, E		

M: m-sequences
W: Hadamard-Walsh (Sylvester)
QR: Quadratic Residue (Paley)
QR2: Quadratic Residue-2 (Paley- 2)
TP: Twin Prime
Q: Quaternion (Williamson)
Q2: Quaternion-2
OD: Orthogona Design
E: Extended

$Q2$. Also, another type of array presented by Hedayat and Wallis [14] is

$$\begin{bmatrix} \mathbf{A} & \mathbf{B} & \mathbf{C} & \mathbf{D} \\ -\mathbf{B} & \mathbf{A} & -\mathbf{E} & \mathbf{F} \\ -\mathbf{C} & \mathbf{E} & \mathbf{A} & \mathbf{G} \\ -\mathbf{D} & -\mathbf{F} & -\mathbf{G} & \mathbf{A} \end{bmatrix}$$

Circulant matrices $\mathbf{A}, \mathbf{B}, .., \mathbf{G}$ of size 47 are used in the construction of a Hadamard matrix of size 188.

Orthogonal Designs

Next we consider orthogonal matrices with entries $0, \pm1, \pm2, \ldots$ known as *orthogonal designs*.

An *Orthogonal Design* (OD) of order n and type $(s_1, s_2, .., s_k)$, s_i positive integers, is defined as an $n \times n$ matrix $\mathbf{Z}$, with entries $\{0, \pm z_1, \pm z_2, .., \pm z_k\}$ (commuting indeterminates) satisfying $\mathbf{ZZ}^T = \left(\sum_{i=1}^{k} s_i z_i^2\right) I_n$. An orthogonal design is then denoted by $OD(n; s_1, s_2, .., s_k)$. Alternatevly, each row of $\mathbf{Z}$ has s_i entries of the type $\pm z_i$, and the distinct rows are orthogonal under the Euclidean inner product. An orthogonal design with no zeros, in which each entry is replaced by $+1$ or -1, is a Hadamard matrix. The OD(4;1,1,1,1) is known as a Williamson array, while the OD(4t;t,t,t,t), known as the Baumert–Hall array, is useful in the construction of Hadamard matrices.

Other orthogonal designs can be derived from orthogonal tranformation. A discrete orthogonal tranformation can be represented by a square orthogonal matrix, $\mathbf{H} = [h_{nm}]$. Examples of such orthogonal matrices are the *Discrete Cosine* orthogonal transformation, for which

$$h_{nm} = \left\{ \frac{1}{\sqrt{2}}, \quad \cos\frac{2n+1}{2N}; \ n = 0, 1, 2, .., M-1, \ m = 1, 2, .., M \right\}$$

and the *Karhunen–Loeve* orthogonal transformation for which

$$h_{nm} = \left\{ \sqrt{\frac{2}{N}} \frac{\sin 2\pi(n/N - m/2)}{2\pi(n/N - m/2)}; \ n, m = 0, 1, 2, .., N-1 \right\}$$

Figure 2.4 shows a plot of Karhunen–Loeve, Hadamard and Fourier orthogonal sequences with size 16.

2.2.5 Extended Orthogonal Sequences

Orthogonal sequences of additional lengths can be constructed using the following proposition:

Proposition 1: Let $\mathbf{G}_x = [g_{i,j}]$ and $\mathbf{H}_y = [h_{i,j}]$ be orthogonal matrices of lengths x and y, respectively; Then the matrix $\mathbf{E}_z = [e_{ij}]$ is formed by substituting $\mathbf{G}_x$ for 1 and $-\mathbf{G}_x$ for -1 in $\mathbf{H}_y$, and is also an orthogonal matrix with size ($z = x \cdot y$).

Each element w_{ij} is then given by $e_{xn+i,xm+j} = h_{nm}g_{ij}$ for $0 \leq n, m < y$ and $0 \leq i, j < x$. This operation is called the *Kronecker product*, and is denoted by $\mathbf{E}_z = \mathbf{G}_x \times \mathbf{H}_y$. The codes generated by the Kronecker product are called *extended orthogonal* codes. The matrix $\mathbf{E}_z$ having size $z = xy$ is generated in the way illustrated below:

$$\mathbf{E}_{xy} = \mathbf{G}_x \times \mathbf{H}_y = \begin{bmatrix} g_{11}\mathbf{H}_y & g_{12}\mathbf{H}_y & \cdots & g_{1x}\mathbf{H}_y \\ g_{21}\mathbf{H}_y & g_{22}\mathbf{H}_y & \cdots & g_{2x}\mathbf{H}_y \\ \cdots & \cdots & \cdots & \cdots \\ g_{x1}\mathbf{H}_y & g_{x2}\mathbf{H}_y & \cdots & g_{xx}\mathbf{H}_y \end{bmatrix}$$

Proof Given that, $\mathbf{G}_x\mathbf{G}_x^T = x\mathbf{I}_x$, $\mathbf{H}_y\mathbf{H}_y^T = y\mathbf{I}_y$ and $(\mathbf{G}_x \times \mathbf{H}_y)^T = \mathbf{G}_x^T \times \mathbf{H}_y^T$ (shown below in Lemma 1), then

$$(\mathbf{G}_x \times \mathbf{H}_y)(\mathbf{G}_x \times \mathbf{H}_y)^T = (\mathbf{G}_x \times \mathbf{H}_y)(\mathbf{G}_x^T \times \mathbf{H}_y^T) = (\mathbf{G}_x\mathbf{G}_X^T) \times (\mathbf{H}_y\mathbf{H}_y^T) = x\mathbf{I}_x \times y\mathbf{I}_y = xy\mathbf{I}_{xy}$$

Lemma 1 If $\mathbf{A}$ and $\mathbf{B}$ are any matrices of size n, then $(\mathbf{A} \times \mathbf{B})^T = \mathbf{A}^T \times \mathbf{B}^T$. If, further, $\mathbf{C}$ and $\mathbf{D}$ are any matrices such that the product $\mathbf{AC}$ and $\mathbf{BD}$ exist, then $(\mathbf{A} \times \mathbf{B})(\mathbf{C} \times \mathbf{D}) = \mathbf{AC} \times \mathbf{BD}$.

	0	1	2	3	4	5	6	7
0	1	1	1	1	1	1	1	1
1	1	0	1	1	0	1	0	0
2	1	1	1	0	1	0	0	0
3	1	1	0	1	0	0	0	1
4	1	0	1	0	0	0	1	1
5	1	1	0	0	0	1	1	0
6	1	0	0	0	1	1	0	1
7	1	0	0	1	1	0	1	0

	0	1	2	3	4	5	6	7	8	9	0	1
0	1	1	1	1	1	1	1	1	1	1	1	1
1	1	0	1	0	1	1	1	0	0	0	1	0
2	1	1	0	1	1	1	0	0	0	1	0	0
3	1	0	1	1	1	0	0	0	1	0	0	1
4	1	1	1	1	0	0	0	1	0	0	1	0
5	1	1	1	0	0	0	1	0	0	1	0	1
6	1	1	0	0	0	1	0	0	1	0	1	1
7	1	0	0	0	1	0	0	1	0	1	1	1
8	1	0	0	1	0	0	1	0	1	1	1	0
9	1	0	1	0	0	1	0	1	1	1	0	0
10	1	1	0	0	1	0	1	1	1	0	0	0
11	1	0	0	1	0	1	1	1	0	0	0	1

	0	1	2	3	4	5	6	7	8	9	0	1	2	3	4	5	6	7	8	9
0	1	0	0	0	0	0	1	1	1	1	0	0	1	1	0	0	1	0	0	1
1	0	1	0	0	0	1	0	1	1	1	0	0	0	1	1	1	0	1	0	0
2	0	0	1	0	0	1	1	0	1	1	1	0	0	0	1	0	1	0	1	0
3	0	0	0	1	0	1	1	1	0	1	1	1	0	0	0	0	0	1	0	1
4	0	0	0	0	1	1	1	1	1	0	0	1	1	0	0	1	0	0	1	0
5	1	0	0	0	0	1	0	0	0	0	0	1	0	0	1	1	1	0	0	1
6	0	1	0	0	0	0	1	0	0	0	1	0	1	0	0	1	1	1	0	0
7	0	0	1	0	0	0	0	1	0	0	0	1	0	1	0	0	1	1	1	0
8	0	0	0	1	0	0	0	0	1	0	0	0	1	0	1	0	0	1	1	1
9	0	0	0	0	1	0	0	0	0	1	1	0	0	1	0	1	0	0	1	1
10	1	1	0	0	1	1	0	1	1	0	1	0	0	0	0	0	1	1	1	1
11	1	1	1	0	0	0	1	0	1	1	0	1	0	0	0	1	0	1	1	1
12	0	1	1	1	0	1	0	1	0	1	0	0	1	0	0	1	1	0	1	1
13	0	0	1	1	1	1	1	0	1	0	0	0	0	1	0	1	1	1	0	1
14	1	0	0	1	1	0	1	1	0	1	0	0	0	0	1	1	1	1	1	0
15	1	0	1	1	0	0	0	1	1	0	1	0	0	0	0	1	0	0	0	0
16	0	1	0	1	1	0	0	0	1	1	0	1	0	0	0	0	1	0	0	0
17	1	0	1	0	1	1	0	0	0	1	0	0	1	0	0	0	0	1	0	0
18	1	1	0	1	0	1	1	0	0	0	0	0	0	1	0	0	0	0	1	0
19	0	1	1	0	1	0	1	1	0	0	0	0	0	0	1	0	0	0	0	1

Figure 2.3 Hadamard matrices of sizes 8, 12 and 20.

Proof By definition we have

$$\begin{bmatrix} \mathbf{X} & \mathbf{Y} \\ \mathbf{Z} & \mathbf{W} \end{bmatrix}^T = \begin{bmatrix} \mathbf{X}^T & \mathbf{Z}^T \\ \mathbf{Y}^T & \mathbf{W}^T \end{bmatrix}$$

Hence, $(\mathbf{A} \times \mathbf{B})^T = [(a_{ij}\mathbf{B})^T] = [a_{ij}\mathbf{B}^T]$. Now, $[x_{ij}][y_{ij}]$ has (i, j) block entry $\sum_k x_{ik}y_{kj}$. So, $(\mathbf{A} \times \mathbf{B})(\mathbf{C} \times \mathbf{D})$ has (i, j) block $\sum_k (a_{ik}B)(c_{kj}\mathbf{D}) = (\sum_k a_{ik}c_{kj})\mathbf{BD}$. However, $\sum_k a_{ik}c_{kj}$ is the (i, j) entry of $\mathbf{AC}$. Hence, $(\mathbf{A} \times \mathbf{B})(\mathbf{C} \times \mathbf{D}) = \mathbf{AC} \times \mathbf{BD}$.

Now, if we apply the above operation repeatedly (k times) with size matrices $z_1, z_2, .., z_k$, the result will be an orthogonal matrix with size z, where

$$z = \prod_{i=1}^{k} z_i = z_1 \cdot z_2 \cdots z_k$$

The orthogonal matrices $\mathbf{H}_z$ can be constructed by any method described above.

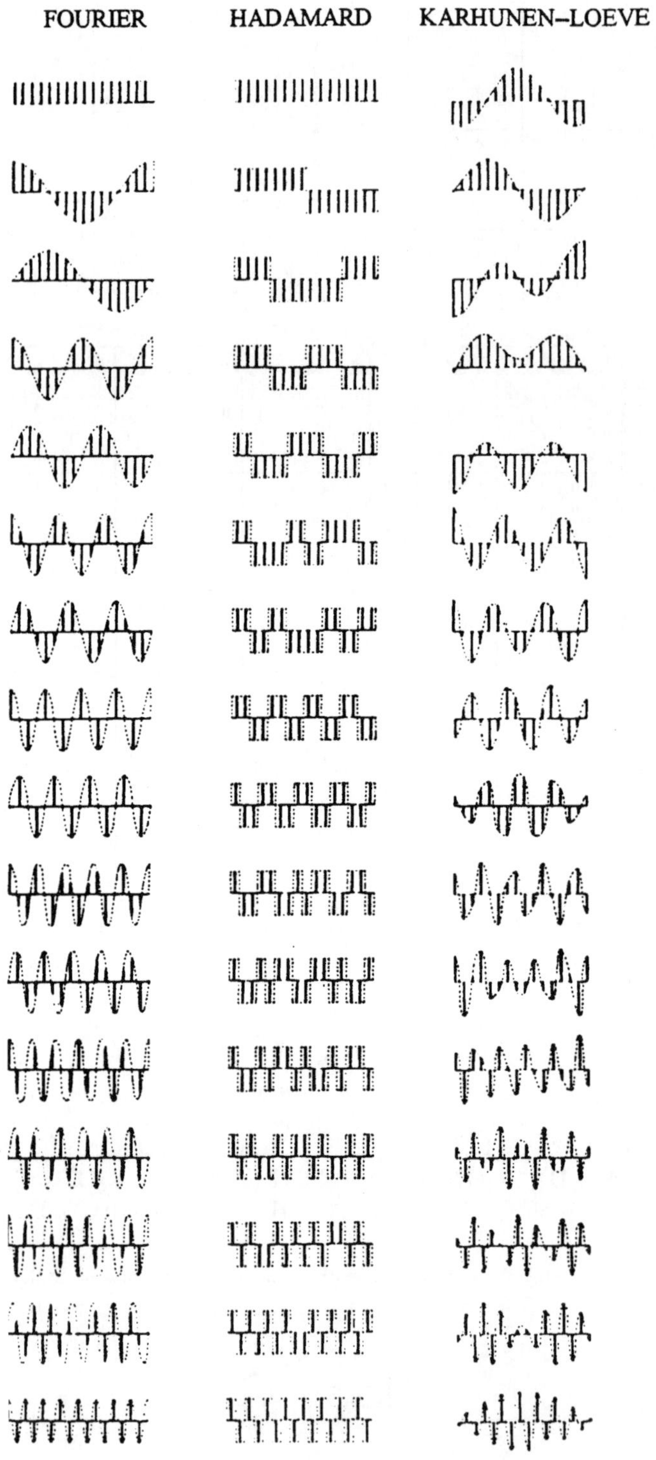

Figure 2.4 Fourier, Hadamard and Karhunen–Loeve orthogonal codes of size-16.

In the special case where all $z_i = 2$ for $i = 1, 2, .., k$, then the generated sizes are $z = 2^k$, and all matrices have the Walsh–Hadamard format.

In another example, if $z_1 = 12$ and $z_i = 2$ for $i = 2, 3, .., k$, then the generated sizes are $z = 12, 24, 48, 96, 192, 384, 768, \ldots$. Among these, matrix sizes 96 and 768 are new sizes, while the rest may also be generated by other basic methods.

In general $\mathbf{G}_z \times \mathbf{G}_w \neq \mathbf{G}_w \times \mathbf{G}_z$. This means that we may generate pairs of distinct orthogonal matrices, each with size zw.

The methods presented above for the construction of orthogonal binary codes can be used to generate all matrices of size $m = 4t \leq 256$ (see reference [15]). In Table 2.3 we present the size and corresponding method(s) of generating them. As we observe, matrix sizes $m = 2^t$ can be generated by both *m–sequences* and *Walsh–Hadamard* (W-H) codes. The Quadratic Residue (QR) method can be used to generate additional sizes of $m = 12, 20, 24, 60, 68, 72, 80, 84, 104, 108, \ldots$ The QR-2 method provides additional sizes $m = 28, 196, 204, 220, \ldots$. The Quaternion (Williamson) method also provides additional sizes $m = 52, 76, 92, 99, 124, 136, \ldots$ Also, Baumert–Hall (Q2) and Hedayat–Wallis (Q3) arrays are used to construct sizes 236 and 188, respectively. The method of extended sequences has also been used to provide many new matrices of the same and additional sizes. Matrices given in Table 2.3 have been constructed and may be supplied to the reader on request. In Figure 2.3 we present orthogonal matrices of sizes 8, 12 and 20 as examples.

2.2.6 Complex Orthogonal Matrices

A matrix $\mathbf{C}$ of order ℓ with elements ± 1, $\pm j$ $(j^2 + 1 = 0)$ that satisfies $\mathbf{CC}^* = \ell \mathbf{I}_\ell$ is called the Complex Orthogonal Matrix, where $\mathbf{C}^*$ denotes the Hermitian conjugate (transpose, complex conjugate) and $\mathbf{I}$ is the unit matrix. A complex Hadamard matrix of order 2ℓ is a complex orthogonal matrix.

An example of a complex Hadamard matrix of size 4 is shown below:

$$\mathbf{W} = \begin{bmatrix} 1 & -j & 1 & -j \\ j & -1 & j & -1 \\ 1 & -j & -1 & j \\ j & -1 & -j & 1 \end{bmatrix}$$

In the above matrix $r_i = c_i^*$, where r_i and c_i are the vectors of row and column i, respectively. Based on the above definition of orthogonality,

$$r_i r_k^* = \delta_{i,k} \equiv \begin{cases} 0 & \text{for } i \neq k \\ n & \text{for } i = k \end{cases}$$

Complex Hadamard matrices were first introduced by Turyn [16].

The above definition of $(\pm 1, \pm j)$ complex orthogonal matrices may be extended to polyphase complex entries. That is, when each entry $w_{ik} = e^{j\phi_{ik}}$ of matrix $\mathbf{W} = [w_{ik}]$ is located on the periphery of a unit circle. Such a matrix is called a *Polyphase Orthogonal Matrix (POM)* if $\mathbf{WW}^* = \mathbf{L}\mathbf{I}_L$, where $\mathbf{L}$ is the size of the matrix, $\mathbf{W}^*$ denotes the Hermitian conjugate (transpose, complex conjugate) and $\mathbf{I}$ is the unit matrix.

$$
H_8 = \begin{pmatrix}
1 & e^{\frac{j\pi}{4}} & j & e^{\frac{j3\pi}{4}} & -1 & e^{\frac{j5\pi}{4}} & -j & e^{\frac{j7\pi}{4}} \\
1 & e^{\frac{j5\pi}{4}} & & e^{\frac{j7\pi}{4}} & -1 & e^{\frac{j\pi}{4}} & -j & e^{\frac{j3\pi}{4}} \\
1 & e^{\frac{j\pi}{4}} & -j & e^{\frac{j7\pi}{4}} & -1 & e^{\frac{j5\pi}{4}} & j & e^{\frac{j3\pi}{4}} \\
1 & e^{\frac{j5}{4}} & -j & e^{\frac{j3\pi}{4}} & -1 & e^{\frac{j\pi}{4}} & j & e^{\frac{j7\pi}{4}} \\
1 & e^{\frac{j}{4}} & j & e^{\frac{j3\pi}{4}} & 1 & e^{\frac{j\pi}{4}} & j & e^{\frac{j3\pi}{4}} \\
1 & e^{\frac{j5\pi}{4}} & j & e^{\frac{j7\pi}{4}} & 1 & e^{\frac{j5\pi}{4}} & j & e^{\frac{j7\pi}{4}} \\
1 & e^{\frac{j\pi}{4}} & -j & e^{\frac{j7\pi}{4}} & 1 & e^{\frac{j3\pi}{4}} & -j & e^{\frac{j7\pi}{4}} \\
1 & e^{\frac{j5\pi}{4}} & -j & e^{\frac{j3\pi}{4}} & 1 & e^{\frac{j5\pi}{4}} & -j & \frac{j3\pi}{4}
\end{pmatrix}
$$

Figure 2.5 A Polyphase Orthogonal Matrix of size 8.

A particular type of POM has the format $w_{l,k} = e^{j(\frac{2\pi m_{lk}}{M})}$, for $m_{l,k} = 0, 1, 2, 3...M-1$, and for $0 \leq l, k \leq L$, where M is the number of phases and L is the size of the matrix. Now we may consider an example of POM having N phases, with $N \geq 4$ and $N \leq L$ (N and L are even numbers), defined as follows; $W = [\mathbf{w_0}, \mathbf{w_1}, .., \mathbf{w_{L-1}}]^T$, where vector $\mathbf{w_n} = \mathbf{h_n a}$, vector $\mathbf{a} = [1, e^{j2\pi/N}, .., e^{j2\pi(KN-1)/N}]$ (for $KN = L$), and the vector $\mathbf{h_n} = [h_{kn}]$ is a real binary Hadamard sequence. If $K = 1$, then the number of phases is the same as the matrix size ($N = L$). We may easily show that $\mathbf{WW}^* = L\mathbf{I}_L$, since,

$$
\mathbf{w_n w_k^*} = \sum_{\ell=0}^{L-1} h_{\ell n} e^{j2\pi\ell/N} h_{\ell k} e^{-j2\pi\ell/N} = \mathbf{h_n} \cdot \mathbf{h_k} = \begin{cases} L & \text{for } n = k \\ 0 & \text{for } n \neq k \end{cases}
$$

Another type of POM for which $N = L$ is based on the Discrete Fourier Transform (DFT). The DFT-POM matrix is given by $\mathbf{W} = [w_{nm}]$, where $w_{nm} = e^{j2\pi nm/N}$ and $n, m = 0, 1, .., N-1$. An example of POM is shown in Figure 2.5.

2.3 Timing Jitter Properties of Sequences

In this section we analyze the effect of time jitter on synchronous CDMA systems employing Quadratic-Residue (Q-R) orthogonal codes and Preferentially Phased Gold (quasi-orthogonal codes or QO) codes. The cross-correlation functions for these codes are derived for time jitter taking values in the interval $[-T_c, T_c]$, where T_c is the chip duration. Our results are derived as functions of the vector of values of the instantaneous time jitter, and they can be used to obtain the average interference with respect to any desirable probability distribution of the time jitter. It is established that time jitter significantly affects fully loaded orthogonal and quasi-orthogonal CDMA systems: the variance of the other-user interference depends on the term $1/N \sum_{k=2}^{K} [\tau_k / T_c]^2$, where τ_k is the time jitter of user k with respect to user 1 (the reference user), N is the spreading factor or processing gain (number of chips per bit), and K is the number of simultaneous users. For full loading ($K = N$), the interference variance assumes a non negligible value of the order $\overline{\tau}^2/T_c^2$ (where $\overline{\tau}$ is the average of the τ_ks) for both the Q-R and QO CDMA systems. For $K = N/2$ (50% loading) and carefully purged Q-R set (eliminate every second shift sequence), the interference variance becomes $1/N^2 \sum_{k=2}^{N/2}[\tau_k/T_c]^2$, which is $N/2$ times smaller than that of the QO code set and practically negligible.

2.3.1 The System Model

In this system model, the kth user signal $b_k(t)$ is a sequence of unit amplitude, positive and negative, rectangular pulses of duration T. This signal represents the data of kth user:

$$b_k(t) = \sum_{l=-\infty}^{\infty} b_{k,l} p_T(t - jT) \qquad \text{where} \quad p_T(t) = \begin{cases} 1: & 0 \leq t < T \\ 0: & \text{otherwise} \end{cases}$$

The kth user is assigned a code waveform $a_k(t)$ which consists of a periodic sequence of period $T = NT_c$ of unit amplitude, positive and negative, rectangular pulses of duration T_c. We can write $a_k(t)$ as $a_k(t) = \sum_{j=-\infty}^{\infty} a_{k,j} p_{T_c}(t - jT_c)$, where $a_{k,j}$ is the code sequence such that $a_{k,j+N} = a_{k,j}$.

The data signal $b_k(t)$ is multiplied by the code, and then modulates a carrier to produce the BPSK CDMA signal $s_k(t)$, which is given by

$$s_k(t) = \sqrt{2P_k} a_k(t) b_k(t) \cos(w_c t + \theta_k)$$

In this analysis we intend to investigate the multi-user interference effect due to time jitter only, so we will neglect the channel noise and distortion. The received signal $r(t)$ is given by

$$r(t) = \sum_{k=1}^{N} \sqrt{2P_k} b_k(t) a_k(t - \tau_k) \cos(w_c(t - \tau_k) + \theta_k)$$

assuming the number of users is equal to the processing gain N and τ_k is the delay of the kth signal.

Without loss of generality, we will consider the receiver for the first user, and will assume that $\tau_1 = 0$ and $\theta_1 = 0$. The received signal is the input to a correlation receiver matched to $s_1(t)$, and the output of the matched filter at time T is

$$\begin{aligned} Z &= \int_0^T r(t) a_1(t) \cos(w_c t) dt \\ &= \int_0^T \sum_{k-1}^{N} \sqrt{2P_k} b_k(t - \tau_k) a_k(t - \tau_k) \cos(w_c(t - \tau_k) + \theta_k) a_1(t) \cos(w_c t) dt \\ &= \sqrt{P_1/2} \left[b_1 T + \sum_{k=2}^{N} \int_0^T \sqrt{\frac{P_k}{P_1}} b_k(t - \tau_k) a_k(t - \tau_k) a_1(t) \cos(w_c \tau_k + \theta_k) dt \right] \end{aligned}$$

So the desired signal component at the output of the correlator is $\sqrt{P_1/2}\ b_1 T$, while the other-user interference is given by

$$I = \sqrt{P_1/2} \sum_{k=2}^{N} \sqrt{\frac{P_k}{P_1}} I_k$$

where $I_k = \cos(w_c \tau_k + \theta_k) \int_0^T b_k(t-\tau_k) a_k(t-\tau_k) a_1(t) dt$.

It is reasonable to assume that $(w_c \tau_k + \theta_k)$ is uniformly distributed in the range $[-\pi, \pi]$:

$$E\{I_k\} = E\{\cos(w_c \tau_k + \theta_k)\} \int_0^T E\{b_k(t-\tau_k)\} E\{a_k(t-\tau_k) a_1(t)\} dt = 0$$

The variance of the interference component of Z is

$$Var(Z) = E\{Z^2\} = \sum_{k=2}^{N} \sum_{j=2}^{N} \sqrt{\frac{P_k}{P_1}} \sqrt{\frac{P_j}{P_1}} E\{I_k I_j\}$$

but I_k and I_j are independent and have zero mean, so for $k \neq j$, $E\{I_k I_j\} = 0$

$$Var(Z) = \sum_{k=2}^{N} \frac{P_k}{P_1} E\{I_k^2\} = \sum_{k=2}^{N} \frac{P_k}{P_1} \left[\int_0^T b_k(t-\tau_k) a_k(t-\tau_k) a_1(t) dt \right]^2 \frac{1}{2}$$

2.3.2 Jitter Impact on QR-Orthogonal Sequences

To calculate $Var(Z)$, we must know the properties of the code used. First we will consider Quadratic Residue orthogonal codes. Calculating the above expression involves calculation of the cross-correlation between the sequences taking into account the possible values that $b_{k,l}$ could take. Again without loss of generality, assume $b_{k,0} =$ '1'. In calculating the integrals we must know the properties of the sequences used. Q-R orthogonal sequences have the following properties:

(1) First chip (code symbol) is a '1'.
(2) Next $N-1$ chips are cyclic shifts of a PN code.

The autocorrelation function of PN codes is shown in Figure 2.6-A. Figure 2.7 shows the generation of QR signals for $N = 8$. We will analyze the cross-correlation $R_{i,j}(\tau)$ between two codes a_i and a_j, given by

$$R_{i,j}(\tau) = \int_0^T b_j(t-\tau) a_j(t-\tau) a_i(t) dt$$

In the analysis of the cross-correlation we must distinguish between two cases for the shift of the codes. As $N-1$ chips of the code are just a cyclic shift of each other, so we could estimate that the cross-correlation between any two sequences will increase very much for a specific delay period. First we will analyze between $R_{i,j}(\tau)$ for two successive sequences $(1,2)$, for shifts $0 \leq \tau < T_c$, which are the values of interest

$$\begin{aligned} R_{1,2}(\tau) &= (N-2)\,\tau - \frac{1}{N-1}\tau - b_{2,-1}\frac{1}{N-1}\tau \\ &= (N-2)\,\tau - \frac{1}{N-1}\tau\,(1+b_{2,-1}) \end{aligned}$$

Secondly, we will consider codes generated by PN code shifted more than two chips, for shifts $0 \leq \tau < T_c$

$$R_{1,k}\,(\tau) = -\tau\left(1 + \frac{1}{N-1}b_{k,-1} + \frac{1}{(N-1)^2}\right)$$

$R_{1,2}(\tau)$ and $R_{1,k}(\tau)$ are shown in Figures 2.6-B and -C, respectively.

Then using the for $Var(Z)$ above, we get

$$E\{I_2^2\} \approx (N-2)^2\tau_2^2\frac{1}{2} \quad \text{and} \quad E\{I_k^2\} \approx \frac{1}{2}\tau_k^2 \quad \text{for} \quad k \neq 1,2$$

Therefore, the total interference power is

$$Var(Z) = \sum_{k=2}^{N} E\left\{I_k^2\right\} = \frac{1}{2}\frac{P_2}{P_1}(N-2)^2\tau_2^2 + \frac{1}{2}\sum_{k=3}^{N}\frac{P_k}{P_1}\tau_k^2$$

It is clear that the first term is much larger than the other terms if P_2 is not much less than P_k. For comparative results we could write

$$Var(Z) \approx \frac{1}{2}\frac{P_2}{P_1}(N-2)^2\tau_2^2$$

The interference power normalized by the processing gain $(Var_n(Z))$ is

$$Var_n(Z) = \frac{P_2}{P_1}\frac{(N-2)^2}{2N^2}\frac{\tau_2^2}{T_c^2} \approx \frac{P_2}{P_1}\frac{\tau_2^2}{2T_c^2}$$

The interference power will have the shape shown in Figure 2.8 for large N. For synchronous systems, good synchronization could happen, so that we could limit τ_k to the interval $[-T_c, T_c]$; in this case it is clear that for every code the previous and the following codes will contribute the biggest interference, so we could decrease the interference by generating only half the number of codes by shifting the PN code by two chips to ensure that the cross-correlation will be minimized in the interval $0 \leq \tau < T_c$, which will result in substantial improvement in performance. In this case the interference power will be

$$Var(Z) = \sum_{k=2}^{\frac{N}{2}}\frac{P_k}{P_1}E\{I_k^2\} = \frac{1}{2}\sum_{k=2}^{\frac{N}{2}}\frac{P_k}{P_1}\tau_k^2$$

and the interference power normalized by the processing gain is

$$Var_n(Z) = \frac{1}{2}\sum_{k=2}^{\frac{N}{2}}\frac{P_2}{P_1}\frac{\tau_k^2}{N^2T_c^2}$$

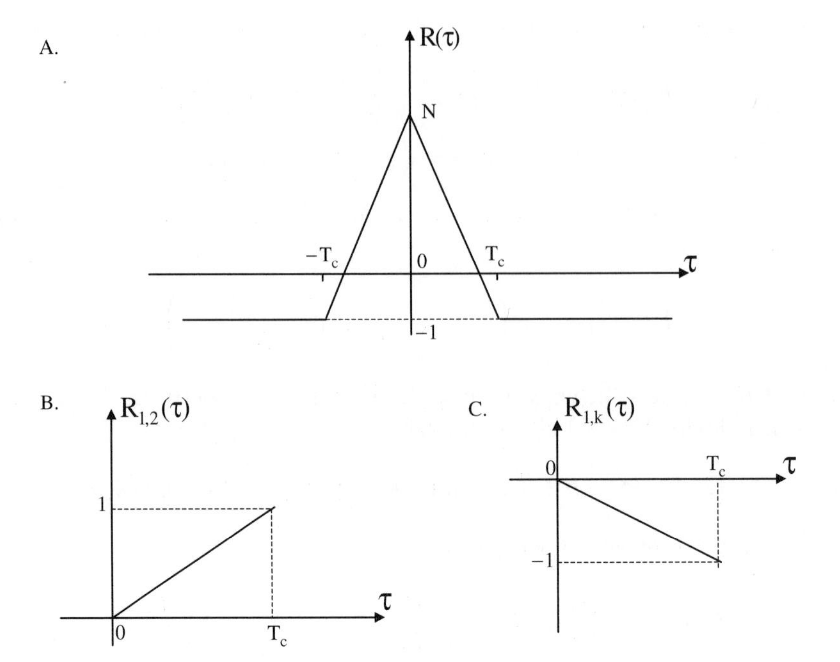

Figure 2.6 A. The autocorrelation function of PN-sequences, B. The worst case cross-correlation, C. Cross-correlation.

So in this case, the interference power will decrease by a factor of N in the interval $-T_c < \tau_k < T_c$. The interference power for different values of τ can be seen in Figure 2.6B.

2.3.3 Jitter Impact on Gold Code Sequences

The cross-correlation properties of the Gold codes of length $N = 2^m - 1$ have been summarized in Table 2.1. As in the case of QR orthogonal sequences (Section 2.3.2), the interference of the previous data bit b_{k-1} does not affect the current one (i.e. for $0 \leq \tau \leq T_c$). Hence, we may write

1. For n even

$$\begin{aligned} Var(Z) &= \sum_{k=2}^{N} \frac{P_k}{P_1} E\left\{I_k^2\right\} \\ &= \frac{1}{2} 4N \sum_{k=2}^{\frac{N}{4}} \frac{P_k}{P_1} \tau_k^2 + \frac{1}{2} \sum_{k=\frac{N}{4}+1}^{N} \frac{P_k}{P_1} \tau_k^2 \end{aligned}$$

and the normalized interference power will be

$$Var_n(Z) = \frac{1}{2} \frac{4}{N} \sum_{k=2}^{\frac{N}{4}} \frac{P_k}{P_1} \frac{\tau_k^2}{T_c^2} + \frac{1}{2} \frac{1}{N^2} \sum_{k=\frac{N}{4}+1}^{N} \frac{P_k}{P_1} \frac{\tau_k^2}{T_c^2}$$

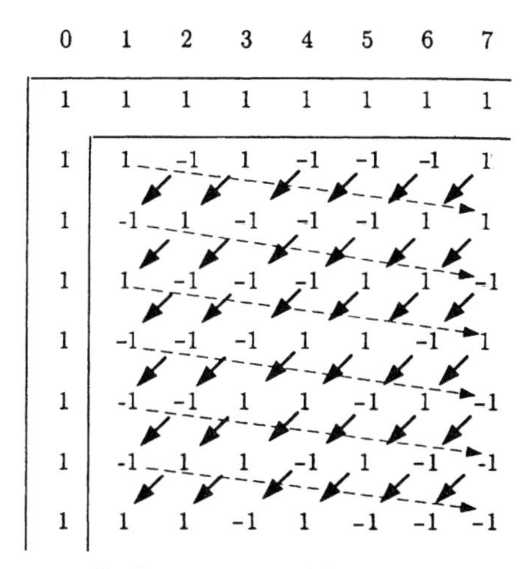

Figure 2.7 Cyclic property of QR orthogonal code set.

2. For n odd

$$Var(Z) = \sum_{k=2}^{N} \frac{P_k}{P_1} E\left\{I_k^2\right\}$$

$$= \frac{P_k}{P_1}\frac{1}{2}2N \sum_{k=2}^{\frac{N}{2}} \tau_k^2 + \frac{P_k}{P_1}\frac{1}{2} \sum_{k=\frac{N}{2}+1}^{N} \frac{P_k}{P_1}\tau_k^2$$

and the normalized interference power will be

$$Var_n(Z) = \frac{1}{2}\frac{2}{N} \sum_{k=2}^{\frac{N}{2}} \frac{P_k}{P_1}\frac{\tau_k^2}{T_c^2} + \frac{1}{2}\frac{1}{N^2} \sum_{k=\frac{N}{2}+1}^{N} \frac{P_k}{P_1}\frac{\tau_k^2}{T_c^2}$$

2.3.4 Jitter Impact on Extended Orthogonal Sequences

In this case we consider extended orthogonal sequences (described in Section 2.2.5), which are composed of two orthogonal codes: an outer code C of length N_1 and an inner code a of length N_2. We assume that the time duration of one inner chip is T_c and the time of each outer chip is $4T_c$, and that the two code sequences are in complete synchronization. We will assume $N_2 = 4$, and the total number of users $N = N_1 N_2 = 4N_1$. The received signal will be given by

$$r(t) = \sum_{k=1}^{N} \sqrt{2P_k} b_k(t-\tau_k) a_k(t-\tau_k) C_k(t-\tau_k) \cos(w_c(t-\tau_k)+\theta_k)$$

Without loss of generality, we will consider the receiver of the first user, and assume that $\tau_1 = 0$ and $\theta_1 = 0$. The received signal is the input to a correlation receiver

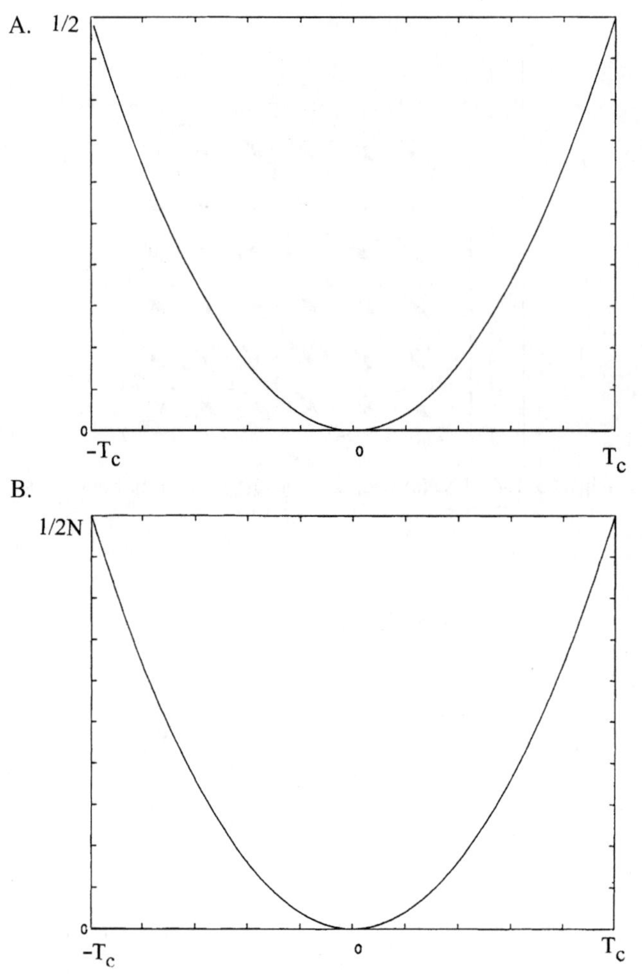

Figure 2.8 The inter-user interference power vs. the time-jitter τ. A. For full QR, set, B. for half QR set.

matched to the first signal, the output of the matched filter at time T is

$$
\begin{aligned}
Z &= \int_0^T r(t) C_1(t) a_1(t) \cos(w_c t) dt \\
&= \int_0^T \sum_{k=1}^{N} \sqrt{2P_k} b_k(t-\tau_k) C_k(t-\tau_k) a_k(t-\tau_k) C_1(t) a_1(t) \\
&\qquad \cos(w_c(t-\tau_k)+\theta_k)\cos(w_c t) dt \\
&= \sqrt{P_1/2}\left[b_1 T + \sum_{k=2}^{N} \int_0^T \sqrt{\frac{P_k}{P_1}} b_k(t-\tau_k) C_k(t-\tau_k) a_k(t-\tau_k) C_1(t) a_1(t) \right. \\
&\qquad \left. \cos(w_c \tau_k + \theta_k) dt \right]
\end{aligned}
$$

As in the previous sections, the desired signal component at the output of the correlator is $\sqrt{P_1/2}b_1T$, while the other-user interference is given by

$$I = \sqrt{P_1/2}\sum_{k=2}^{N}\sqrt{\frac{P_k}{P_1}}I_k$$

where $I_k = \cos(w_c\tau_k+\theta_k)\int_0^T b_k(t-\tau_k)C_k(t-\tau_k)a_k(t-\tau_k)C_1(t)a_1(t)dt$ and $E\{I_k\} = 0$. The variance of the interference component of Z is

$$Var(Z) = \sum_{k=2}^{N}\frac{P_k}{P_1}E\left\{I_k^2\right\}$$

where

$$\begin{aligned}
E\left\{I_k^2\right\} &= \frac{1}{2}E\left\{\left[\int_0^T b_k(t-\tau_k)C_k(t-\tau_k)a_k(t-\tau_k)C_1(t)a_1(t)dt\right]^2\right\} \\
&= \frac{1}{2}E\left\{\sum_{j=1}^{N_1}\left[C_{k,j}C_{1,j}b_{k,0}\int_{\tau_k}^{4T_c} a_k(t-\tau_k)a_1(t)dt\right.\right. \\
&\quad \left.\left. + \; C_{k,j}C_{1,j-1}b_{k,-1}\int_0^{\tau_k} a_k(t-\tau_k)a_1(t)dt\right]^2\right\} \\
&= \frac{1}{2}\sum_{j=1}^{N_1}\left[C_{k,j}C_{1,j}b_{k,0}\hat{R}_{1,k}(\tau_k) + C_{k,j-1}C_{1,j}R_{1,k}(\tau_k)b_{k,-1}\right]^2 \\
&= \frac{1}{2}\left[b_{k,0}\hat{R}_{1,k}(\tau_k)\sum_{j=1}^{N_1}(C_{k,j}C_{1,j}) + R_{1,k}(\tau_k)b_{k,-1}\sum_{j=1}^{N_1}(C_{k,j-1}C_{1,j})\right]^2
\end{aligned}$$

where $\hat{R}_{1,k}(\tau_k) = \int_{\tau_k}^{4T_c} a_k(t-\tau_k)a_1(t)dt$ and $R_{1,k}(\tau_k) = \int_0^{\tau_k} a_k(t-\tau_k)a_1(t)dt$

$$E\left\{I_k^2\right\} = \frac{1}{2}\hat{R}_{1,k}^2(\tau_k)\left[\sum_{j=1}^{N_1}(C_{k,j}C_{1,j})\right]^2 + \frac{1}{2}R_{1,k}^2(\tau_k)\left[\sum_{j=1}^{N_1}(C_{k,j-1}C_{1,j})\right]^2$$

If we assume that the outer code is Q-R orthogonal code, then we will have the following possibilities:

Case 1: Same or different a, different C
$N-4$ sequences, $\sum_{j=1}^{N_1}C_{k,j}C_{1,j} = 0$, and for $N-8$ users $\sum_{j=1}^{N_1}C_{k,j-1}C_{1,j} \approx -1$, and for four sequences (with outer sequence number 2) $\sum_{j=1}^{N_1}C_{2,j-1}C_{1,j} \approx N_1$.

Case 2: Different a, Same C
$\sum_{j=1}^{N_1} C_{1,j}C_{1,j} = N_1$ and $\sum_{j=1}^{N_1} C_{1,j-1}C_{1,j} \approx -1$. For any choice of the inner codes, we will have $R_{1,k}(\tau_k) = \pm\tau_k$ for all k, and the interference power will be

$$Var(Z) = \frac{1}{2}\sum_{k=1}^{N-8} \frac{P_k}{P_1}\tau_k^2 + \frac{1}{2}\sum_{k=N-7}^{N-4} \tau_k^2 \frac{P_k}{P_1} N_1^2 + \frac{1}{2}\sum_{k=N-3}^{N-1} \frac{P_k}{P_1}\left(\tau_k^2 + N_1^2 \hat{R}_{1,k}^2(\tau_k)\right)$$

and the normalized variance will be

$$Var_n(Z) = \frac{1}{2}\sum_{k=1}^{N-8} \frac{P_k}{P_1}\frac{\tau_k^2}{T_c^2}\frac{1}{N^2} + \frac{1}{2}\sum_{k=N-7}^{N-4} \frac{\tau_k^2}{T_c^2}\frac{P_k}{P_1}\frac{N_1^2}{N^2}$$
$$+ \frac{1}{2}\sum_{k=N-3}^{N-1} \frac{P_k}{P_1}\left(\frac{\tau_k^2}{T_c^2}\frac{1}{N^2} + \frac{N_1^2}{N^2}\hat{R}_{1,k}^2(\tau_k)\right)$$

For all practical purposes, the above expression could be approximated for large N by

$$Var_n(Z) = \frac{1}{2}\sum_{k=N-7}^{N-4} \frac{\tau_k^2}{T_c^2}\frac{P_k}{P_1}\frac{1}{4^2}$$

2.3.5 Results of Time-Jitter Interference

The results are summarized in Table 2.4 for the case of signals with equal received power.

Similarly, for the case of signals with unequal received power, the resulting interference is shown in Table 2.5.

Figure 2.6-B shows the interference power for the QR and the Gold systems, assuming all the τ_ks are equal and all the P_ks are equal. For the proposed 1/2 QR system the interference power is shown in Figure 2.6-B. We can see the huge improvement in performance with half the number of users.

2.4 Interference Impact of Band-limited Pulse-Shapes

In previous sections we assumed that the pulse-shape of the code symbols or chips is time-limited between nT_c and $(n+1)T_c$, such as square time-pulses. This, however, requires infinite bandwidth. Practical systems use band-limited and not time-limited

Table 2.4 Variance of other-user interference $Var_n(Z)$ for Q-R and Preferred Phase Gold Codes.

Q-R Code Set	$\frac{1}{2}$ Q-R Code Set	Preferred Gold Code Set
$K = N$	$K = \frac{N}{2}$	$K = N$
$\frac{\tau_2^2}{2T_c^2} + \frac{1}{2N^2}\sum_{k=3}^{N}\frac{\tau_k^2}{T_c^2}$	$\frac{1}{2N^2}\sum_{k=2}^{\frac{N}{2}}\frac{\tau_k^2}{T_c^2}$	$\frac{1}{2}\frac{2}{N}\sum_{k=2}^{\frac{N}{2}}\frac{\tau_k^2}{T_c^2} + \frac{1}{2}\frac{1}{N^2}\sum_{k=\frac{N}{2}+1}^{N}\frac{\tau_k^2}{T_c^2}$

Table 2.5 Variance of other-user interference $Var_n(Z)$ for Q-R and Preferred Phase Gold Codes.

Q-R	$\frac{P_2}{P_1}\frac{\tau_2^2}{2T_c^2}$
$\frac{1}{2}$*Q-R*	$\frac{1}{2}\sum_{k=2}^{\frac{N}{2}}\frac{P_k}{P_1}\frac{\tau_k^2}{N^2T_c^2}$
Gold	$\frac{1}{2}\frac{2}{N}\sum_{k=2}^{\frac{N}{2}}\frac{P_k}{P_1}\frac{\tau_k^2}{T_c^2}+\frac{1}{2}\frac{1}{N^2}\sum_{k=\frac{N}{2}+1}^{N}\frac{P_k}{P_1}\frac{\tau_k^2}{T_c^2}$
Q-R outer, Walsh inner	$\frac{1}{2}\sum_{k=1}^{N-8}\frac{P_k}{P_1}\frac{\tau_k^2}{T_c^2}\frac{1}{N^2}$ $+\frac{1}{2}\sum_{k=N-7}^{N-4}\frac{\tau_k^2}{T_c^2}\frac{P_k}{P_1}\frac{N_1^2}{N^2}$ $+\frac{1}{2}\sum_{k=N-3}^{N-1}\frac{P_k}{P_1}\left(\frac{\tau_k^2}{T_c^2}\frac{1}{N^2}+\frac{N_1^2}{N^2}\hat{R}_{1,k}^2(\tau_k)\right)$

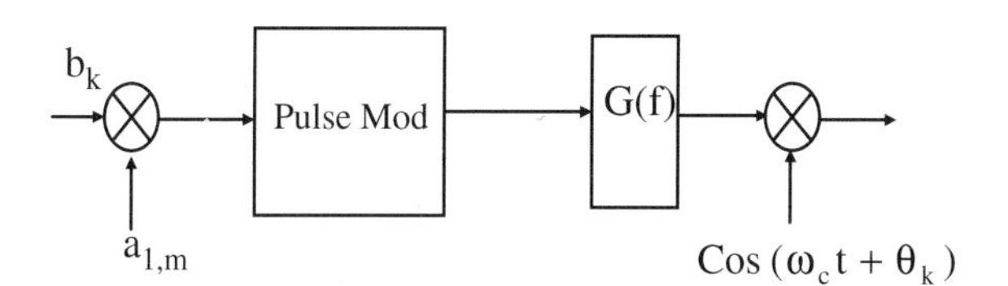

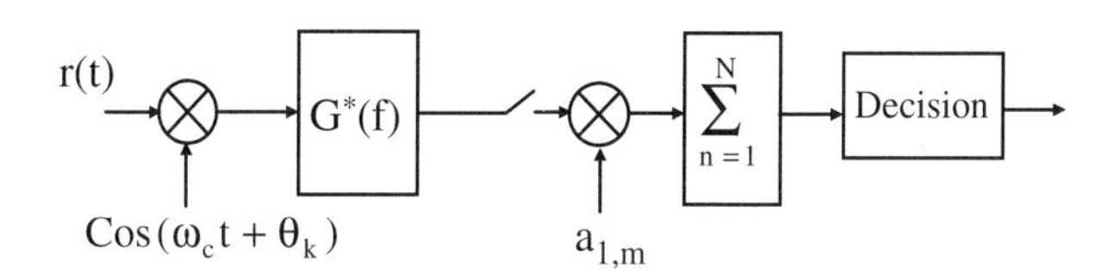

Figure 2.9 Block diagram of the band-limited system.

pulse-shape waveforms. In this section we investigate the effect of chip pulse-shape on orthogonality.

The data signal for the kth user is

$$s_k(t) = 2\sum_{m=-\infty}^{\infty} a_{k,m}b_k g(t - mT_c)\cos(w_c t + \theta_k)$$

where $a_{k,m+N} = a_{k,m}$. For the system shown in Figure 2.9, we define

$$H(f) = G(f)G^*(f)$$

As shown the receiver is a matched filter to the input chip pulse-shape. The output signal $y_k(n)$ will then have three components:

1. The desired output due to $a_{k,n}$.
2. The interchip interference (ICI) component, which depends only on $a_{k,n+m}$, $m \neq 0$.
3. The component due to other user interference, which depends on $a_{i,m}$, for all m.

We will neglect the effect of the data bit stream $b(t)$, because as shown for synchronous systems, its effect is negligible. Since this system is linear we could use superposition.

Let $y_1^s(n)$ denote the first component, assuming that the signal of interest is the signal of the first user. For $y(n)$, we sample at the output of the matched filter at time nT_c. Then, the convolution for the mth component evaluated at $t = nT_c$ will be

$$\int_{-\infty}^{\infty} g(\tau - mT_c)g(nT_c - \tau)d\tau = h((m-n)T_c)$$

The discrete time-response to the signals of the first user will be

$$y_1(n) = a_{1,n} \sum_{m=-\infty}^{\infty} a_{1,m}h((n-m)T_c).$$

Then, $y_1^s(n) = h(0)$.

Let $y_1^i(n)$ denote the interchip interference component. Then

$$y_1^i(n) = a_{1,n} \sum_{m=-\infty, m\neq n}^{\infty} a_{1,m}h((n-m)T_c)$$

Generally, $h(lT_c) \neq 0$, only for $|l| \leq M$, hence

$$y_1^i(n) = a_{1,n} \sum_{m=n-M, m\neq n}^{n+M} a_{1,m}h((n-m)T_c)$$

Now let us consider the other user interference, denoted by $y_1^o(n)$:

$$y_1^o(n) = a_{1,n} \sum_{i=2}^{N} \sum_{m-\infty}^{\infty} b_i a_{i,m} h((n-m)T_c)\cos(\phi_i)$$

The decision at the output of the receiver is based on $\sum_{n=1}^{N} y(n)$, which is defined as Z. The signal component of Z is $Nh(0)$. The output due to self-interference is

$$\begin{aligned}\sum_{n=1}^{N} \sum_{m=n-M, m\neq n}^{n+M} a_{1,n}a_{1,m}h((m-n)T_c) &= \sum_{n=1}^{N} \sum_{l=-M, l\neq 0}^{M} a_{1,n}a_{1,n+l}h(lT_c) \\ &= \sum_{l=-M, l\neq 0}^{M} h(lT_c) \sum_{n=1}^{N} a_{1,n}a_{1,l+n}\end{aligned}$$

For other users

$$\sum_{i=2}^{N}\sum_{n=1}^{N}\sum_{m=n-M}^{n+M} b_i a_{1,n} a_{i,m} h((n-m)T_c)\cos(\phi_i)$$
$$=\sum_{i=2}^{N}\sum_{l=-M}^{M} h(lT_c)\sum_{n=1}^{N} b_i a_{1,n} a_{i,l+n}\cos(\phi_i)$$

In the case of time-jitter, the difference is that the output due to other user interference will not be at lT_c instants of time, it will be at $lT_c+\tau_i$, where τ_i is the time shift between the first user and the ith user. The interchip interference will not depend on the shift, and the other user interference will be given by

$$\sum_{i=2}^{N}\sum_{n=1}^{N}\sum_{m=n-M}^{n+M} a_{1,n} a_{i,m} h((n-m)T_c+\tau_i)\cos(\phi_i)$$
$$=\sum_{i=2}^{N}\sum_{l=-M}^{M} h(lT_c+\tau_i)\sum_{n=1}^{N} a_{1,n} a_{i,l+n}\cos(\phi_i)$$

The pulses usually used are Nyquist pulses, defined in the frequency domain by

$$H(f)=\begin{cases} T, & 0\le \|f\|<\frac{1-\alpha}{2T} \\ \frac{T}{2}\left[1-\sin\left(\frac{2\pi\|f\|T-\pi}{2\alpha}\right)\right], & \frac{1-\alpha}{2T}\le\|f\|<\frac{1+\alpha}{2T} \\ 0, & \|f\|>\frac{1-\alpha}{2T} \end{cases}$$

The corresponding impulse response is

$$h(t)=\frac{\sin(\pi t/T}{\pi t/T}\frac{\cos(\alpha\pi t/T)}{(1-4\alpha^2t^2/T^2}$$

All of these signals have the property that $h(lT_c)=0$, for $l\neq 0$, therefore no interchip interference will occur, and the interference will only be from other users.

Orthogonal Code Sequences

Under complete synchronization conditions the other-user interference will be given by

$$\sum_{i=2}^{N} h(0)\sum_{n=1}^{N} a_{1,n}a_{i,n}$$

and *for orthogonal sequences, the above will be zero.* In general, when there is time-jitter, interference is

$$\sum_{i=2}^{N}\sum_{l=-M}^{M} h^2(lT_c+\tau_i)\ \frac{1}{2}\left[\sum_{n=1}^{N} a_{1,n}a_{i,n+l}\right]^2$$

For Q-R orthogonal sequences we find one sequence at each shift that will have the maximum cross-correlation that we described previously, also assuming the data will have equal probability of '1' and '−1', and we get

$$\frac{N^2}{2} \sum_{l=-M, l\neq 0}^{M} h^2(lT_c + \tau_l)$$

where τ_l is the delay of the code that will have maximum cross-correlation at shift l.

PN-Sequences

Following the same procedure as above and assuming completely synchronous systems, the output due to self-interference will be

$$\sum_{l=-M, l\neq 0}^{M} h(lT_c) \sum_{n=1}^{N} a_{1,n} a_{1,l+n}$$

which is zero for Nyquist pulses.

With other users present, the other-user interference will be given by

$$\sum_{i=2}^{N} h(0) \sum_{n=1}^{N} a_{1,n} b_i a_{i,n}$$

and the interference power will be given by

$$\sum_{i=2}^{N} h^2(0) \left[\sum_{n=1}^{N} a_{1,n} a_{i,n} \right]^2$$

For PN sequences the mean square value of $\sum_{n=1}^{N} a_{1,n} a_{i,n}$ is N. Therefore, the interference power is

$$h^2(0)N(N-1)$$

or more generally, when $K \neq N$ (partial loading)

$$h^2(0)N(K-1)$$

For the general case where there is time-jitter, the interchip interference power will be zero for Nyquist pulses and the other-user interference power will be given by

$$\sum_{i=2}^{N} \sum_{l=-M}^{M} h^2(lT_c + \tau_i) \frac{1}{2} N$$

2.5 Conclusion

In this chapter we have studied a wide variety of spreading sequences. Based on their cross-correlation values, code sequences are categorized as pseudo-orthogonal, quasi-orthogonal and orthogonal. Pseudo-orthogonal PN-sequences have been used in traditional asynchronous CDMA applications. Quasi-orthogonal sequences, such as the preferentially-phased Gold codes, can achieve lower maximum cross-correlation values than pseudo-orthogonal sequences, while they are more tolerant to timing jitter in synchronous CDMA applications than the orthogonal sequences. Orthogonal code sequences, on the other hand, always have zero cross-correlation, but they are required to have precise synchronization.

We have examined and presented a wide variety of methods to construct orthogonal binary Hadamard matrices. In particular, we have presented methods for the construction of Hadamard matrices of all possible length up to 256. Then we focused our attention on extended orthogonal sequences, which are based on Kronecker products orthogonal Hadamard matrices and are used in the design of composite orthogonal CDMA systems. We have also presented complex, polyphase orthogonal codes and other orthogonal code designs.

Next, we investigated the timing jitter properties of orthogonal and quasi-orthogonal sequences in synchronous CDMA applications. The impact of timing jitter in orthogonal sequences is greater than in quasi-orthogonal ones. For example, while the preferentially-phased Gold codes can tolerate a timing jitter of 50% of the chip length, quadratic residue orthogonal codes can only tolerate up to 10% of the chip length. Also, as we have shown, the power of interuser interference increases parabolically with the timing jitter, as a percentage of the chip length, in quadratic residue orthogonal sequences.

Finally, we investigated the impact of band-limited pulse-phape waveforms on interference, i.e. when the code-symbol or chip waveform is the result of frequency band-limit filtering. In this case we have evaluated the interchip and interuser interference when there is an equalizing matched filter at the receiver. The interference is evaluated when we have orthogonal or PN-sequences and under synchronous and asynchronous conditions. As we have shown, the interference power (both interchip and interuser interference) depends on the filter impulse response and the size of the system.

References

[1] D.V. Sarwate and M.B. Pursley 'Crosscorrelation Properties of Pseudorandom and Related Sequences' *Proceedings of the IEEE*, Vol. 68, No. 5, May 1980, pp. 593–619.

[2] R.E. Paley 'On Orthogonal Matrices' *Journal of Mathematics and Physics*, Vol. 12(1933), pp. 311–20.

[3] M. Hall 'A Survey of Difference Sets' *Proceeding of the American Mathematical Society*, Vol. 7 (1956), pp. 975–986.

[4] R. Gold 'Maximal Recursive Sequences with 3-valued Recursive Cross-Correlation Functions' *IEEE Trans. Information Theory*, January 1968, pp. 154–156.

[5] J.L. Massey and T. Mittelholzer 'Technical Assistance for the CDMA communication system analysis' Institute for Signal and Information Processing, ETH, Zurich, Switzerland, European Space Agency (ESA-ESTEC) Contract 8696/89/NL/US, May 1992.

[6] L.R. Welch 'Lower Bounds on the Maximum Cross Correlation of Signals' *IEEE Trans. Information Theory*, Vol. IT-20, May 1974, pp. 397–399.

[7] X.D. Lin and K.H. Chang 'Optimal PN Sequence Design for Quasisynchronous CDMA Communication System' *IEEE Trans. on Commun.*, Vol. 45, No. 2, February 1997, pp. 221–226.

[8] F.J. MacWilliams and N.J.A. Sloane 'The Theory of Error-Correcting Codes' North-Holland Mathematical Library, Vol. 16, 1977.

[9] J. Hadamard 'Resulation d'une question relative aux determinants' *Bull. Sciences Math.*, Vol. 17, (1893), pp. 240–246.

[10] A.M. Brauer 'On a New Class of Hadamard Determinants' *Mathematicsche Zeitschrifty*, Vol. 58 (1953), pp. 219–25.

[11] J.L. Walsh 'A Closed Set of Normal Orthogonal Functions' *Am. Journal of Mathematics*, Vol. 55 (1923), pp. 5–24.

[12] J.J. Sylvester 'Thoughts on inverse orthogonal matrices, simultaneous sign successions, and tesselated pavements in two or more colours' *Phil. Mag.* Vol. 34 (1867), pp. 461–475.

[13] J. Williamson 'Hadamard's Determinant Theorem and the Sum of 4 Squares' *Duke Journal of Mathematics*, Vol. 11 (1944), pp. 65–81.

[14] A.S. Hedayat and W.D. Wallis 'Hadamard matrices and their applications' *Ann. Statist.*, Vol. 6 (1978), pp. 1184–1238.

[15] D. Gerakoulis and S. Ghassemzadeh 'Extended Orthogonal Code Designs with Applications in CDMA' AT&T Labs-Research Technical Memorandum, HA1360000-990730-01TM.

[16] R.I. Turyn 'Complex Hadamard Matrices' in *Combinatorial Structures and Their Applications*'. Gordon & Breach, New York, 1970, pp. 435–437.

3

Switched CDMA Networks

3.1 Overview

Code Division Multiple Access (CDMA) has been widely accepted and used for wireless access in terrestrial and satellite applications. These applications often require switching of the CDMA traffic channels in order to establish connectivity between end users. In existing terrestrial wireless networks, while CDMA is used for access, connectivity and routing is achieved via the Public Switched Telephone Network (PSTN). It is often desirable, however, that access and switching is performed within the same network in many applications. An example of such an application is the Satellite Switched CDMA (SS/CDMA) system presented in [1]. The SS/CDMA network is comprised of a multibeam satellite and a large population of ground users, as illustrated in Figure 3.1-A. Ground users within each beam access the satellite by CDMA. The satellite is equipped with an on-board switch for routing inter- or intra-beam calls. The SS/CDMA network is described in detail in Section 3.2. Similar satellite systems based on TDMA, called Satellite Switched TDMA (SS/TDMA), are presented elsewhere [2], [3] and [4].

As in the satellite example, CDMA switching may also be used in terrestrial applications. These applications include wireless and cable networks that have CDMA as their access method. An example of such a network, called Base-station Switched CDMA (BS/CDMA), is illustrated in Figure 3.1-B. The BS/CDMA is comprised of a CDMA exchange node connected to a number of Radio Distribution Points (RDPs) via distribution lines which carry the CDMA signal. The exchange node in this case provides the switching capability for establishing connectivity between the wireless users. This wireless network may be used for fixed or mobile services. Similar systems based on TDMA have also been proposed (see [5] and [6]). Reference [5] presents a wireless TDMA switching system which provides connectivity between mobile users in a community of interest, while reference [6] presents another TDMA switching system for fixed service wireless metropolitan area networks. In addition to wireless applications, CDMA has been proposed for standardization in coax-cable networks for providing upstream voice, data and video services (see reference [7]). In this case, a switching CDMA device at the exchange node will provide an efficient mechanism for routing CDMA channels between cable users. Such an application is called Cable-Switched CDMA (CS/CDMA), and is illustrated in Figure 3.1-C. The above applications, both satellite and terrestrial, are refered to by the term switched CDMA (SW/CDMA) networks.

A. The satellite switched CDMA (SS/CDMA)

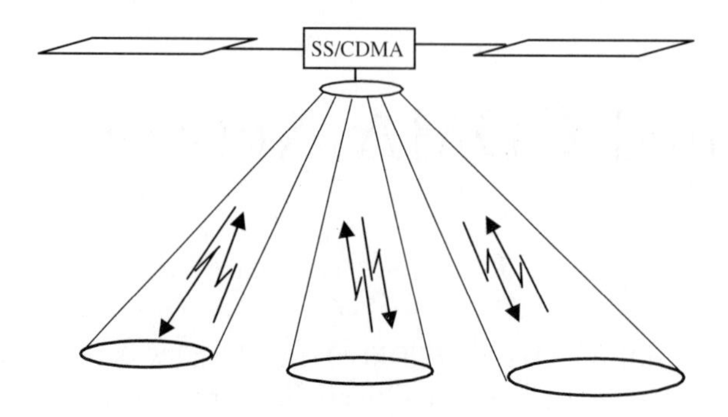

B. The base station switched CDMA (BS/CDMA)

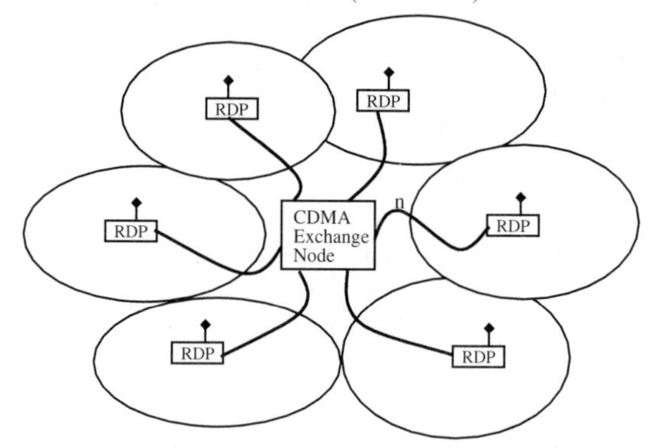

RDP: Radio Distribution Point

C. The cable switched CDMA (CS/CDMA)

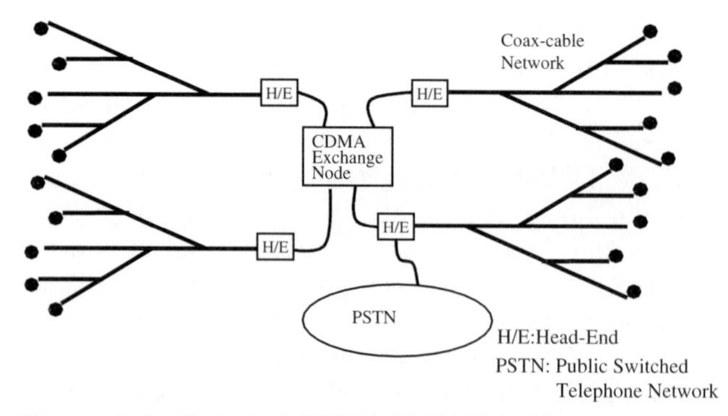

Figure 3.1 Switched CDMA (SW/CDMA) networks.

In this chapter we focus our attention on the satellite switched CDMA system. We present the network architecture, the access method and switching mechanism, and describe the design of its system units. We also examine the network operation and control algorithm.

3.2 Satellite Switched CDMA (SS/CDMA)

The service needs for future geostationary satellite systems demand direct two-way communication between end satellite users having Ultra Small Aperture Terminals (USAT) (antenna dish 26″ in diameter). The requirement for this type of service is the capability of call routing on-board the satellite. That is, the satellite will operate not only as a repeater, but also as a switching center in space. Such services, however, can only become economically feasible if the satellite communication capacity and throughput is sufficiently high while its service quality is comparable to the quality of wireline service. For this reason the system has to provide higher spectral efficiency, but also more efficient utilization of the available mass and power of the spacecraft. Higher spectral efficiency is achieved by using multibeam satellite antennas which allow resuse of the available spectrum. Also, the power needs of the transceiver units can be reduced by introducing new access and modulation methods operating at a very low signal-to-noise ratio in order to allow the use of USAT. Also, higher throughput can be achieved with a demand assignment control mechanism, which allows the distribution of system functionalities between the satellite and end users.

The system proposed to meet the above needs is the Satellite Switched Code Division Multiple Access (SS/CDMA). The SS/CDMA resolves both the multiple access and the satellite switching problems. The uplink access method is based on CDMA, the downlink on Code Division Multiplexing (CDM) and the on-board switching on compatible technology which is also code division (CDS). The system operates with demand assignment control for both access and switching. That is, service bandwidth and switch connections are assigned only upon a user request. The SS/CDMA can achieve higher spectral efficiency by allowing frequency reuse, i.e. reuse of the available spectrum in every beam of a multibeam satellite. In addition, it provides an efficient switching mechanism by establishing a direct end-to-end route with minimal on-board signal processing and no on-board buffering. The access and switching problems are resolved in one step by the demand assignment control mechanism. This approach also allows system optimization by using an assignment control algorithm to maximize throughput and to integrate the traffic of circuit calls and data packets. A large population of end users may then access the geostationary satellite which provides the routing of calls and packets between them. The system may offer fixed services for circuit switched calls (voice, data and video) and packet switched data.

A related method based on Time Division Multiple Access (TDMA), called Satellite Switched TDMA (SS/TDMA), has been proposed in the past for packet switched data services, [2], [3]. In SS/TDMA the access method is TDMA and the switching is based on time multiplexing (TMS). A similar TDMA demand assignment system is also used in the ACTS satellite for low burst rate traffic [4]. The TDMA approach, however, requires frequency reuse of 1/4 or 1/7 (depending on the beamwidth), while its switch implementation and algorithm control may be more complex for large switch sizes.

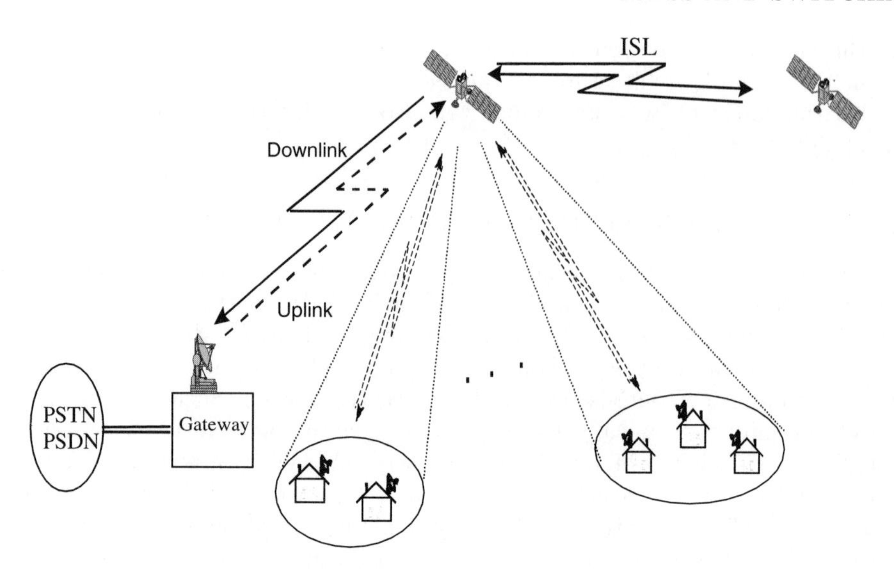

Figure 3.2 The Satellite Switched CDMA (SS/CDMA).

The SS/CDMA system has been developed for AT&T's VoiceSpan satellite project and Ka-band application filling (the VoiceSpan project has not been realized). In the following section we present the system description, in Section 3.2.2 the satellite switching mechanism, in Section 3.2.3 the description of transmitter and the receiver units, and in Section 3.2.4 the network operation and control.

3.2.1 System Description

The Satellite Switched Code Division Multiple Access (SS/CDMA) is the underlying communication system proposed for a network of satellites. This network is comprised of the space segment containing a number of geostationary satellites and the ground segment containing the Customer Premises Equipment (CPE) and gateway offices to the Public Switched Telephone and Data Networks (PSTN and PSDN). The geostationary satellites are equipped with multibeam antennas, on-board processing and switching for providing fixed service communications. The network configuration is shown in Figure 3.2.

Transmission Rates and Services

The main objective of the satellite network is to provide services with a direct connection to each subscriber. The services offered are both circuit switched and packet switched. The circuit switched services are for voice, video and data, while the packet switched services are only for data. The transmission bit rates, the source bit rates and the quality of each circuit switched service are shown in Table 3.1. The transmission rate in each channel type includes a source rate, a subrate, framing bits, and a frame quality indicator (CRC). The offered rates for voice services are: 16, 32, and 64 Kbps;

Table 3.1 Transmission and source bit rates and the corresponding services.

Channel Type	Source Rate(kb/s)	Transmiss. Rate(kb/s)	Service Offer	Required BER
I	64	76.8	Voice/Data	10^{-6}
II	32	38.4	Voice/Data	10^{-6}
III	16	19.2	Voice/Data	10^{-6}
IV	144	153.6	ISDN(2B+D)	10^{-6}
V	384	460.8	Video	10^{-8}
VI	1544	2304	T1	10^{-8}
VII	2048	2304	E1	10^{-8}

while the offered rates for data are: 16, 64 and basic ISDN 144 kbps (2B+D). The system also offers video services with rate of 384 Kbps and 4.608 Mbps, and T1 or (E1) carriers with rates of 1544 (or 2048) Kbps. Each transmission rate is the result of multiplexing the source data with the frame quality indicator, signaling data and/or other information data. Each channel Type (I, II) corresponds to a required Bit Error Rate (BER). The SS/CDMA system will also offer packet switched services for bursty data.

Multiple Access

The SS/CDMA provides both multiple access and switching to the multibeam satellite. The multiple access problem is resolved by space, frequency and code division. The space division multiple access is achieved by multibeam antennas in order to reuse the available spectrum in each beam. The frequency division multiple access is achieved by segmenting the available spectrum into frequency bands, each having a convenient size of 10 MHz (see Figure 3.3). The Code Division Multiple Access (CDMA) will then provide access for each user within each frequency band and in each beam. The CDMA will spread the user data over the bandwidth of 10 MHz.

The satellite also performs the switch function. That is, user traffic channels will be switched from any uplink to any downlink beam. This is done with an on-board code division switch which performs the switching of the CDMA codes (identifying traffic channels) from any uplink CDMA channel in beam-i to any downlink CDMA channel in beam-j. The SS/CDMA system architecture, shown in the block diagram of Figure 3.4, is comprised of a satellite and the Customer Premises Equipment (CPE). The CPE contains the Subscriber Unit (SU) and the Terminal Equipment (TE). Each SU is comprised of the Transceiver Unit (TU) and the Call Control Unit (CCU). The

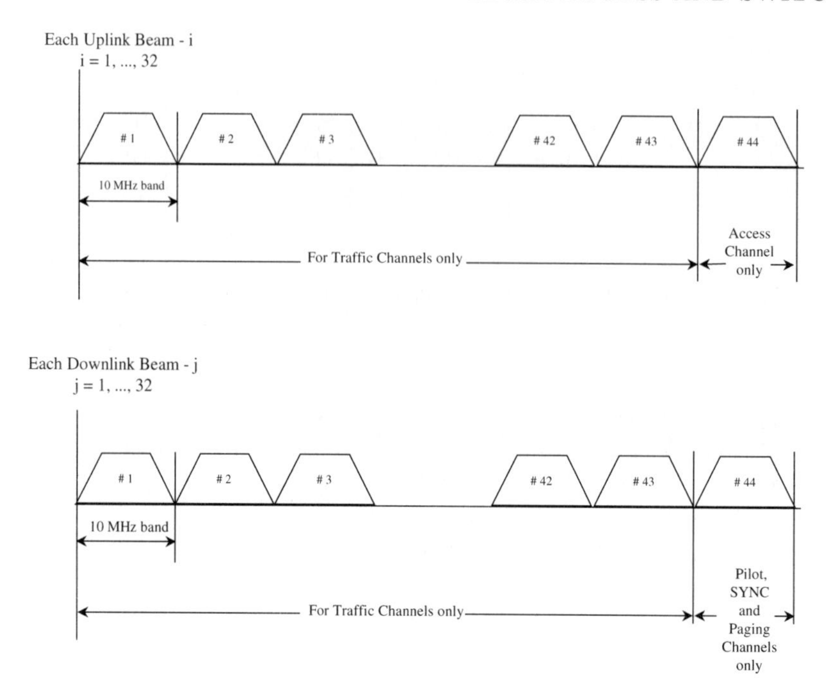

Figure 3.3 Frequency band assignments for the SS/CDMA.

TU includes the transmitter units for the Access and the Traffic channels (ACTU and TCTU) on the uplink and the receiver units for Synchronization and Paging (S&PRU) as well as Traffic channels (TCRU) on the downlink. The on-board system architecture has as its basic functional blocks the Code Division Switch (CDS), the Control Unit (CU) and the receiver and transmitter for the Access (ACRU) and Satellite Broadcast channels (SBTU).

Common Air Interface

The Common Air Interface (CAI) is defined as the interface between the space and the earth segments of the system, i.e. between the satellite and the subscriber units or gateway offices. The CAI provides the Control and the Traffic channels. The Control channels are: the Access in the uplink, and the Pilot, SYNC and Paging in the downlink. These channels operate on an assigned frequency band (see Figure 3.3). The Pilot and the SYNC provide timing and synchronization to the system while the Access and Paging channels deliver signaling messages to and from the satellite. The Traffic channels, on the other hand, carry voice, data and signaling information between the end subscriber units. The multiple access and modulation of the Traffic Channel is based on the Spectrally Efficient Code Division Multiple Access (SE-CDMA) scheme presented in Chapter 6. The SE-CDMA provides orthogonal separation of Traffic channels within each beam, as well as between beams. On-board the satellite, the Traffic channels are simply switched from an uplink to a downlink beam without any data decoding or buffering.

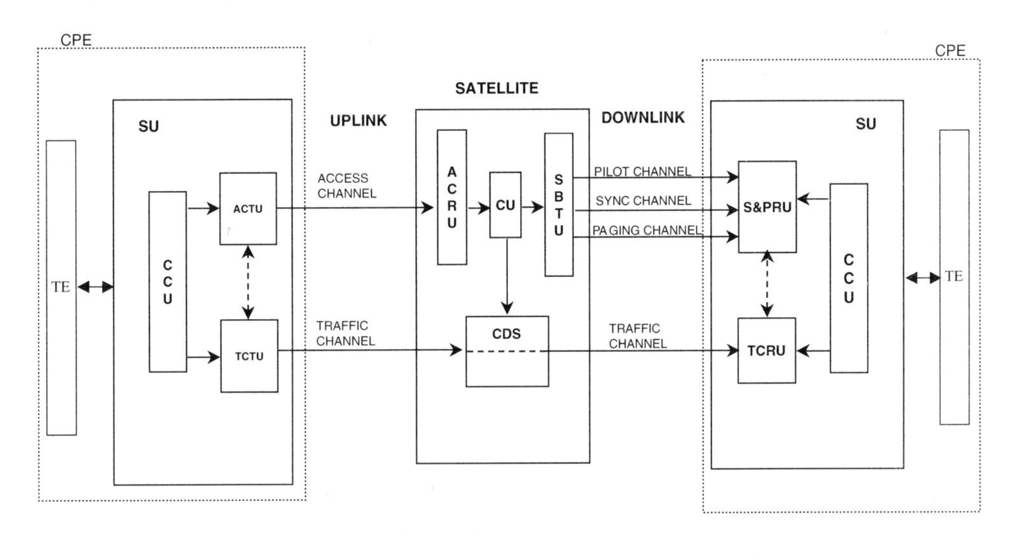

ACRU:	Access Channel Receiver Unit	SBTU:	Satellite Broadcast Transmitter Unit
ACTU:	Access Channel Transmitter Unit	S&PRU:	SYNC & Paging Receiver Unit
CCU:	Call Control Unit	SU:	Subscriber Unit
CDS:	Code Division Switch	TCRU:	Traffic Channel Receiver Unit
CU:	Control Unit	TCTU:	Traffic Channel Transmitter Unit
CPE:	Customer Premises Equipment	TE:	Terminal Equipment

Figure 3.4 The SS/CDMA system architecture.

3.2.2 Satellite Switching

The SS/CDMA system has an on-board switching mechanism which routes the Traffic channel data from any uplink beam-i to any downlink beam-j. The on-board system architecture provides the Access Channel Receiver Unit (ACRU), the Satellite Broadcast Transmitter Unit (SBTU), the Control Unit (CU) and the Code Division Switch (CDS), as shown in Figure 3.4. The ACRU and SBTU handle the signaling messages to or from the CU, while the CDS routes the Traffic channels. The satellite switching system design is based on code division technology, while its operation is based on the Demand Assignment method.

Code Division Switch

Code Division Switching allows the implementation of a nonblocking switch fabric of low complexity (linear to the size of the switch) without any channel decoding/encoding or buffering on-board, while it maintains compatibility with the SE-CDMA Common Air Interface (CAI). The proposed switching system consists of Code Division Switch (CDS) modules. Each CDS module routes calls between N uplink and N downlink beams, where each beam contains of a single frequency band W ($W = 10$ MHz). The size of the CDS module then is $(NL \times NL)$, where L is the number of Traffic channels in the SE-CDMA band. (In a particular implementation, $N = 32$ and $L \leq 60$.) The basic design idea in a CDS module is to combine the input port Traffic channels into a bus by spreading them with the orthogonal code of their destination port. This bus is called a Code Division Bus (CDB). All Traffic channels in the CDB are orthogonally separated, and can be routed to the destination output by despreading with the orthogonal code of the particular output port. The detailed system architectures of the CDS modules are presented in Chapter 4. The CDS fabric has been shown to be a nonblocking switch fabric. Also, routing via the CDS fabric will cause no additional interference to the Traffic channels other than the interference introduced at the input satellite link. A complexity analysis and performance assessment of the CDS is also presented in Chapter 4.

Demand Assignment Control

The demand assignment process provides access and switching to the Subscriber Unit (SU) in the SS/CDMA system. That is, the CDMA frequency band and Traffic channel allocations for circuit or packet switched services are made upon a user request. Message requests and assignments are sent via the signaling control channels (Access in the uplink and Paging in the downlink), while the information data are transmitted via the Traffic channels. The demand assignment approach allows the establishment of a direct route between the end SUs via the Code Division Switch (CDS) without any buffering or header processing on board the satellite. It also allows dynamic sharing of system resources for different services while maximizing the system throughput. A basic description of the Demand Assignment Control process is the following: each SU initiates a call by sending a message request to the on-board Control Unit (CU) via the Access channel. The CU will assign (if available) a Traffic channel for the duration of the call by allocating uplink–downlink frequency bands and CDMA codes identifying the Traffic channel. The CU will then send the assigned Traffic channel information to the end SUs via the Paging channels, while the switch makes the appropriate connection for it. The end-SU will then begin transmitting on this channel. A detailed description of this process is given in Section 3.2.4.

As described above, the switching system consists of CDS modules. Each CDS module performs intra-band switching by routing the traffic between beams within a single pair of uplink and downlink frequency bands. There is a number of uplink–downlink pairs of frequency bands allocated for Traffic channels (see Figure 3.3), and an equal number of CDS modules corresponding to these pairs. The demand assignment algorithm will also be used to handle the inter-module or inter-band routing of traffic. This is done by the following procedure:
upon the arrival of a call, the SU sends a message request via the Access channel to the on-board Control Unit which assigns an uplink–downlink pair of frequency bands and sends back the assignment data via the Paging channel to the SUs. The SUs then tune up on the assigned frequency bands and use the corresponding CDS module to switch its traffic. The frequency bands for the Access and Paging channels are pre-assigned to each SU. Also, this approach requires that each SU is capable of tuning its transceivers (TCTU and TCRU) to the assigned RF frequency upon arrival of a call. (No frequency band assignment can be made to TCTU and TCRU before any call request.)

The proposed method of frequency band assignments for inter-module routing avoids the need for additional hardware on board the satellite, while providing a balance of the traffic load among the available frequency bands. The number of CDS modules will be equal to the number of uplink or downlink frequency bands. For reliability purposes, a spare module is added for use in case one fails. Also, the demand assignment algorithm will further optimize system performance by extending the size of the Traffic channel pool beyond the single frequency band.

In addition, the demand assignment operation is utilized to integrate circuit and packet switched services, and maximizes the utilization of the available switching resources. The proposed method is based on the *Movable Boundary*, and is described as follows.

Given a pool of K orthogonal Traffic channels, K_c out of K will be allocated for circuit switched calls and K_p for packet switched data. Then $K = K_c + K_p$. (The total number of Traffic channels K is $K = qL$, where q is the number of frequency bands and L is the number of Traffic channels per frequency band.) Any unused circuit traffic channel may be assigned momentarily for packets. Traffic channels allocated for packet services are not assigned for circuits.

Let k_c be the number of active circuit calls and k_p the number of packets in transmission at a given time instant, then the Traffic channel assignment rules will be based on conditions (a) $k_c \leq K_c$, and (b) $k_c + k_p \leq K_c$. Condition (a) indicates that no more than K_c circuit calls may be routed to any uplink beam i and downlink beam j. Similarly, condition (b) indicates that the total number of circuits and packets admitted in the uplink beam-i and downlink beam-j, respectively, cannot exceed the beam capacity K. If condition (a) does not hold true after the arrival of any new circuit call, the call will be blocked. Similarly, if condition (b) does not hold true after the arrival of a new data packet, the packet will remain buffered in the SU. Given conditions (a) and (b), scheduling algorithms have been designed to maximize the switch throughput (see Chapter 5).

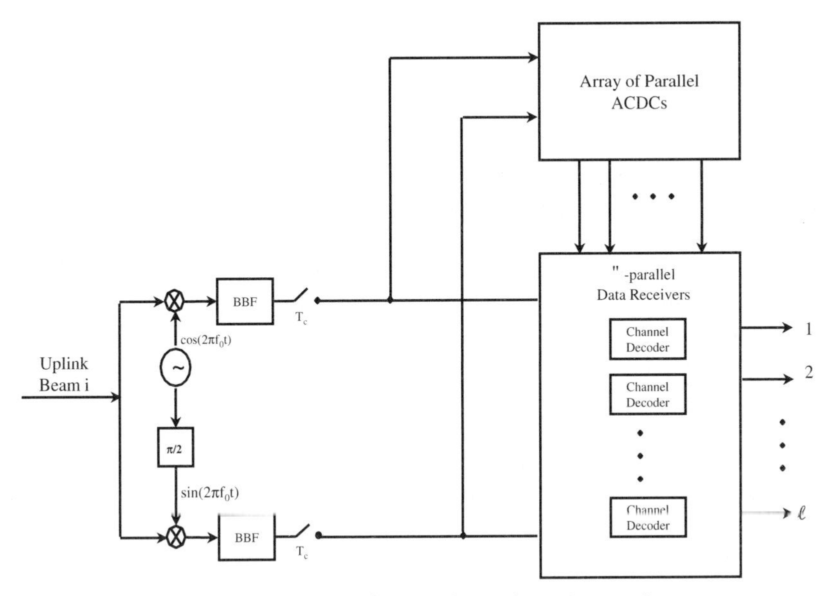

Figure 3.5 The Access channel receiver unit.

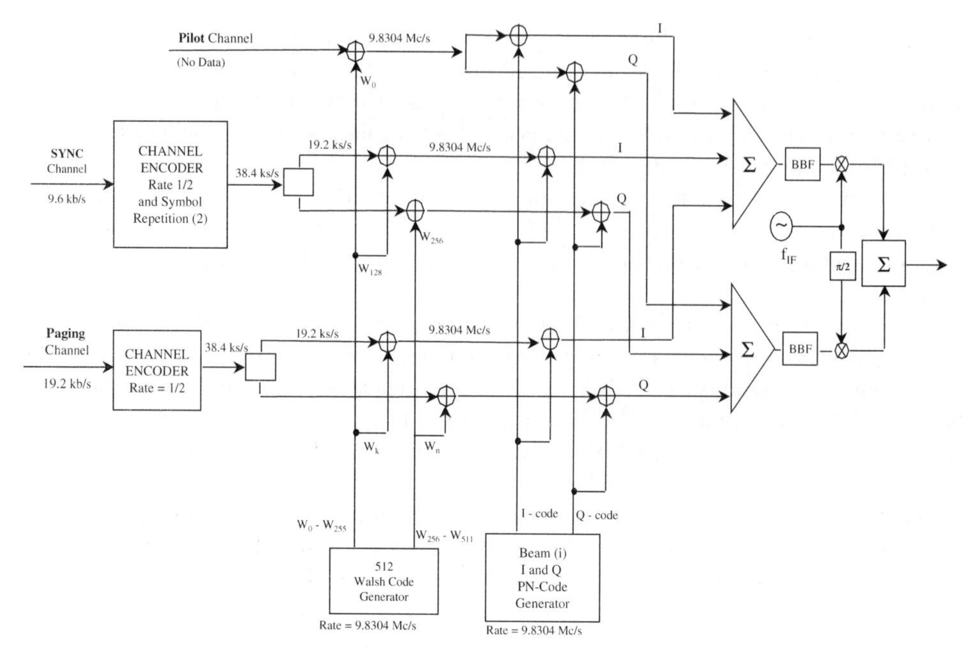

Figure 3.6 The satellite broadcast transmitter unit.

3.2.3 Transmitter and Receiver Units

Access Channel

The Access channel operates on the assigned uplink frequency band or bands. The basic structure of the Access Channel Transmitter Unit (ACTU) provides a channel encoder followed by the spreader and a quadrature modulator. The channel encoder has a rate 1/2 and may be convolutional or turbo. Data are then spread by a PN code g_i. The PN codes g_i have a length of L ($L = 2^{10} - 1$) chips. The spreading chip rate is $R_c (R_c = 9.8304$ Mc/s), and the CDMA channel nominal bandwidth is W ($W \approx 10$ MHz).

Transmissions over the Access channel obey the *Spread Spectrum Random Access* (SSRA) protocol. The SSRA protocol assumes that the Access channel transmissions are Asynchronous or Unslotted. According to SSRA protocol, there is a unique PN code $g_i(t)$ assigned to each beam i. Since each ACTU may begin its transmission randomly at any time instant (continuous time), the phase offset of the PN code at the receiver i.e. $g_i(t - nT_c)$. On the receiver side there will be a set of parallel Access Channel Detection Circuits (ACDC) in order to detect and despread the arrived signal at any phase offset. Signals that arrive at the receiver with a phase offset of more than one chip will be distinguished and received. Unsuccessful message transmissions will be retransmitted after a random delay, while messages that are successfully received will be acknowledged. All responses to the accesses made on an Access channel will be received on a corresponding Paging channel. A detailed description of the SSRA protocol and its throughtput performance is presented in Chapter 7.

The Access channel message has a preamble and an information data field. The preamble contains no data and is used to aquire the phase offset of its PN-code. Depending on the number of parallel ACDCs on the receiver, the preample length will vary, but will not exceed τ_{aq} ($\tau_{aq} \leq 5$ msec). The Access channel, in addition of delivering access messages, will also be used for synchronization of the Traffic channel. On board the satellite is the Access Channel Receiver Unit (ACRU), shown in Figure 3.5. The ACRU consists of a noncoherent demodulator, an array of parallel Access Channel Detection Circuits (ACDC) and a pool of k data decoders. The array of parallel ACDCs provide a combination of parallel with serial aquisition circuits. Each ACDC searches for synchronization of the message by correlating over a window of w chips. Given L chips the length of the PN code g_i, and k the number of ACDCs, the window size will then be, $w = L/K$. (For example, if $L = 1094$ chips and $P = 16$, then $w = 64$ chips.) The correlation process takes place during the message preamble using the serial search (double dwell) approach. The design parameters of the serial search circuit (such as the lengths of the dwell times and the corresponding thresholds) are determined so that it meets the requirement for the false alarm and detection probabilities. This analysis is presented in Chapter 7.

Considering the long round trip satellite propagation delay (200 ms), the main performance requirement of the Access channel is to provide a high probability of success at the first transmission attempt. The probability of message success depends (a) on the successful PN-code acquisition during the message preample, (b) on the probability of collision, and (c) on the probability of no bit errors in the message after channel decoding. The design requirement for successful aquisition with a probability of $(1-10^{-4})$ or higher is to have the preamble length two standard deviations above the mean aquisition time. The successful retention of the message (no bit errors) requires that the message has an optimum length. In Chapter 7 we also provide an estimate of the optimum mumber of ACRU receivers given the total number of Traffic channels in the system.

Satellite Broadcast Channels

The satellite broadcast channels have assigned downlink frequency bands of bandwidth W ($W \approx 10$ MHz) in each beam. Figure 3.6 shows the basic structure of the Satellite Broadcast Transmitter Unit (SBTU). Each SBTU transmits one Pilot, one Sync and a number of Paging channels. Each broadcast channel is identified by two orthogonal codes (W_k for I and W_n for Q components) and a beam PN-code g_j. The I and Q components have different orthogonal and PN-codes. All channels within a beam are 'orthogonally' separated, while the beams are separated only by PN-codes (semi-orthogonal implementation). After spreading, the satellite broadcast channels are digitally combined, then modulated and filtered. The Pilot channel is transmitting at all times and contains no data. Each satellite beam is identified by the Pilot's PN sequence. The Sync channel transmits system information for synchronizing and receiving a Paging channel or transmitting on an Access channel. The Paging channel is used by the satellite for transmitting paging information and for responding to Access channel requests.

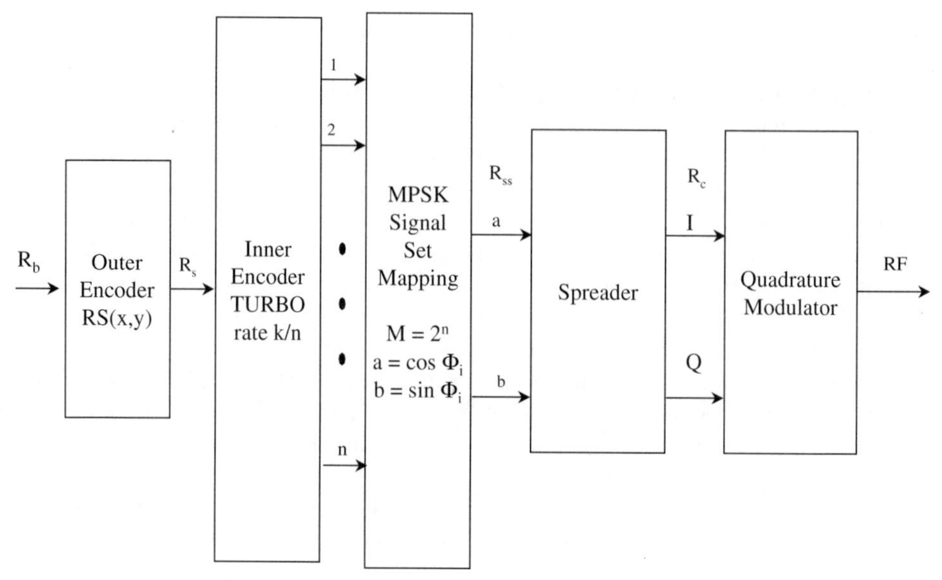

Figure 3.7 The SE-CDMA modulation process.

Traffic Channels

The Traffic channels provide a direct connection between the end subscriber units. Their paths consists of three segments: the uplink, the switching and the downlink. The Traffic channel multiple access and modulation procedures are based on the *Spectrally Efficient Code Division Multiple Access* (SE-CDMA) scheme. The SE-CDMA scheme has the following characteristics:

1. It is an orthogonal CDMA scheme which utilizes an optimized concatenation of error correcting codes and bandwidth efficient modulation. The orthogonal code of length L chips will span over the entire length of a symbol.
2. The concatenated codes are: Reed–Solomon $RS(x, y)$ with a rate x/y as the outer code and Turbo with a rate of k/n as the inner code. (Turbo codes we refer to a general class of codes that use serial or parallel concatenation of convolutional codes linked by an interleaf. One such class uses two parallel recursive systematic convolutional codes linked by an interleaver.) The input bits, after framing, first enter the Reed–Solomon code, then the Turbo code, and are then spread and modulated using M-ary Phase Shift Keying (M-PSK) ($M = 2^n$) (see Figure 3.7). The spreading of the orthogonal sequence will span over the length of the M-ary symbol at the input of the spreader.
3. The SE-CDMA provides orthogonal separation of all Traffic channels within the CDMA bandwidth W ($W \approx 10$ MHz). This is achieved by assigning orthogonal codes to each Traffic channel. In addition, orthogonal and/or PN-codes are used for separating the satellite beams (beam codes).
4. The SE-CDMA can be implemented as Fully Orthogonal (FO), Mostly Orthogonal (MO) or Semi-Orthogonal (SO). All of these implementations provide orthogonal separation of all of the Traffic channels within each beam. In addition, the FO/SE-CDMA provides orthogonal separation of the first

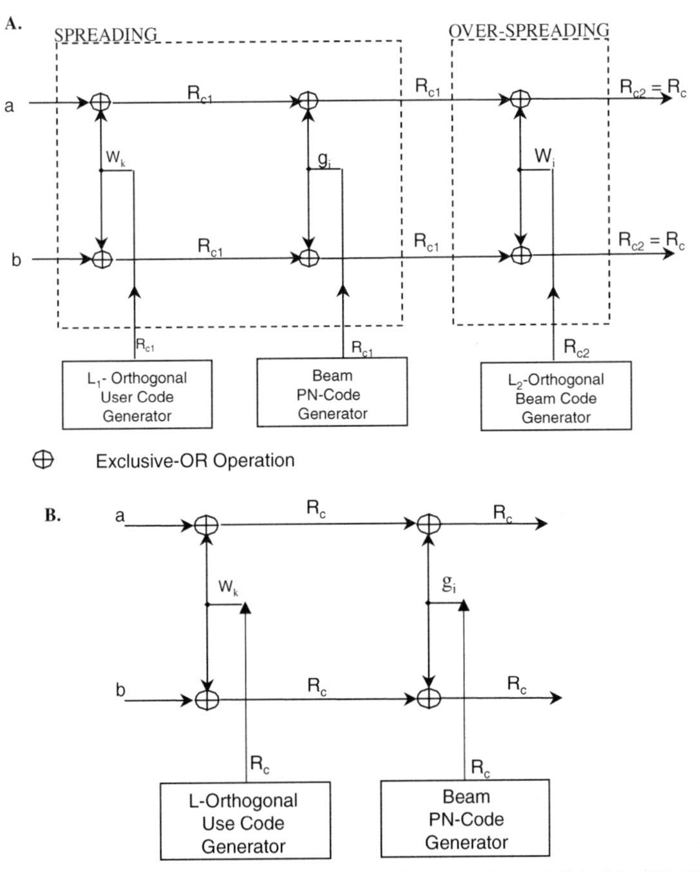

Figure 3.8 The spreading operation for (a) FO, MO and (b) SO, SE-CDMA.

tier of the satellite beams (four beams). The MO/SE-CDMA has two beams in the first orthogonal tier, while the SO/SE-CDMA has all of its beams separated by PN-codes.

5. The spreading operation for the FO and MO SE-CDMA is shown in Figure 3.8-A, while for the SO/SE-COMA it is shown in Figure 3.8-B. Spreading takes place in two steps, the first at a rate R_{c1} and the second at a rate R_{c2} (overspreading). The FO/SE-CDMA has $R_{c2} = 4 \times R_{c1}$, while the MO/SE-CDMA has $R_{c2} = 2 \times R_{c1}$. Also, an (I, Q) PN-code generator is used to isolate the interference from the second tier of beams. Its rate is R_{c1}. In the SO/SE-COMA, spreading has an orthogonal user code and a PN-beam code. In this case, the satellite beams are only separated by the PN-code, which has the same rate R_c as the orthogonal code.
6. The SE-CDMA scheme requires synchronization. That is, the codes from all users must be perfectly aligned at the satellite despreaders.

The SE-CDMA has been designed to optimize the SS/CDMA system performance. That is, to maximize the system capacity and spectral efficiency, while achieving very low E_b/N_o (3 – 5 dB) at a very low bit error rate (10^{-6} to 10^{-10}). The intra-beam or

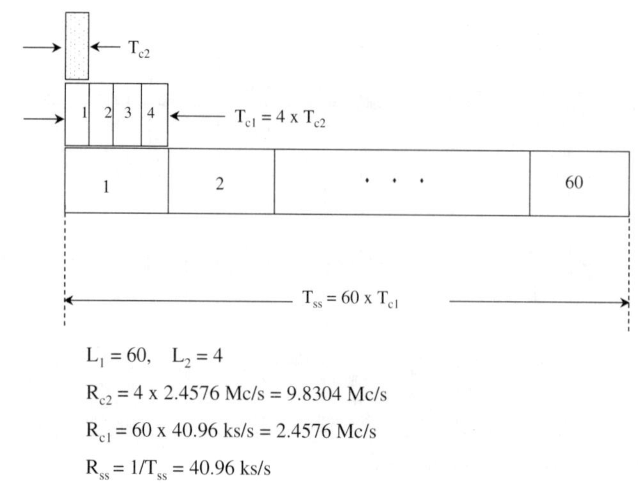

Figure 3.9 The spreading and overspreading operations for FO/SE-CDMA.

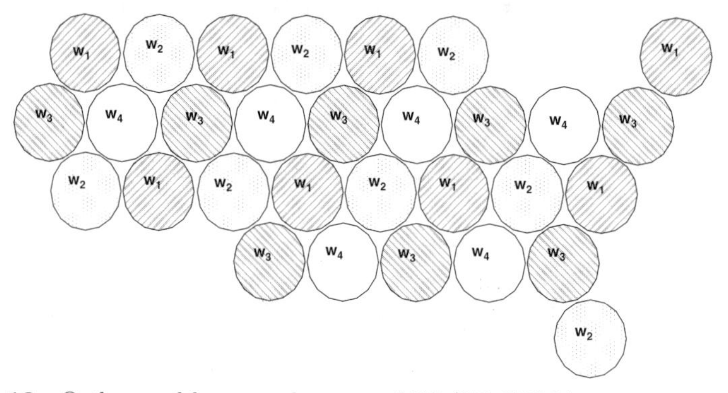

Figure 3.10 Orthogonal beam-code reuse of FO/SE-CDMA over continental USA.

other user interference is eliminated with the use of orthogonal codes. The other-beam interference is minimized by isolating beams with orthogonal and/or PN-codes, while providing a frequency reuse beam. The concatenation of Reed–Solomon and Turbo codes is optimized in order to provide a very low BER, required for better service quality, and very low E_b/N_o to allow a sufficient margin to mitigate the Output Back-Off (OBO) problem at the power amplifier (TWT). (The concatenated of the RS-Turbo scheme is used for rejecting the noise floor that appears at BERs of 10^{-4} or 10^{-5} in Turbo codes: see Chapter 6.) The modulation load (QPSK, 8-PSK) is optimally chosen so that it maximizes the spectral efficiency, and also succeeds in achieving a very low BER at low E_b/N_o.

The generalized block diagram of the SE-CDMA is shown in Figure 3.7. The system parameters of each implementation (FO, MO and SO) are given in Table 3.2, and the system bit, symbol and chip rates in Table 3.3. The choice of the specific SE-CDMA implementation will be based on the service type and the required BER-E_b/N_o. The SE-CDMA utilizes Aid Symbols for nearly coherent detection, as described in Chapter 8. After framing, the bit stream enters the Reed–Solomon (RS) outer code RS(x,y) (rate y/x), resulting in a symbol rate R_s.

Table 3.2 SE-CDMA selected implementations.

SE-CDMA IMPLEM.	OUTER ENCODER	INNER ENCODER	MPSK SCHEME	BEAM CODE REUSE
FO-1	$RS(16\lambda, 15\lambda)$	Turbo, 2/3	8-PSK	1/4
MO-1	$RS(16\lambda, 15\lambda)$	Turbo, 1/2	QPSK	1/2
SO-1	$RS(16\lambda, 15\lambda)$	Turbo, 1/3	QPSK	1

Table 3.3 Bit, symbol and chip rates for each SE-CDMA implementation.

RATE	FO-1	MO-1	SO-1
R (kb/s)	64.0	64.0	64.0
R_b(kb/s)	76.8	76.8	76.8
R_s(ks/s)	81.92	81.92	81.92
R_{ss}(ks/s)	40.69	81.92	122.88
R_{c1}(Mc/s)	2.4576	4.9152	$R_{c1} = R_{c2}$
$R_c = R_{c2}$(Mc/s)	9.8304	9.8304	9.8304
R_{c1}/R_{ss}	60.0	60.0	80.0
R_{c2}/R_{c1}	4.0	2.0	1.0

Following the outer RS encoder is the inner Turbo encoder with a rate of k/n. The Turbo code rates for FO-1, MO-1 and SO-1 SE-CDMA are 2/3, 1/2 and 1/3, respectively. The Turbo encoder output generates n (parallel) symbols which are mapped into the M-ary PSK signal set $M = 2^n$. The MO and SO/SE-CDMA use QPSK, while the FO/SE-CDMA uses 8-PSK. The signal phases ϕ_i ($i = 1, 2, \ldots$) are then mapped into the inphase and quadrature components (a, b). $\phi_i \rightarrow (a, b)$ ($a = \cos\phi_i$, $b = \sin\phi_i$). The modulated signal will then spread over a bandwidth W ($W \approx 10$ MHz).

The spreading operations for the FO and MO SE-CDMA are shown in Figure 3.8-A and for the SO/SE-CDMA in Figure 3.8-B. The inphase and quadrature components are spread by the same orthogonal and PN codes. The chipping rates for the FO and MO implementations are shown in Table 3.3. The FO spreading and overspreading rates are illustrated in Figure 3.9. FO implementation requires 60 orthogonal codes for user Traffic channels having a chip rate of $R_{c1} = 2.4576$ Mc/s. Then, overspreading by a factor of 4 will raise the chip rate to $R_c = 9.8304$ Mc/s. The overspreading will provide four orthogonal codes for separating the satellite beams. Figure 3.10 shows the re-use patern of the four orthogonal beam codes. The resulting pattern has all beams orthogonal in the first tier, while in the second tier beams are separated

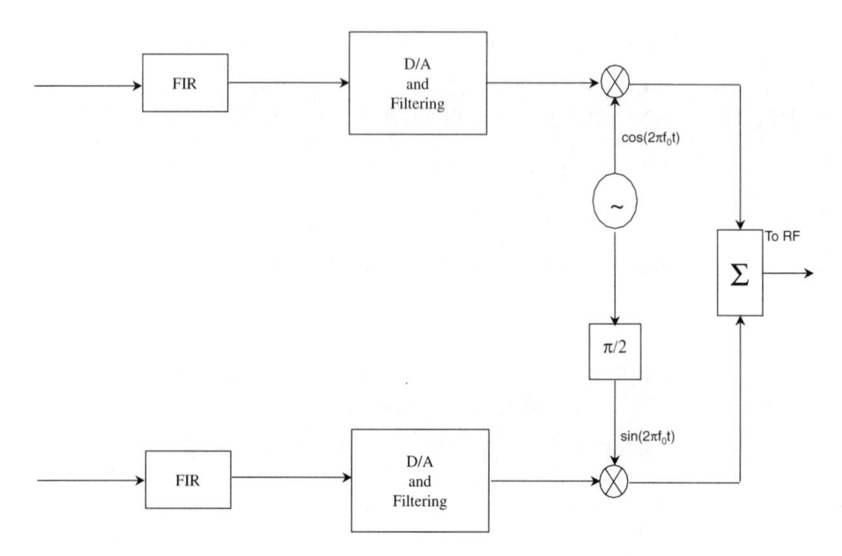

Figure 3.11 The baseband quadradure modulator.

by PN-codes. The MO implementation requires 60 orthogonal codes for user Traffic channels which spread at a rate of $R_c = 4.9152$ Mc/s. The overspreading will provide two orthogonal codes for beam isolation. In addition, cross-polarization may be used in this case for further reduction of the other beam interference. In the resulting pattern four out of six beams in the first tier are orthogonally isolated and two by cross-polarization and PN-codes only. In the SO/SE-CDMA the spreading operation consists of the user orthogonal code and the beam I and Q PN-codes. All codes have the same rate $R_c = 9.8304$ Mc/s. Beams are only separated by PN-codes. For additional protection against other beams' interference, cross-polarization is also needed in this case. (Orthogonal codes will be generated using Hadamard–Walsh if the required length is $L = 2^k$. If L is not a power of 2 then we use the Quadratic Residue method, or any other method presented in Chapter 2.) Following the spreading operation, the resulting I and Q waveforms will be band-limited by a digital FIR filter. The digital FIR filter can be a raised cosine filter with a roll-off factor of 0.15 to 0.2. following the digital filter, the signal is converted into analog form and modulated by a quadratic modulator, as shown in Figure 3.11. The resulting IF signal bandwidth will be about 10 MHz. The Traffic Channel Receiver Unit (TCRU) contains a quadrature demodulator, a despreader and a channel decoder. The despreading operation for the FO and MO SE-CDMA is shown in Figure 3.12-A and for the SO SE-CDMA in Figure 3.12-B. The channel decoding for the Reed–Solomon and Turbo codes will only take place at the Subscriber's Unit (SU).

Synchronization and Timing

The SE-CDMA is a synchronous CDMA system. All uplink traffic channels are required to arrive synchronously at the satellite despreaders in order to maintain the orthogonality between those channels within the same beam, as well as between those in other beams. That is, the starting time of all beam PN-codes $g_i(t)$

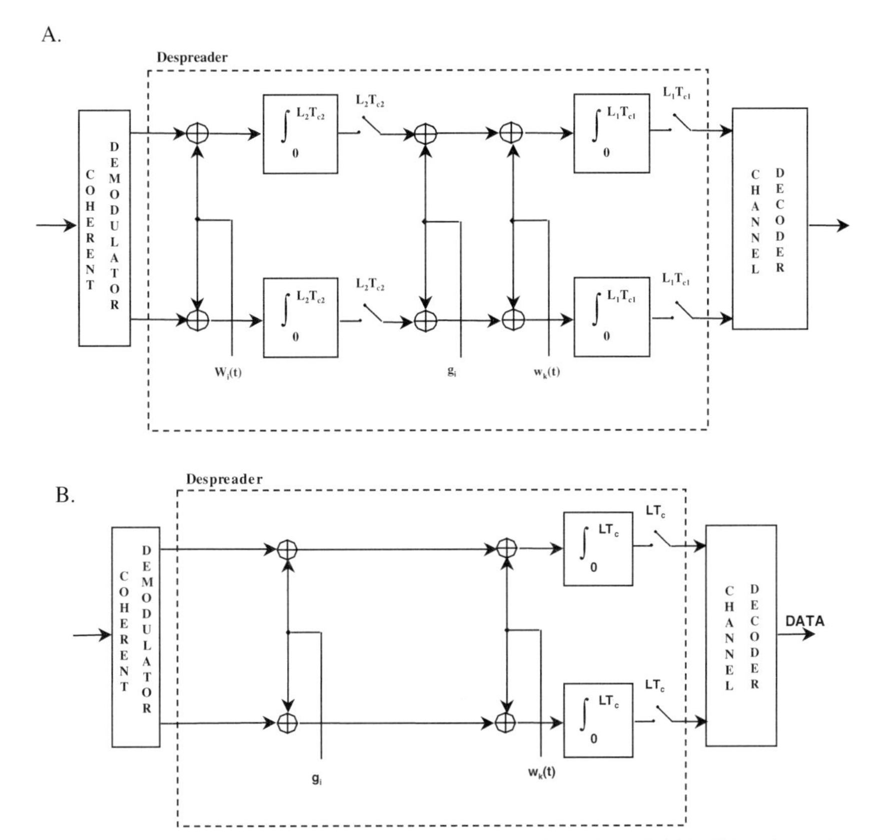

Figure 3.12 The despreading operation for A. FO, MO/SE-CDMA, and B. SO/SE-CDMA.

$(i = 1, 2, 3 \ldots)$, all user orthogonal codes W_k $(k = 1, 2, 3 \ldots)$ and beam orthogonal codes W_i $(i = 1, 2, 3 \ldots)$ should be aligned upon arriving at the satellite. The synchronization procedure that leads to this global code alignment has the following steps:

1. Upon power-on the SU acquires synchronization to the pilot PN-sequence using the serial search acquisition circuit in the S&PRU. This leads to the acquisition of a Paging channel (in the downlink), which provides the PN-code of the corresponding Access channel (in the uplink).
2. The SU establishes coarse synchronization to the satellite reference time by transmitting a successful message over the Access channel and receiving (via the Paging channel) the relative (to the reference time) message arrival time.
3. The SU calibrates the code clock (by advancing or delaying its starting point) and begins transmitting on the Traffic channel.
4. Once transmission on the Traffic channel begins, the Tracking circuit will provide fine alignment with the reference arrival time at the satellite despreaders. (The Tracking and Sync control circuit is presented and analysed in Chapter 7.)

5. In the last step, the synchronization system retains the fine Sync achieved in step 4 by using the downlink (Traffic channel) Tracking circuit and the uplink SYNC control circuit.

In Chapter 7 we present a detailed description and performance analysis of the synchronization procedures.

SE-CDMA capacity

The SE-CDMA capacity is defined as the maximum number of users or Traffic channels the system can supply in each 10 MHz CDMA band. The system capacity for the FO and MO implementations is equal to the ratio R_{c1}/R_{ss}, while the SO SE-CDMA capacity is limited by the ratio R_c/R_{ss}. The SE-CDMA capacity and the corresponding BER-E_b/N_o for the FO, MO and SO implementations is shown in Table 3.4. These results are derived analytically in Chapter 6, and are based on the assumption that coherent detection has been used. Also, in evaluating E_b/N_o for MO and SO, the assumption is made that half of the surrounding beams have a cross-polarization isolation of -6 db. The capacity indicated in Table 3.4 corresponds to 100% loading for the FO and MO and 75% loading for the SO implementation. As shown, implementation of MO achieves the lowest E_b/N_o. This is about 4 dB at a BER of 10^{-10}. This allows a sufficient margin for mitigating the output back-off at the power applifier (Traveling Wave Tube, TWT). The problem of nonlinear applification and its impact on the BER performance has been examined in Chapter 9.

The impact of system loading from 1 to 60 users on the E_b/N_o at a BER of 10^{-6} is shown in Table 3.5 for FO, MO and SO. As shown, the impact of 100% loading for the FO is only 0.15 dB, on the MO it is 1.5 dB, while for the SO it is about 11 dB for 75% loading. This reflects the fact that the MO, and particularly the SO, implementations are limited by the other beam interference. Therefore, the use of voice activity with the SO implementation will provide a significant increase in the S0 SE-CDMA capacity. As shown in Table 3.4, the required E_b/N_o at a BER of 10^{-6} with voice activity utilization is only 4.3 dB. The impact on E_b/N_o performance with symbol-aided demodulation has been evaluated analytically in Chapter 8, and is found to be between 0.8 and 1 dB below the coherent demodulation. Another factor that will impact on the E_b/N_o performance is the synchronization time-jitter, which has been examined in Chapters 2 and 7.

3.2.4 Network Architecture

The satellite system described above provides direct connectivity between the Customer Premises Equipment (CPE) and gateway offices via one or more satellites which have switching or routing capabilities. In particular, the satellite network will have the following capabilities:

(a) *Line switching* of circuit calls between CPEs or between a CPE and a PSTN gateway.
(b) *Data packet routing* between CPEs or between CPEs and a PSDN gateway.
(c) *Trunk switching* between PSTN gateways.

Table 3.4 The capacity of each channel type and the corresponding (BER, E_b/N_o) for each SE-CDMA implementation.

Channel Type	Capac. C*	BER	E_b/N_o(dB)			
			FO	MO	SO	SO**
I	60	10^{-6}	4.95	3.8	11.65	4.3
II	120	10^{-6}	4.95	3.8	11.65	4.3
III	240	10^{-4}	4.90	3.65	11.55	4.0
IV	30	10^{-6}	4.98	3.8	11.65	4.3
V	10	10^{-8}	4.98	3.9	11.85	4.3
VI	2	10^{-10}	5.05	4.05	12.0	4.52
VII	2	10^{-10}	5.05	4.05	12.0	4.52

* C is the capacity in number of users per 10 MHz channel.
** We assume voice activity is utilized.

Table 3.5 The impact of system loading on the attainable E_b/N_o for each SE-CDMA implementation.

User Loading	E_b/N_o(dB) at 10^{-6}		
	FO	MO	SO
1	4.8	2.3	0.38
30	4.85	3.05	3.4
60	4.95	3.8	11.65

(*Line* is a link leading to the end user. *Trunk* is a link between central offices.) The main use of this satellite network, though, is to provide the switching of lines for circuit calls and the routing of data packets to the end users (CPE). The satellite network will also provide signaling for lines and trunks which is specifically designed to meet the needs of the system. The satellite will not provide on-board multiplexing of lines routed to a gateway office. Instead, the multiplexing of lines into trunks will take place at the gateway.

Figure 3.13-A shows the main system components, as well as the routes between them. The CPE is comprised of the Subscriber Unit (SU) and the Terminal Equipment (TE). Similarly, a gateway office will consist of a SU to interface the satellite and the central office multiplexers and switching equipment. The CAI signaling messages will be carried by the Access/Paging channels only if they are addressing the satellite, otherwise they will be carried by the Traffic channel. The CAI signaling is specifically designed to meet the needs of the satellite system. Therefore, any external signaling

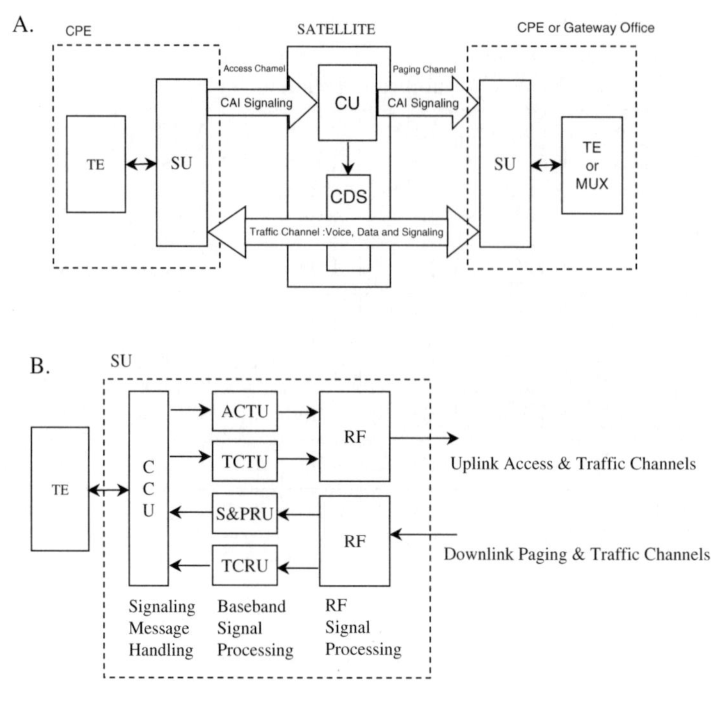

CPE	Customer Premise Equipment
TE:	Terminal Equipment
SU:	Subscribe Unit

Figure 3.13 The SS/CDMA network interfaces.

system (for lines or trunks) interfacing the SU will be 'covered' by a CAI signaling overhead. In particular, all messages with their destination as the satellite or an end-SU will be converted in to CAI signaling messages, while messages routed via the Traffic channel to or from external points will be treated as data, and will remain unconverted.

The SU system design is shown in Figure 3.13-B. It is comprised of the RF, the baseband signal processors and the Call Control Unit (CCU). The RF processors are common between the Traffic and the Signaling control channels (Access/Paging), while the baseband transceiver units are dedicated to each particular channel. Each SU, while in the traffic channel state (see call control operation), will have the capability to send and receive signaling messages to either the end SU or the satellite. This is done by multiplexing the additional traffic into the voice or data frames (frame design makes provision for such traffic). The CCU will provide all the necessary software for call control. The CCU signaling interfaces are shown in Figure 3.14. The CCU will interface the user's Terminal Equipment (TE) at one end and the CAI signaling network at the other. It will also make the message translation (protocol conversion) between the TE signaling (such as ISDN Q.931) and the CAI signaling, and perform the CAI call control procedures. The CAI signaling is particularly designed for the satellite environment, so that the call set-up time does not exceed 1 sec in a single

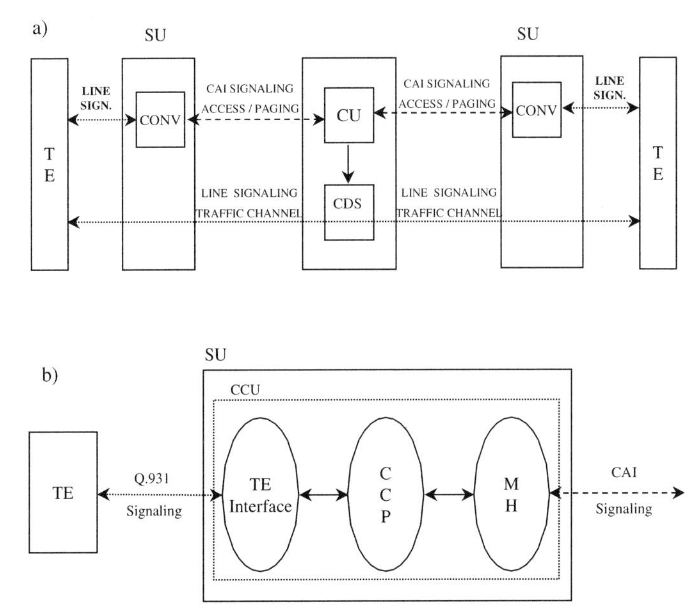

Figure 3.14 The SS/CDMA signalling interfaces.

hop satellite network. For this purpose, the CCU will maintain all dynamic data (a traffic matrix of active calls) on board the satellite. The static database (user data), however, will be on the ground, and will be accessed via a dedicated link.

The layer structure of the CAI has Physical, Link and Network layers, as shown in Figure 3.15. The Network layer defines the CAI call control procedures and the signaling messages. The Link layer defines the Traffic channel, the signaling channels (the Access in the uplink and the Paging in the downlink) and the Pilot and SYNC channels (in the downlink). The Physical layer contains the radio interfaces of the Traffic and signaling control channels, which are defined in Section 3.2.3.

3.2.5 Network Control System

Demand Assignment Algorithm

Network control is based on the demand assignment algorithm, which may then be described as follows:

1. Upon the arrival of a circuit call or a data packet, the SU sends a message request via the Access channel to the on-board control unit.
2. The control unit assigns an end-to-end Traffic channel for the circuit or packet by searching the uplink and the downlink pool of traffic channels for an available one, i.e. which meets the scheduling conditions. If there is one available the control unit sends the assignment information to the end SUs via the Paging channels. If there is none, the circuit call will be blocked while the data packet will be buffered until one becomes available.

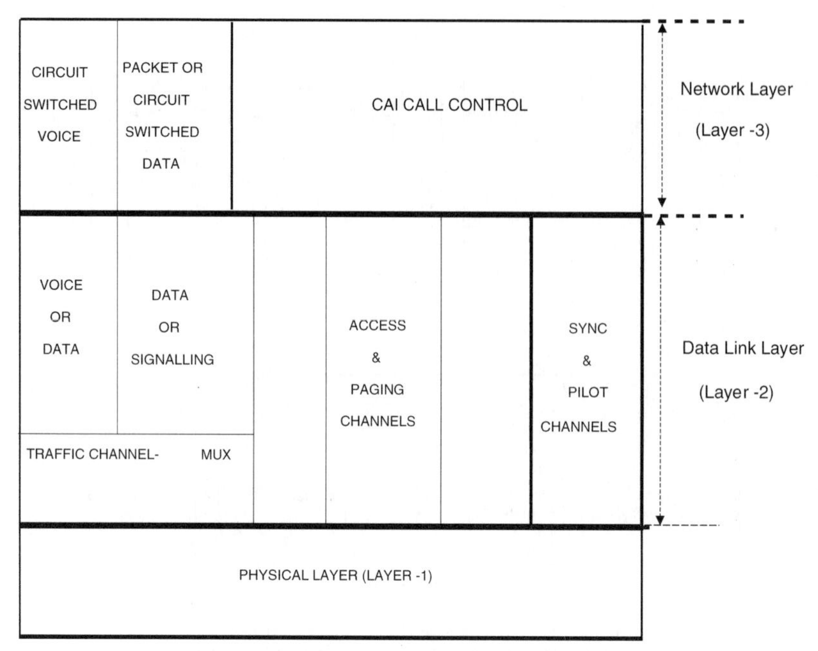

Figure 3.15 The CAI layer structure.

3. Each end-SU then transmits or receives on the assigned Traffic channel (i.e. frequency band and codes) while the control unit sets up the connection in the assigned CDS module.
4. Traffic channels will be reserved for the duration of the circuit call or packet transmission. After the termination of the circuit call or packet transmission, the SU sends an indication to the control unit via the Access channel that this Traffic channel has became available.

On-board the satellite the control unit makes the Traffic channel assignments by using a scheduling algorithm. A scheduling algorithm may be optimum, sub-optimum or random. Such algorithms are presented and evaluated in Chapter 5.

Call Control Operation

The control operation described here is used for circuit switched calls between CPEs, as well as for calls between a CPE and a PSTN gateway office. The call processing design will allow portability of the TE and/or the SU. A satellite user may also access the system from different CPEs. For this purpose, the call control will provide procedures for registration, authentication and identification. Each call requires the mapping of the dial number to the SU-ID and the SU to a user location (beam number). The above parameters are contained in the ground database, which will be accessed by the satellite upon call set-up.

The Call Processing States: the call processing is described in terms of the SU's functional condition, called State. As shown in Figure 3.16-A, the SU has the following call-states:

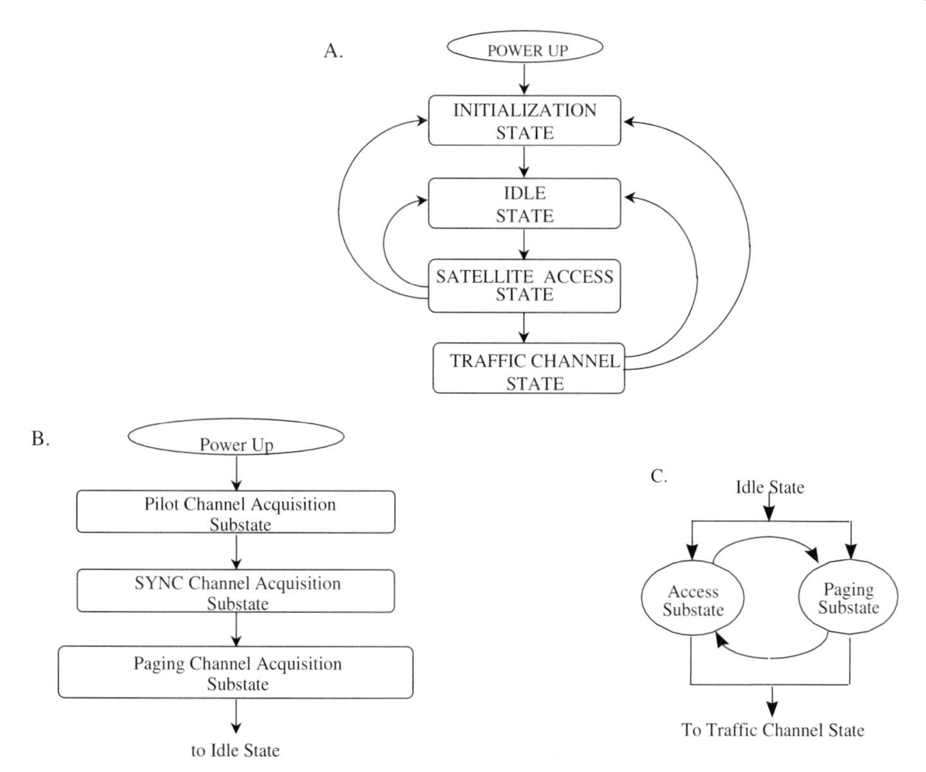

Figure 3.16 SS/CDMA call processing flow diagrams.

The Initialization State: upon powering up, the SU will enter the initialization state in order to acquire the system. This state consists of the following sub-states (see Figure 3.16-B); The Pilot Acquisition sub-state, in which the SU searches and acquires the strongest beam Pilot signal; the SYNC Channel Acquisition sub-state, in which the SU receives the system timing and identification information; and the Paging Channel Acquisition sub-state, in which the SU acquires an assigned Paging channel and then enters the Idle state.

The Idle State: in the Idle state the SU monitors the Paging channel. It can receive messages, receive an incoming call or initiate a call.

The Satellite Access State: in this state the SU may either transmit messages to the satellite via the Access channel (Access Channel sub-state), or receive messages from the satellite on the Paging channel (Paging Channel sub-state) (see Figure 3.16-C). Responses to the accesses made on an Access channel will be transmitted by the satellite on an associated Paging channel. SU transmissions on the Access channel will obey the Random Access Protocol (see Chapter 7). The main task the SU performs in this state is to initiate a call or to receive a call which leads to the Traffic Channel State with the assignment of a Traffic channel. The Traffic channel assignments are made at the satellite scheduling algorithm. Another task that takes place in this state is the SU registration and authentication. The SU will perform the initial registration automatically after its initialization. This is done with an initial transmission over an Access channel and confirmation response over its associated Paging channel. The SU in the Access channel sub-state may perform the following tasks: (1) Initiate a call by sending a call set-up message via the Access channel. This message will contain the dialled digits which will be mapped (at the satellite) to the destination address of the called SU (ID and location); (2) Respond to a call set-up message by sending a Call Proceeding or Connect messages. Respond to any other Page message requiring a response; (3) Initiate a registration procedure by sending a Registration message. The SU in the Paging channel sub-state may perform the following tasks: (1) receive a Call Set-Up, Alerting, Connect or a Connect Acknowledgment message; (2) receive an Authentication request message or any other Page message.

The Traffic Channel State: in this state the end SUs communicate directly with each other via the Traffic channel. While in this state, the SU may return to the Satellite Access State if there is a message to or from the satellite.

3.3 Conclusion

In this chapter we have given an overview of switched CDMA networks, and presented the Satellite Switched CDMA (SS/CDMA) as a case study for such networks. The SS/CDMA system illustrates how we can apply CDMA for both access and switching. We have presented the SS/CDMA network architecture and the design of each system component, and examined the network operation and control. The switching mechanism is based on code division technology, which is examined in detail in Chapter 4. The SS/CDMA access and switching requires a demand assignment control mechanism, which provides efficient routing of circuit calls and data packets upon user request. The switch control algorithms and the evaluation of the network throughput is presented in Chapter 5. The satellite multiple access is based on a synchronous CDMA scheme utilizing a concatenation of Reed–Solomom with Turbo codes. Such a scheme is called Spectrally Efficient CDMA (SE-CDMA), and is analyzed in Chapter 6. The SE-CDMA requires code synchronization of all users in the network. The satellite spread-spectrum random access and network synchronization procedures are presented in Chapter 7. The SE-CDMA carrier recovery utilizes a symbol-aided demodulation scheme which has been analyzed in Chapter 8. Finally, the impact of the nonlinear amplification of the SE-CDMA signal by the on-board TWT and the required 'back-off' is presented in Chapter 9.

References

[1] D. Gerakoulis, E. Geraniotis, R.R. Miller and S. Ghassemzadeh 'A satellite Switched CDMA System Architecture for Fixed Service Communications' *IEEE Commun. Magazine*, July 1999, pp. 86–92.

[2] T. Inukai 'An efficient SS/TDMA time-slot assignment algorithm' *IEEE Trans. Commun.*, Vol. 27, No. 10, October 1979, pp. 1449–1455.

[3] T. Scarcella and R.V. Abbott 'Orbital Efficiency Through Satellite Digital Switching' *IEEE Commun. Magazine*, May 1983, pp. 38–46.

[4] L.C. Palmer and L.W. White 'Demand Assignment in the ACTS LBR System' *IEEE Trans. Commun.*, Vol. 38, May 1990.

[5] D. Gerakoulis and E. Drakopoulos 'A Demand Assignment System for Mobile Users in a Community of Interest' *IEEE Trans. Vehic. Tech.*, Vol. 44, No. 3, August 1995, pp. 430–442.

[6] A.S. Acampora, T-S. Chu, C. Dragone and M.J. Gans 'A Metropolitan Area Radio System Using Scanning Pencil Beams' *IEEE Trans. on Commun.*, Vol. 39, No. 1, January 1991, pp. 141–151.

[7] M. Grimwood and P. Richardon *Terayon Communications Systems.* 'S-CDMA as a High-Capacity Upstream Physical Layer' *IEEE802.14a/98-016*, June 15 1998.

4

Code Division Switching

4.1 Overview

In this chapter we present and analyze the switching architecture of the exchange node for switched CDMA networks. As we have discussed in the previous chapter, in such a network CDMA traffic channels will be routed by the exchange node from any input to any output link.

If we consider a traditional switching approach, the exchange node can be implemented as shown in Figure 4.1. In this case we assume that Time Multiplexed Switching (TMS) is used to provide the switch functions. As shown, after the despreading operation, all CDMA user channels are time multiplexed, then routed to the destination output port by the TMS, demultiplexed, spread again, and then combined for the output CDMA channel. The TMS approach, however, introduces additional complexities, because the switch input and output ports require time multiplexing, while the incoming and outgoing signal is based on code multiplexing. (In traditional switching methods such as time slot interchangers or space switching, traffic channels are time multiplexed in each input or output port.) Also, the complexity for a strictly nonblocking TMS switch fabric is significant. This means that in applications such as SS/CDMA where the available power and mass at the satellite are limited, TMS may not be an efficient switching approach.

Therefore, we propose an alternative switching method which is based on code division. That is, the signals in the switch are distinguished and routed according to their spreading codes. This method is directly applicable in all switched CDMA networks such as SS/CDMA, BS/CDMA or CS/CDMA. In this chapter we provide illustrative Code Division Switch (CDS) architectures, performance and complexity evaluation analysis and comparisons with traditional switching methods. As shown, the proposed CDS architecture is nonblocking and its hardware complexity and speed is proportional to the size of the switch. Also, the CDS routes the CDMA user channels without introducing interference. The switch performance evaluation includes the amplitude distribution of the combined signal in the CDS bus and the interference evaluation of the end-to-end link in the proposed network applications. The code division switch performance evaluation will utilize the satellite switching (SS/CDMA) as a basis for study. This work was originally presented in references [1] and [2].

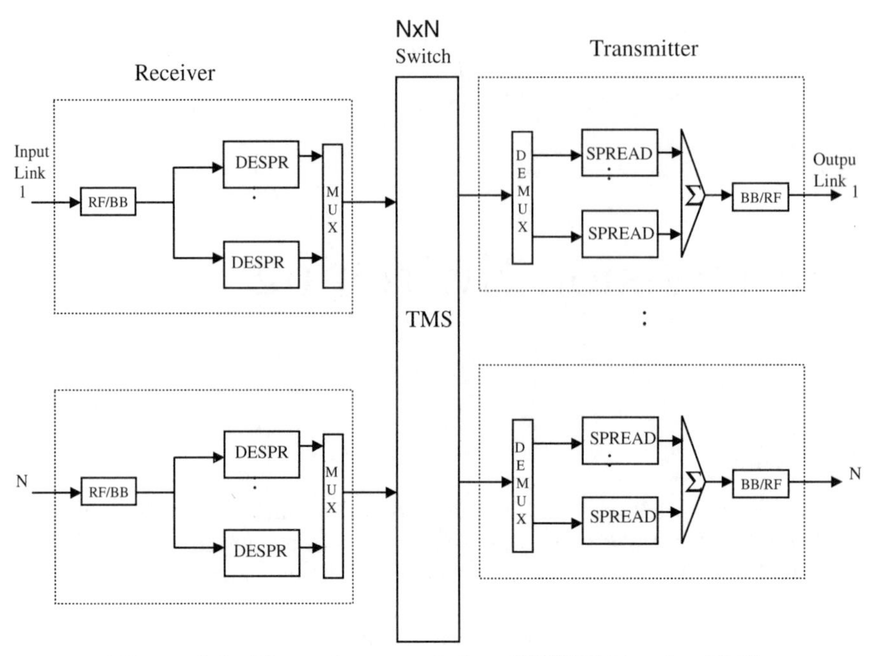

Figure 4.1 The exchange node in a SW/CDMA using TMS.

4.2 Switched CDMA (SW/CDMA) Architectures

In this section we examine the network and switch architectures in SS/CDMA and SW/CDMA for terrestrial wireless and cable applications. We also examine traditional switch architectures (such as the TMS) for routing CDMA channels, and present a CDS method for routing time multiplexed channels.

4.2.1 Satellite Switched CDMA (SS/CDMA) System

As we have described in the previous chapter, the on-board design of a SS/CDMA system provides the CDS modules, the switch control unit and the transceivers of the control channels (Access and Broadcast). The switching and control architecture at the exchange node on board the satellite is illustrated in Figure 4.2.

Traffic channels are routed from uplink to downlink beams via the switch modules without data decoding on board the satellite. The Traffic channel modulation and spreading processes are based on the Spectrally Efficient CDMA (SE-CDMA) which are illustrated in Figures 3.27 and 3.28 of Chapter 3. The SE-CDMA spreading process requires the following codes: (1) a set of orthogonal codes w_k having a chip rate R_{c1} assigned to satellite users $k = 1, 2, .., L_u$ within each beam; (2) pseudo-random (PN) codes c_i with a chip rate R_{c1} assigned to satellite beams $i = 1, 2, \ldots N$; and (3) a set of orthogonal codes w_i with a chip rate R_{c2} for orthogonal isolation of L_b satellite beams, $i = 1, 2, \ldots, L_b$.

The PN-codes spreading rate R_{c1} is the same as the rate of the user orthogonal codes w_k. The orthogonal codes w_i, however, require a higher spreading rate $R_{c2} = L_b R_{c1}$. The process of spreading a previously spread signal at a higher rate is called

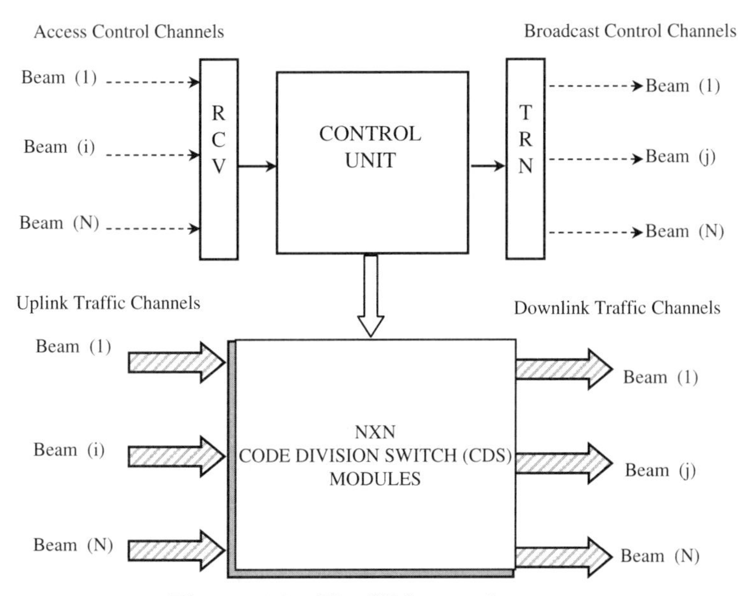

Figure 4.2 The CDS control system.

overspreading (see Chapter 1, Section 1.4.2). When $L_b = 4$ the system is called a Fully Orthogonal (FO), when $L_b = 2$, a Mostly Orthogonal (MO), and when $L_b = 1$ (i.e. $R_{c1} = R_{c2} = R_c$) is called it Semi-Orthogonal (SO) SE-CDMA. Hence, the SE-CDMA will eliminate the interference between users within each beam, as well as between the L_b beams in the cluster, while it allows a frequency reuse of one.

In a particular implementation, presented in Appendix 4A, $R_{c2} = 9.8304$ Mc/s and $L_u = 60$. Also, the orthogonal codes can be either Quadratic Residue (QR) codes or Walsh codes when the length $L = 2^k$.

The Code Division Switch (CDS)

The proposed CDS architecture is shown in Figure 4.3. Each uplink CDMA channel is first converted into an Intermediate Frequency (IF) and then into baseband (BB) without demodulating the incoming signal (switching at IF has also been considered). After that, the signal is despread by the uplink orthogonal beam code w_i and the PN beam code c_i (see Figure 4.4-A). Each particular user signal is then recovered by the Traffic Channel Recovery Circuit (TCRC) shown in Figure 4.5. This is achieved by despreading with the user's uplink orthogonal code w_k. The signal will then be respread with the user (w_m) and beam (c_j, w_j) downlink codes.

Finally, the signal will be overspread again by an orthogonal (switch) code w_n ($n = 1, 2, .., L_s$), having a chip rate $R_{c3} = L_s R_{c2}$. This step of overspreading will achieve orthogonal separation of all user Traffic channels in the system, and thus can be combined (summed up) into a common bus. The number of w_n codes, L_s, is equal to the number N of switch ports ($L_s = N$), if no prior orthogonal separation between uplink beams exists. In such a case the rate is $R_{c3} = N \cdot R_{c1}$. The SE-CDMA scheme, however (shown in Figures 3.27 and 3.28), has the L_b beams already orthogonalized. Hence, $L_s = N/L_b$ and $R_{c3} = (N/L_b) \cdot R_{c2}$. Each uplink beam in the cluster will

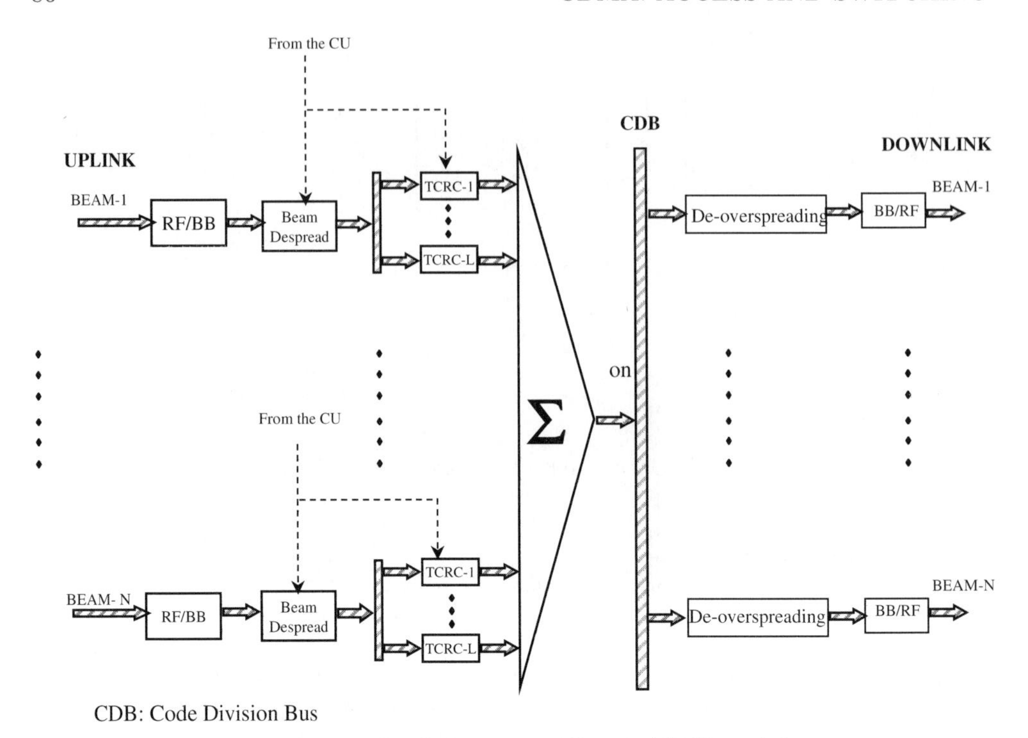

Figure 4.3 The Code Division Switch (CDS) module.

then be overspread by the same w_n orthogonal code ($n = 1, 2 \ldots, N/L_b$). For $L_b = 4$ (FO/SE-CDMA), $N = 32$ and $L_s = 8$, the chip rate is $R_{c3} = 78.6432$ Mc/s. (See the example presented in Appendix 4A.) The I and Q components are combined (summed-up) in parallel by two separate adders (in the case where both I and Q are summed, the rate will be $R_{c3} = 2N \cdot R_{c1}$). The steps of overspreading, the codes involved, and the corresponding chip rates for this application are shown in Figure 4.6.

After overspreading, all incoming (I or Q) signals are combined (summed up) into a (I or Q) bit stream called a Code Division Bus (CDB). The CDB then contains all Traffic channels spread by their corresponding downlink user and beam destination codes. Hence, each downlink beam may be recovered by the de-overspreading circuit shown in Figure 4.4-B, and routed to its destination port. The signal will then be converted into an IF, and subsequently into an RF frequency for downlink transmission. The set of all codes in the TCRCs for routing the Traffic channels to their destinations are supplied by a Control Unit (CU). The number of TCRCs required in each beam is L_u, and is equal to the number of Traffic channels per beam (beam capacity), so that no blocking occurs in the switch. Also, uplink orthogonal codes, w_k and w_i, require synchronization in order to maintain orthogonality. This is achieved by a synchronization mechanism which adjusts the transmission time of each user so that all codes are perfectly aligned upon reception at the TCRC despreaders.

An equivalent functional arrangement of the code division switch is shown in Figure 4.7. The corresponding circuits for Traffic channel recovery and respreading are shown in Figure 4.8. In this architecture the incoming signal, after conversion to baseband, is despread by the uplink beam orthogonal code (beam recovery), and

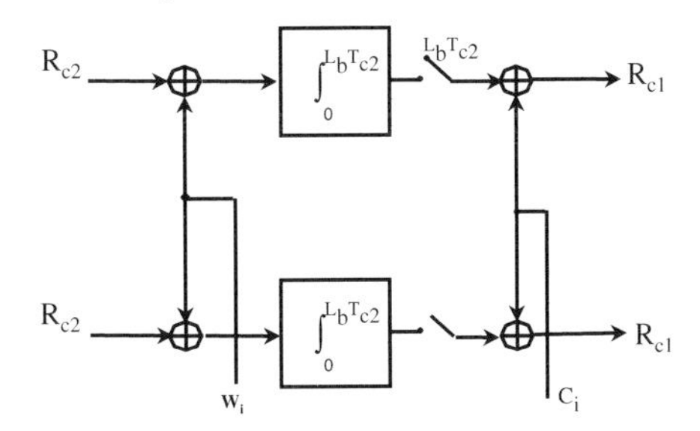

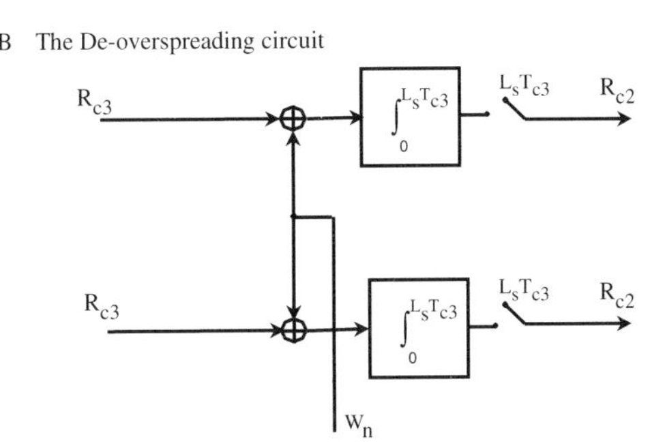

Figure 4.4 The beam-despreading and the de-overspreading circuits.

then overspread so that it can be combined (summed up) into the Code Division Bus (CDB). Overspreading by the switch codes w_n allows orthogonal separation in the CDB between all uplink beams or incoming switch inputs. The beam recovery and overspreading (BR&OS) operation is illustrated in Figure 4.8-A. A Traffic channel recovery and respreading (TCR&RS) circuit recovers the desired Traffic channel from the CDB by de-overspreading its signal with the corresponding switch orthogonal code ($w_n, n = 1, .., n$), and then despreading it with the uplink user code w_k. After recovery, Traffic channels are routed to the desired downlink beam (output port) by respreading them with the corresponding destination user (w_m) and beam (c_j, w_j) codes. The TCR&RS circuit is shown in Figure 4.8-B. At the output, all TCR&RS circuits having the same destination beam will be combined (summed up) and converted into the RF carrier for downlink transmission. Each output beam requires L_u TCR&RS circuits equal to the maximum number of Traffic channels per beam.

Comparing the two architectures presented above (Figures 4.3 and 4.7), we observe that both of them perform the same functions, but in a different order. In the first configuration (Figure 4.3), Traffic Channel Recovery (TCR) takes place before channels are combined into the CDB, while in the alternative configuration (Figure 4.7), TCR takes place after the CDB. In the alternative configuration, only beam recovery takes place before the CDB to the rate $R_{c1} = L_u R_s$. In both cases, the CDB has the same rate which is R_{c3} ($R_{c3} = NR_{c1} = L_s R_{c2}$ and $L_s = N/L_b$). The relation between chip rates is shown in Figure 4.6. Performance comparisons between the above CDS configurations are provided in Section 4.3.

In the above CDS architectures, the baseband signal (i.e. the output of the RF to baseband converter for any M-ary PSK scheme, $M \geq 4$), has two components, I

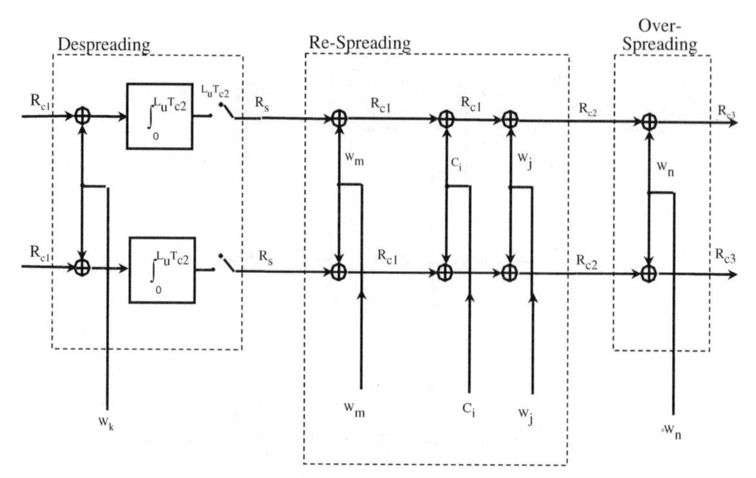

Figure 4.5 The Traffic Channel Recovery Circuit (TCRC).

and Q. The I and Q outputs are not orthogonal in baseband. Hence, either the I and Q components must be switched separately (using I and Q signal combiners), or if a single combiner is used, the speed of overspreading must be doubled (using twice as many orthogonal codes). Here, we consider the first case in which there is space separation between the I and Q components as in Figure 4.7.

Time Multiplexed Switching (TMS) of CDMA Channels

In SS/CDMA we may also use Time and/or Space Division switching for routing the code multiplexed signals. In these cases, the incoming signal is first downconverted to baseband and despread. Data symbols are then time multiplexed and time slots will be routed via a Time Slot Interchanger (TSI) or a Space Division Switch (SDS). Figure 4.9 illustrates a Time Division Code Switch (TDCS) consisting of a TSI between the input despreader and the output respreader. Similarly, a Space Division Code Switch (SDCS) would consist of despreaders, followed by a space switch, followed by respreaders. The TSI in the TDCS rearranges the time slots in each frame, while the SDS in the SDCS provides physical connections during the period of the time slot. The size of a TSI is limited by practical speed and memory. In space switching, on the other hand, the limiting factor is the number of cross point connections (N^2 for a nonblocking cross-bar switch fabric) which may be constrainted by the volume available within the spacecraft. For large switch sizes, a multi-stage switching network is generally used. Such a network may consist of TSIs interconnected with a space switch (known as the Time-Space-Time architecture). The complexity of this approach, however, may be excessive in satellite switching applications. An implementation example of time multiplexed switching CDMA channels is given in reference [3].

4.2.2 SW/CDMA Applications in Terrestrial Networks

Terrestrial SW/CDMA applications include wireless CDMA networks for mobile and fixed services, called Base Station Switched CDMA (BS/CDMA), and coax-cable

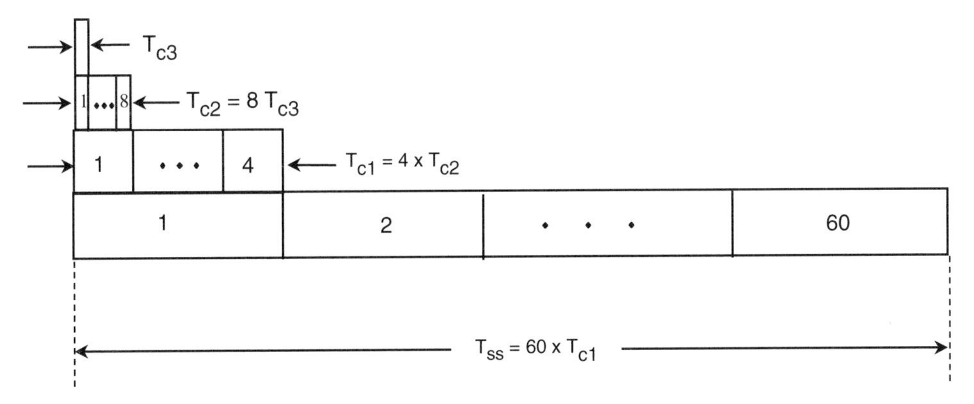

$$R_{c3} = \frac{1}{T_{c3}} = 8 \times 9.8304 = 78.6432 \text{ Mc/s (Orthogonal Separation of Beams in the Switch)}$$

$$R_{c2} = \frac{1}{T_{c2}} = 4 \times 2.4576 = 9.8304 \text{ Mc/s (Cluster Beams Orthogonal Separation)}$$

$$R_{c1} = \frac{1}{T_{c1}} = 60 \times 40.96 = 2.4576 \text{ Mc/s (User Traffic Channel Orthogonal Separation)}$$

$$R_{ss} = \frac{1}{T_{ss}} = 40.96 \text{ ks/s}$$

	Unspread	Orth. User Code	PN Beam Code	Orth. Beam Code	Orth.Switch Code
Uplink Codes	--	W_k	g_i	W_i	W_n
Downlink Codes	--	W_m	c_j	W_j	
Code Rates	R_{ss}	R_{c1}	R_{c1}	R_{c2}	R_{c3}

Figure 4.6 The overspreading relations in the CDS module.

networks having CDMA access for two-way multimedia services called Cable Switched CDMA (CS/CDMA) (see Chapter 3, Section 3.1).

Base Station Switched CDMA (BS/CDMA)

In BS/CDMA we consider the cases of mobile and fixed service applications: see references [4] and [5]. In the case of mobile service, we assume that the uplink spreading consists of a user code g_k and a cell or cell-sector cover-code c_i, where both of them are PN-codes having the same chip rate (as, for example, in the TIA/IS-95 standard). In the downlink, there are orthogonal user codes W_m and PN cover-codes c_j. The code division switch design in this case is then similar to that in Figures 4.3 or 4.7, but without the beam codes W_i and W_j, while the uplink user code W_k is replaced with the PN-code g_k.

In fixed service applications (such as wireless local loop), we may use PN-codes as in the mobile case, or orthogonal codes as in SS/CDMA (since synchronization is possible for nonmobile service), depending on the network application or the propagation

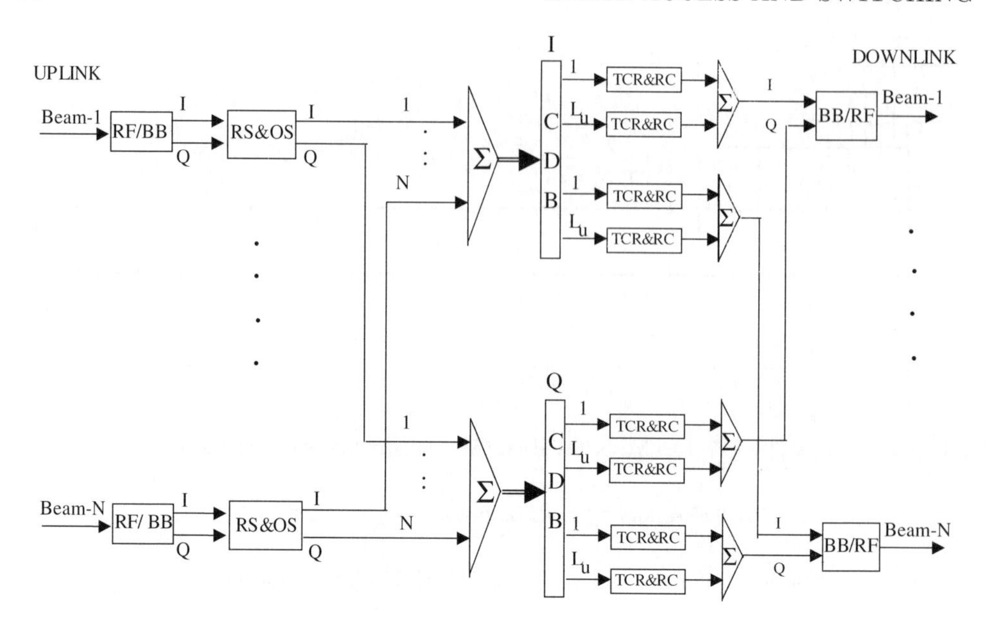

RF/BB: RF to Baseband converter
BR&OS: Beam Recovery and Overspreading
CDB: Code Division Bus
TCR&RS: Traffic Channel Recovery and Respreading

Figure 4.7 An alternative Code Division Switch (CDS) architecture.

characteristics. If we use orthogonal codes, the CDMA spreading design may be based on the Mostly Orthogonal (MO/SE-CDMA) implementation described in Chapter 3. In this case, considering multi-sector cells, we use two orthogonal sector-codes for rejecting the interference from the adjacent sectors. Then, assuming the spreading circuit of Figure 3.28, the rate $R_c = R_{c2} = 2R_{c1}$. The code division switch design in this case will be the same as in Figures 4.3 or 4.7. Based on the end-to-end interference analysis presented in Section 4.3, it is recommended that in the BS/CDMA the CDS also includes both the demodulation/remodulation process and channel decoding and re-encoding.

Cable Switched CDMA (CS/CDMA)

In CS/CDMA the upstream access is based on a synchronized orthogonal CDMA as described in reference [6]. The upstream spreading process, unlike SS/CDMA or BS/CDMA, does not require orthogonal beam or cell-codes, for the reason that CDMA channels (operating in the same frequency band) are in different coax-cables, and are thus completely isolated from each other. Upstream user (code) channels within the cable are then isolated by orthogonal user codes W_k, while CDMA channels in different cables do not interfere with each other. Similarly, for the downstream we only use orthogonal user codes W_m. The code division switch design in this case will be as in

A. The beam recovery and overspreading (BR&OS)

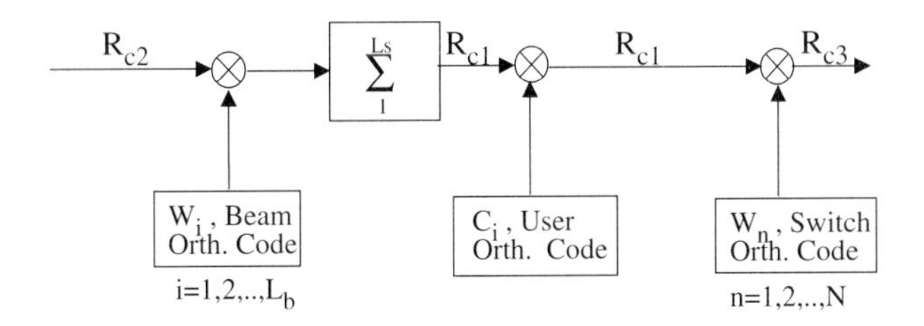

B. The Traffic channel recovery and respreading (TCR&RS)

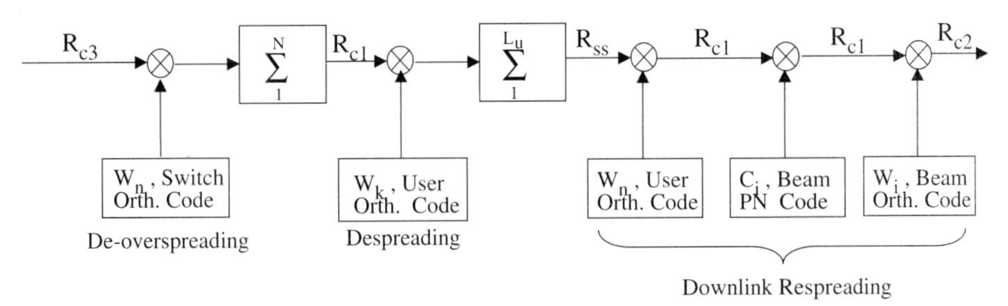

Figure 4.8 The BR&OS and TCR&RS circuits for the alternative CDS module architecture.

Figures 4.3 or 4.7, but without beam codes W_i and c_i in the uplink and W_j and c_j in the downlink.

Based on the end-to-end interference analysis presented in Section 4.3.3, the CDS in the CS/CDMA application may take full advantage of direct connectivity between end users, since no demodulation/remodulation, no channel or source decoding/encoding, and no data buffering are required at the exchange node.

Code Division Switching of Time Multiplexed Channels

Code division switching may also used in systems where Traffic channels at the input or output links of the exchange node are Time Division Multiplexed (TDM). In this case the TSI can be replaced by a Code Division Switch. The CDS architecture in this case is shown in Figure 4.10. The input signals first are spread with orthogonal code W_m of the destination port m ($m = 1,..,N$) of the current time slot k ($k = 1,..,L$), and then are combined (summed up) into a code division bus (CDB). Each output port signal then is recovered from the CDB by despreading with the output code W_m in time slot k. All signals in the CDB are orthogonal in time and code. The speed of the signal in the CDB is NR, where R kb/s is the bit rate at the input or output ports. Orthogonal codes W_m are supplied by the control unit on a time-slot by time-slot basis.

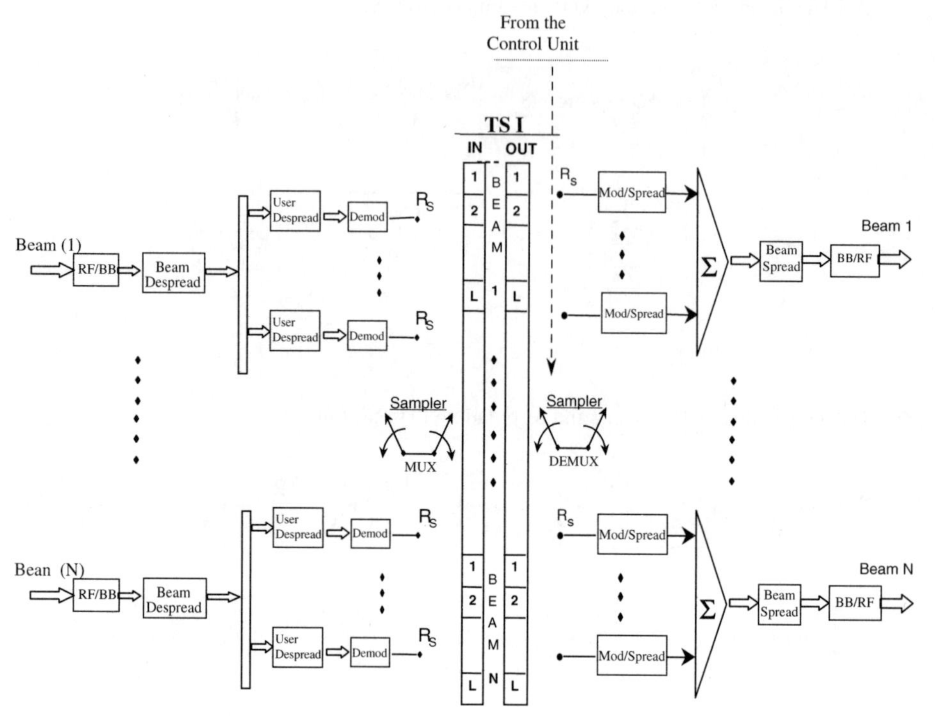

Figure 4.9 The Time Division Code Switch (TDCS).

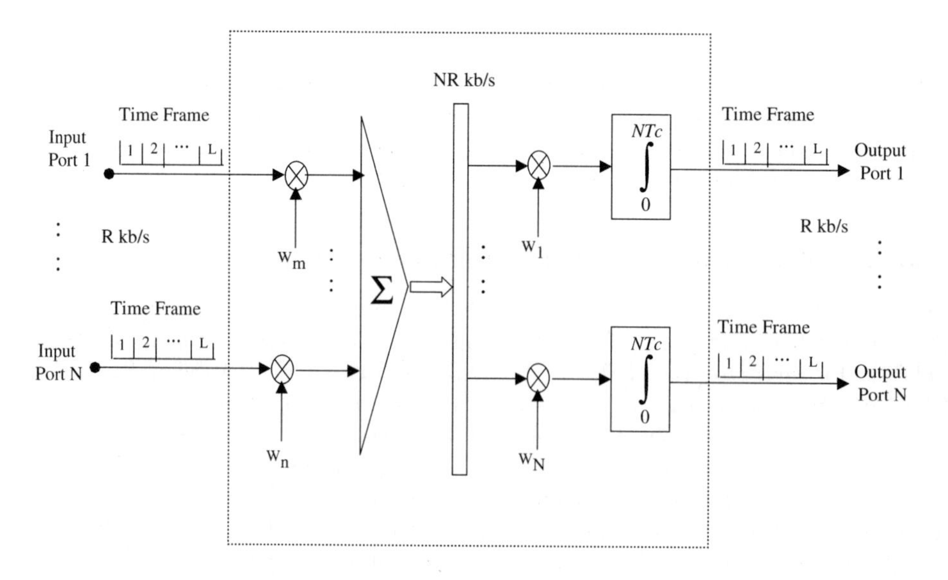

w_n,w_m: Orthogonal codes with rate NR kb/s destined for ports m and n, respectively.

Figure 4.10 A CDS architecture for time multiplexed channels.

4.3 Performance Evaluation of Code Division Switching

In this section we evaluate the interference or noise caused by the switch during the routing process (in Section 4.3.1), the instantaneous signal amplitude in the code division bus as a function of the user load (in Section 4.3.2), and the end-to-end interference for each SW/CDMA application (in Section 4.3.3).

4.3.1 Evaluation of the Switch Interference

Let us consider the SS/CDMA application with the CDS architecture of Figure 4.3, having an $N \times N$ CDS switch module with N input and N output ports. Also, let $s_I^{(n)}[l]$ and $s_Q^{(n)}[l]$ denote the I and Q signal samples at times lT_{c1} at the nth input port of the switch: $1 \le n \le N$ and $l = \ldots, -2, -1, 0, 1, 2, \ldots$. The chip duration is $T_{c1} = 1/R_{c1}$ (see Figure 4.6).

Let $w_I^{(n)}[m]$ and $w_Q^{(n)}[m]$ for $m = 1, 2, \ldots, N$ be the overspreading codes used in the I and Q subports of port n. The result of overspreading is that the mth overspreading chip of the lth chip I and Q components, is equal to

$$\left(\sum_{n=1}^{N} s_I^{(n)}[l] w_I^{(n)}[m], \ \sum_{n=1}^{N} s_Q^{(n)}[l] w_Q^{(n)}[m] \right) = (\bar{s}_I[l,m], \ \bar{s}_Q[l,m]) \equiv \bar{s}[l,m]$$

where $m = 1, 2, \ldots, N$. During de-overspreading at an output port of interest, and for the signal I of the $n = 1$ input port, we have:

$$\begin{aligned}
\sum_{m=1}^{N} \bar{s}_I[l,m] w_I^{(1)}[m] &= \sum_{m=1}^{N} \left\{ \sum_{n=1}^{N} \left[s_I^{(n)}[l] w_I^{(n)}[m] \right] \right\} w_I^{(1)}[m] \\
&= \sum_{n=1}^{N} s_I^{(n)}[l] \left[\sum_{m=1}^{N} w_I^{(n)}[m] w_I^{(1)}[m] \right] = \sum_{n=1}^{N} s_I^{(n)}[l] \delta_I[n,1] = s_I^{(1)}[l]
\end{aligned}$$

where $\delta_I[n,1] = \sum_{m=1}^{N} w_I^{(n)}[m] w_I^{(1)}[m]$, and thus $\delta_I[n,1] = 1$ if $n = 1$ and 0 otherwise. Similarly, we obtain $\sum_{m=1}^{N} \bar{s}_Q[l,m] w_Q^{(1)}[m] = s_Q^{(1)}[l]$, and thus both the I and Q components of the signal of interest ($n = 1$) are recovered at this output port.

Critical to the above derivation is the assumption that the signals at all the input ports $s_I^{(n)}[l]$ and $s_Q^{(n)}[l]$ remain constant (unchanged) over the duration of one chip: T_c. It is thus assumed that the N samples of the overspreading codes $w_I^{(n)}[m]$ and $w_Q^{(n)}[m]$ ($m = 1, 2, \ldots, N$) are multiplied (modulo-2 added) by the same value (single sample) of the signals $s_I^{(n)}[l]$ or $s_Q^{(n)}[l]$ for all n (and l). To guarantee that these chip samples do not change value within the chip duration, there must be no chip waveform shaping taking place in the switch. The chip waveform (raised cosine chip filter) is, of course, used at the input matched filters and at the output of the switch before the signal is transmitted over the downlink.

Provided that there is no time variation within the duration of the chip, there is *no interference* of any type introduced by the CDS. Of course, whatever interference is already included in the soft inputs (real numbers) $s_I^{(n)}[l]$ and $s_Q^{(n)}[l]$ at l^{th} chip time of the nth input port, is transferred intact to the output port that uses the orthogonal codes $w_I^{(n)}[m]$ and $w_Q^{(n)}[m]$ for de-overspreading. The phenomenon of interference transfer is examined in Section 4.3.3.

In the alternative CDS architecture shown in Figure 4.7, overspreading takes place after beam despreading, but before user despreading and re-spreading. In this case the value of the chip amplitude is fixed for the duration of the chip (T_{c2}), since the signal is taken at the output of the matched filter (interchip interference filter). Hence, neither of the alternative CDS configurations introduce any interference to the Traffic channels during the switching process.

4.3.2 Signal Amplitude Distribution

As shown in section 4.3.1, the signal samples at the input ports are first modulo-2 added to the overspreading codes and then added (real addition) together. The resultant total signal at the mth overspread chip time of the l-th (regular) chip time is

$$\bar{s}[l,m] \equiv (\bar{s}_I[l,m],\ \bar{s}_Q[l,m]) = \left(\sum_{n=1}^{N} s_I^{(n)}[l]w_I^{(n)}[m],\ \sum_{n=1}^{N} s_Q^{(n)}[l]w_Q^{(n)}[m]\right)$$

Assuming that

$$s_I^{(n)}[l] = b_I^{(n)}[i] + \bar{I}_I^{(n)}[l] \quad and \quad s_Q^{(n)}[l] = b_Q^{(n)}[i] + \bar{I}_Q^{(n)}[l]$$

$\left(b_I^{(n)}[i], b_Q^{(n)}[i]\right)$ represents the in-phase and quadrature components of the ith M-ary symbol of the nth user; i changes every $T_s = 1/R_{ss}$ secs, while l changes every $T_{c2} = 1/R_{c2}$.

$b_I^{(n)}[i] = \cos\phi^{(n)}[i]$ and $b_Q^{(n)}[i] = \sin\phi^{(n)}[i]$, where $\phi^{(n)}[i]$ denotes the phase angle of the ith M-ary symbol of the nth (user) signal; and they take values in the sets:

$$b_I^{(n)}[i] \in \left\{\cos\left[\frac{(2j-1)\pi}{M}\right], j = 1, 2, \ldots, M\right\}$$
$$b_Q^{(n)}[i] \in \left\{\sin\left[\frac{(2j-1)\pi}{M}\right], j = 1, 2, \ldots, M\right\}$$

It is assumed that the sequences of phase angles (symbols) $\phi^{(n)}[i]$ of the $n = 1, 2, \ldots, N$, signals are i.i.d. That is, there is independence for different j (symbols), and for different n (signals/users/ports) and are also identically distributed. With respect to the latter it is assumed that the phase angle $\phi^{(n)}[i]$ of the ith symbol of the nth signal is uniformly distributed in the set $\{\pi/M, 3\pi/M, \ldots, (2M-1)\pi/M\}$ and subsequently the inphase and quadrature components $b_I^{(n)}[i]$ and $b_Q^{(n)}[i]$ are i.i.d (for different n and i) and uniformly distributed (take each value with equal probability $1/M$) in the above sets. For the same n and i, $b_I^{(n)}[i]$ and $b_Q^{(n)}[i]$ are not independent of each other but are uncorrelated, since we can easily show that the expected value over the above sets results in

$$E\left\{b_I^{(n)}[i]\right\} = E\left\{b_Q^{(n)}[i]\right\} = 0, \quad E\left\{b_I^{(n)}[i]b_Q^{(n)}[i]\right\} = 0 \ \text{ and}$$

$$E\left\{\left[b_I^{(n)}[i]\right]^2\right\} = E\left\{\left[b_Q^{(n)}[i]\right]^2\right\} = \frac{1}{2}$$

Moreover, the terms $\bar{I}_I^{(n)}[l]$ and $\bar{I}_Q^{(n)}[l]$ represent the interference (from other users and AWGN or other channel) present during the lth chip. Observing that during respreading each chip is multiplied by $+1$ or -1 and thus its variance does not change, and that overspreading follows despreading and respreading (see Figures 4.4 and 4.5), the mean of these interference terms is zero and their variances are $var(\bar{I}_I^{(n)}[l]) = var(\bar{I}_Q^{(n)}[l])$. These variances are also equal to the normalized (with respect to the received power of the desired signal) variance of the interference at the output of the despreaders (before respreading and overspreading) at input port n of the switch.

In the SS/CDMA system the normalized power of interference (not including AWGN) at the uplink (after despreading) is denoted by $\bar{I}_{0,t}^u$. Therefore

$$\begin{aligned}\sigma_I^2 = var(\bar{I}_I^{(n)}[l]) = var(\bar{I}_Q^{(n)}[l]) &= \bar{I}_{0,t}^u + \left(\frac{2(\log_2 M)E_b^u}{N_0}\right)^{-1} \\ &= \frac{K}{L_u}(2\bar{I}_2^u) + \left(\frac{2(\log_2 M)E_b^u}{N_0}\right)^{-1}\end{aligned}$$

where K is the number of users per beam, L_u is the spreading gain of the orthogonal user codes, $\bar{I}_2^u$ denotes the normalized interfering power from a single user in the first-tier of beams (averaged with respect to the interfering user's location), E_b^u is the uplink bit energy, and N_0 is the one-sided power spectral density of AGWN (under clear-sky SATCOM channel conditions). Assuming FO/SE-CDMA for link access (see Section 3.2.3 in Chapter 3), $\bar{I}_2^u = 1.226 \times 10^{-3}$ and $L_u = 60$; while the beam capacity is $K \leq 60$ users transmitting at 64 kbps and bandwidth $W \approx 10$ MHz.

Provided that N is sufficiently large ($N \geq 8$), we can apply the Central Limit Theorem (CLT) on each of the asymptotically Gaussian random variables $\bar{s}[l,m]$ for $m = 1, 2, \ldots, N$ above. Notice that they are equal to the sum of N random variables all with (unconditional) mean 0 and variance $1 + \sigma_I^2$. The (unconditional) mean and variance of $\bar{s}_I[l,m]$ (or $\bar{s}_Q[l,m]$) then are:

$$E\{\bar{s}_i[l,m]\} = 0,\ Var\{\bar{s}_i[l,m]\} = N(1 + \sigma_I^2)$$

where $i = I$ *or* Q. The conditional mean value is given by

$$E\{\bar{s}_i[l,m] \mid b_i^{(n)}[l], n = 1, 2, \ldots, N\} = \sum_{n=1}^{N} b_i^{(n)}[l] w_i^{(n)}[m]$$

and the conditional variance is given by

$$Var\{\bar{s}_i[l,m] \mid b_i^{(n)}[l], n = 1, 2, \ldots, N\} = N\sigma_I^2, \quad \text{where } i = I \text{ or } Q$$

Thus, the dynamic amplitude range of the sum-signal in the CDB of the switch, when no interference is present, is given by

$$[-NA_i(M), NA_i(M)]$$

$$For \quad i = I, \quad A_I(M) = \max_{k=1,2,\ldots,M} \left\{ \cos\left(\frac{(2k-1)\pi}{M}\right) \right\} = \cos\left(\frac{\pi}{M}\right) \quad \text{and}$$

$$For \quad i = Q \quad A_Q(M) = \max_{k=1,2,\ldots,M} \left\{ \sin\left(\frac{(2k-1)\pi}{M}\right) \right\} = \sin\left(\frac{(2k^*-1)\pi}{M}\right)$$

The above dynamic range corresponds to the worst case in which the first chip of all the orthogonal codes $w_I^{(n)}[m], n = 1, 2, \ldots, N$ (total of N orthogonal codes) is positive and all N I-type input ports carry the same symbol $\cos(\pi/M)$ (which has the maximum positive value), where $k^* = [(M/4) + (1/2)]$ and $[x]$ denotes the integer part of x.

When interference is taken into account, we must adjust the above expression by adding a multiple of the noise variance. This adjustment should be $2\sqrt{N}\sigma_I$ (for 95.44% confidence) and or $3\sqrt{N}\sigma_I$ (for 99.74% confidence). Therefore, for 99.74% confidence and under sufficiently large N for the CLT approximation to be valid, the normalized dynamic range of the sum-signal in the CDB is given by

$$\left[-NA_i(M) - 3\sqrt{N}\sigma_I, NA_i(M) + 3\sqrt{N}\sigma_I\right]$$

where $i = I, Q$ and σ_I^2 depends on the E_b^u/N_0, the number of users per beam K, the spreading gain L_u, the type of system (beam isolation technique used), and the power control scheme as discussed earlier in this section. This should be compared with a range of $[-1, 1]$ for bipolar samples (again normalized to the received desired signal power at the input of the switch) for non-CDS switches.

4.3.3 Switch Interference Coupling

As we have described in Chapter 3, we consider the following two options in the system design of the CDS: (i) baseband despreading/respreading without demodulation (i.e. no phase detection and symbol recovery) and without channel decoding at the switch site; and (ii) baseband despreading and demodulation (i.e. phase detection and symbol recovery) followed by remodulation and respreading, but without channel decoding at the switch site (see Figure 4.11).

In case (i) we switch the baseband signal at the output of the matched filter (after the A/D converter). This is actually a sampled signal, where each sample represents the

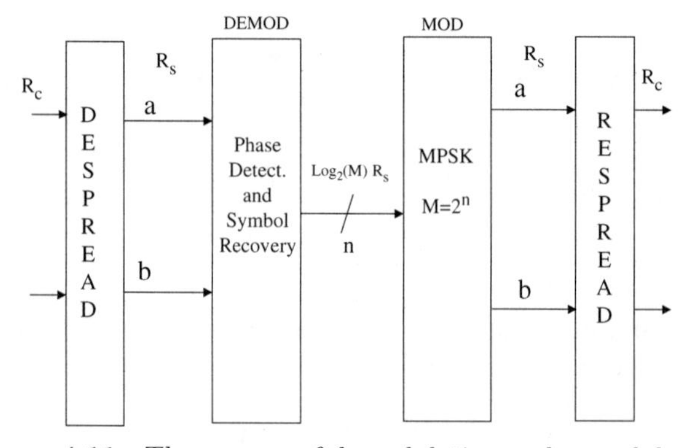

Figure 4.11 The process of demodulation and remodulation.

quantization levels (real numbers), but in the digital domain with an 8-bit (or more) A/D resolution. The process is illustrated in Figure 4.12 in Appendix 4A. This process is followed by the despreading/respreading operation, which routes these samples via the switch as shown in Figures 4.3, 4.4 and 4.5. In this case the required sampling rate is $R_{c3} = L_s R_{c2} = N R_{c1} = N L_u R_{ss}$, where R_{c2} is the total spread-bandwidth of the signal to be switched and $L_s = N/L_b$ (see Figure 4.6). Therefore, the required switch bandwidth or speed is $N L_u R_{ss}$, where R_{ss} is the symbol rate of each (unspread) Traffic channel and L_u is the number of the Traffic channels per beam.

In Case (ii) we switch the recovered symbols or demodulated data but without channel decoding. That is, after demodulation the M-ary symbols are mapped back to 0s (−1) and 1s and are then switched. (The symbols are recovered by detecting the m-PSK phase, i.e. comparing it with a given threshold and making a hard decision.) This process takes place after despreading the signal (in the sampled format), but before respreading it. The required switch speed in this case must be increased by a factor of $\log_2 M$ (that is, $[\log_2 M] N R_{c1}$).

In this section we examine the phenomenon of link coupling or interference transfer from the input to output link, which takes place in case (i). The end-to-end interfence power $\hat{I}^e$ normalized by the power of the desired signal has been derived in Chapter 6 and the results are summarized below.

$\hat{I}^e$ is expessed in terms of the ratio of powers $\frac{P_k}{P_i}$ of interfering user k over the desired user i in both input (uplink) and output (downlink) ports, the interference power between users $\bar{I}_s$ and the AWGN-to-signal ratio $\left(\frac{2E_s^d}{N_0}\right)^{-1}$ in both input and output ports:

$$\hat{I}^e \approx \left[\bar{I}_s \sum_{k \in \mathcal{I}^u(i)} \frac{P_k^u}{P_i^u}\right] + \left[\bar{I}_s \sum_{k' \in \mathcal{I}^d(i)} \frac{P_{k'}^d}{P_i^d}\right] + \left[(2\bar{I}_s) \sum_{k' \in \mathcal{I}^d(i)} \frac{P_{k'}^d}{P_i^d}\right] \cdot \left[\bar{I}_s \sum_{k \in \mathcal{I}^u(k')} \frac{P_k^u}{P_{k'}^u}\right] + \left(\frac{2E_s^u}{N_0}\right)^{-1} \cdot \left[(2\bar{I}_s) \sum_{k' \in \mathcal{I}^d(i)} \frac{P_{k'}^d}{P_i^d}\right] + \left(\frac{2E_s^u}{N_0}\right)^{-1} + \left(\frac{2E_s^d}{N_0}\right)^{-1}$$

This expression has the following terms (in order of appearance): the input (uplink) MAI (Multiple-Access Interference), the output (downlink) MAI, the cross-product of output MAI and input MAI, the cross-product of output MAI and input AWGN, the input AWGN, and the output AWGN. There are six terms of interference and noise instead of the typical two terms (MAI and AWGN) involved in a single-hop transmission system. Next, we evaluate the above expression for each SW/CDMA application.

The Power of End-to-End Interference in SS/CDMA

The total other-user interference power $\bar{I}_{0,t}^u$ is given by

$$\bar{I}_{0,t}^u = K\bar{I}_s \left[\sum_{j=1}^{6} Var\{I_1^{u,(j)}\} + \sum_{j=1}^{12} Var\{I_2^{u,(j)}\}\right] = K\bar{I}_s \left(6\bar{I}_1^u + 12\bar{I}_2^u\right)$$

where K is the total number of users within the beam. The first summation term represents the interference from the first tier of beams (six beams), and the second one from the second tier of beams (12 beams). $\bar{I}_s$ represents the interference power from a single user being at the same power level with the user of interest (it reflects the cross-correlation properties of the spreading codes). $\bar{I}_1^u$ and $\bar{I}_2^u$ represent the average relative power of users in the first and second tiers of beams with respect to the power of the user of interest. That is,

$$\bar{I}_1^u = Var\{I_1^{u,(j)}\} = E\left\{\frac{P_k^u}{P_i^u}\right\} \quad \text{and} \quad \bar{I}_2^u = Var\{I_2^{u,(j)}\} = E\left\{\frac{P_k^u}{P_i^u}\right\}$$

for a user k belonging to the 1st-tier beam j interfering with user i of beam 0 and for a user k belonging to the 2nd-tier beam j interfering with user i of beam 0, respectively.

Now, we consider the FO/SE-CDMA implementation for the SS/CDMA link design (described Chapter 3) under fully synchronous conditions. In this case we get no interference from within the same beam, and no interference from the adjacent first tier beams, since all users within the beam, as well as between beams in the first tier, are isolated by orthogonal codes. Beams in the second tier are isolated by PN-code. Hence there is interference from four beams (out of the 12) of the second tier which is suppressed by the processing gain of the PN-code L_u. Then $\bar{I}_s = 1/2L_u$. The total uplink interference for the FO/SE-CDMA implementation is $\bar{I}_{0,t}^u = \frac{K}{2L_u}\left(4\bar{I}_2^u\right)$. In this expession $\bar{I}_2^u$ represents the average uplink interference from one user in a second tier beam interfering with the user of beam 0 (as if there was no spreading; the spreading is reflected in $\bar{I}_s$). The detailed evaluation of $\bar{I}_2^u$ is presented Chapter 6. Similarly, for the downlink $\bar{I}_{0,t}^u = \frac{K}{2L_u}\left(4\bar{I}_2^d\right)$, where $\bar{I}_2^d$ represents the average downlink interference from one user in a second tier beam interfering with the user of beam 0.

Replacing these values into the expressions for the end-to-end interference power $\hat{I}^e$, we obtain

$$\begin{aligned}\hat{I}^e \approx\ & \frac{K}{2L_u}\left(4\bar{I}_2^u\right) + \frac{K}{2L_u}\left(4\bar{I}_2^d\right) + \frac{K}{2L_u}\left(4\bar{I}_2^u\right)\cdot\frac{K}{L_u}\left(4\bar{I}_2^d\right) + \\ & \left(\frac{2E_s^u}{N_0}\right)^{-1}\cdot\frac{K}{L_u}\left(4\bar{I}_2^d\right) + \left(\frac{2E_s^u}{N_0}\right)^{-1} + \left(\frac{2E_s^d}{N_0}\right)^{-1}\end{aligned}$$

In the above expression the values of $\bar{I}_2^u$ and $\bar{I}_2^d$ are very small, since they represent only the interference from the second tier of beams. Therefore, the end-to-end link performance for the case (i) (baseband despreading/respreading) will be about the same as in case (ii) (baseband despreading and demodulation). In case (ii) the interference for the uplink is $\frac{K}{2L_u}\left(4\bar{I}_2^u\right)$, while for the downlink, it is $\frac{K}{2L_u}\left(4\bar{I}_2^d\right)$. Therefore, the implementation of the CDS without demodulation/remodulation (as in case (i)), is feasible, since it can provide acceptable end-to-end performance.

The Power of End-to-End Interference in BS/CDMA

In the case of BS/CDMA for wireless (fixed or mobile) networks, the computation of the end-to-end interference is similar to that of the previous sections. There are, however, some major differences.

In particular, due to the frequency reuse factor being equal to one, all cells use the same frequency. Thus, there is no isolation between adjacent cells other than the interference suppression provided by the cell PN-code. The implication is that interference is now present from all six cells of the first tier (surrounding the cell of interest) as well as from the twelve cells of the second tier. Therefore, both the terms $\bar{I}_1^u$ and $\bar{I}_2^u$ representing(average) interference from one user of a first tier beam and a second tier beam are now present. Clearly, $\bar{I}_1^u > \bar{I}_2^u$ and these terms depend on cell geometry, antenna gains, the propagation law and the channel fading (which is more severe than for the GEO satellite links of the previous section). Moreover, there is now interference from other users within the cell (even if orthogonal uplink CDMA is used, multipath fading will degrade the code orthogonality). Therefore, a term of $K\bar{I}_s\bar{I}_0^u$ should now be added to the total interference, resulting in the representation $I_{0,t}^u = K\bar{I}_s\left(\bar{I}_0^u + 6\bar{I}_1^u + 12\bar{I}_2^u\right)$, where K is the number of users in each cell, $\bar{I}_s = 1/(2L)$, L is the spreading gain (number of chips per bit), and $\bar{I}_0^u$ is evaluated in the same manner as $\bar{I}_1^u$ or $\bar{I}_2^u$ but, of course, for interfering users in the same cell as the user of interest. Clearly, on average, we have $\bar{I}_0^u > \bar{I}_1^u > \bar{I}_2^u$. Similar development can be followed for the downlink, but all individual terms need be recomputed to reflect the differences in channel characteristics between the two links.

Then, for case (i) (no demodulation/remodulation), the total end-to-end interference is

$$\hat{I}^e \approx \bar{I}_{0,t}^u + \bar{I}_{0,t}^d + \bar{I}_{0,t}^u \cdot \bar{I}_{0,t}^d + \left(\frac{2E_s^u}{N_0}\right)^{-1} \bar{I}_{0,t}^d + \left(\frac{2E_s^u}{N_0}\right)^{-1} + \left(\frac{2E_s^d}{N_0}\right)^{-1}$$

In conclusion, for BS/CDMA applications, the total end-to-end coupled interference power is much higher than for SS/CDMA, since the in uplink $I_{0,t}^u$ and downlink $I_{0,t}^d$ interference power in this case is significant. Therefore, it is suggested that the operation of code division switching should be preceded by despreading demodulation and decoding and followed by re-encoding, remodulation and repsreading. This process increases complexity but eliminates noise coupling in the switch. It should be noted that the same processing would be required to support time-switched systems.

The Power of End-to-End Interference in CS/CDMA

This case corresponds to the opposite extreme than the BS/CDMA (wireless) case. Specifically, in the coax-cable network case the interference from other cables is insignificant, and the interference within the same cable is very small, considering a Synchronous CDMA (S-CDMA) similar to that proposed in S-CDMA [5] (the S-CDMA in [5] is recommended for upstream only). The interference from other users in the same cable can be neglected if the maximum time-jitter (τ) after synchronization is bounded to only a fraction of the chip duration T_c (say $\tau < T_c/10$). If this is not the case, the interference from other users in the same cable is upper-bounded (actually approximated) by $(K-1)/L \cdot (\tau/T_c)^2$, where L is again the spreading gain.

Thus, typically we only have the terms due to the uplink and downlink AWGN noise present in the expression for the power of end-to-end interference:

$$\hat{I}^e \approx \left(\frac{2E_s^u}{N_0}\right)^{-1} + \left(\frac{2E_s^d}{N_0}\right)^{-1}$$

In this expression we only assume AWGN. However, ingress-noise (non-white) may also be present. The above implies that if the signal-to-AWGN noise ratios of the upstream and downstream are the same, the noise ratios of the end-to-end link (without demodulation/remodulation) is twice the noise ratios of either the upstream or downstream links individually (decoupled links, case (ii)). Therefore, it is feasible to implement the CDS as in case (i) (without demodulation/remodulation) in this case, and have acceptable end-to-end performance.

In general, the phenomenon of interference transfer may be avoided when either (a) all input traffic channels within each port of the switch (or beam), as well as between ports (or beams), are orthogonal to or isolated from each other, or (b) when the switching process includes demodulation and symbol recovery after despreading which will effectively decouple the incoming and outgoing links.

4.3.4 Switch Control and Optimization

The proposed CDS system also includes a control unit which makes switch assignments based on signaling information received during the call establishment process. The CU collects all requests received during a time frame and applies a Traffic Channel Assignment (TCA) algorithm which assigns the incoming and outgoing Traffic channels. The traffic record of call requests at the control unit is organized into a traffic matrix format, each entry t_{ij} of which represents the number of calls from input port i to output port j. Given the traffic matrices of the ongoing calls $T_o(k-1)$ and newly-arrived call requests $T_a(k-1)$ in frame $(k-1)$, the Traffic Channel Assignment (TCA) algorithm derives the traffic matrices $T_o(k)$ of the ongoing calls (including the newly-assigned) and the blocked calls $T_b(k)$ in frame k, so that the following assignment conditions are met:

$$\sum_{j=1}^{N} t_{ij}(T_o) \leq L_u \quad \text{and} \quad \sum_{i=1}^{N} t_{ij}(T_o) \leq L_u$$

where L_u is the capacity of each input or output link. Then, the relation between the above matrices is given by the flow equation:

$$T_r(k-1) + T_a(k-1) = T_o(k) + T_b(k)$$

where $T_r(k-1) = T_o(k-1) - T_e(k-1)$ and $T_e(k-1)$ is the matrix of calls ending in frame $(k-1)$. The switch TCA may be optimum or random. An optimum TCA algorithm will minimize the number of blocked calls (matrix T_b). Optimum, sub-optimum and random TCA algorithms are presented and analyzed in Chapter 5.

The CDS maximum size N is limited by the speed of the available electronics. The maximum switch size is also related to the bandwidth of the CDMA channel, and to the number of samples per chip. In the SS/CDMA implementation example presented in Appendix 4A, the satellite can switch traffic between $N = 32$ beams, where each beam has $L_u = 60$ traffic channels. The onboard switching system consists of CDS modules with size (32×32). Each switch module provides inter- and inta-beam routing between the 32 beams, but within a specific pair of uplink-downlink frequency bands. The satellite system operates over S uplink and S downlink frequency bands of bandwidth $W \approx 10$ MHz each (S=40 at Ka band). Hence, there are S switching modules, each corresponding to a pair of uplink-downlink-frequency bands. The demand assignment approach may then be used to achieve modular growth, and also provide routing calls at different frequency bands. The proposed method can be described as follows: Each user sends a request via the common access control channel to the control unit which assigns the CDS module and the corresponding pair of (uplink, downlink) frequency bands for switching this particular call. The assignment information is sent back to the user via the downlink control channel. Then, each user transceiver is tuned to the assigned frequency band to transmit and receive information. In this method, the load of all users (SNL_u) is shared among all S switch modules. In addition, this approach can provide fault tolerance. That is, if one switch module fails, the load can be shared among the rest.

4.4 Switch Capacity and Complexity Assessment

Let us consider a SW/CDMA application which has N input or output switch ports (or beams) and L_u CDMA Traffic channels per port or beam. The CDS will then switch between a total of NL_u input or output channels, providing a *capacity* of $NL_u \times NL_u$ simultaneous connections. This capacity is achieved when the switch fabric has NL_u TCRCs which are assigned on demand to the desired input/output connections. That is, the TCRCs' input/output (despreading/respreading) codes are not fixed, but are assigned upon a call request. Given the above assumptions, the code division switch fabric is *nonblocking* for any incoming call to an input port; there is always a connection available to a destination output having available traffic channels. The number of active calls at any input i or output j must be less than L_u, i.e. $\sum_{j=1}^{N} t_{ij} \leq L_u$ and $\sum_{i=1}^{N} t_{ij} \leq L_u$, where t_{ij} is the number of calls between i and j. Then, a call may be blocked only by the input and/or output port capacity limit L_u. In addition, the number of TCRCs required in a CDS, NL_u, is *linearly proportional* to the switch size N. (This may be compared to a crossbar switch which has N^2 crosspoints.)

Hence, based on the analysis presented above, the proposed Code Division Switch (CDS) has the following features:

1. The code division switch fabric is nonblocking.
2. The hardware complexity of the CDS is linearly proportional to the switch size N. That is, the number of TCR circuits used by the switch is equal to the number of input channels.
3. The CDS does not introduce any interference or noise to the traffic channels while it performs the switching process.
4. Any existing interference or noise in the input channel will be transfered to the output, assuming that no demodulation or channel decoding takes place at the switch. In this case, the total interference at the receiver of an end-to-end link is the sum of the interferences of the input link, the output link and their cross-product.
5. Although the CDS does not require demodulation/re-modulation and channel decoding/re-encoding at the switch node in order to perform the switching function, in certain applications, demodulation/re-modulation and decoding/re-encoding at the switch node is needed in order to provide satisfactory end-to-end link performance. An example of such an application

is the BS/CDMA for wireless terrestrial networks. On the other hand, SS/CDMA and CS/CDMA applications do not require demodulation/remodulation in the CDS. The CDS also does not require source decoding/re-encoding and data buffering physically located at the switch node in any of the proposed applications.

6. The speed or clock rate of the CDS is at most N times the rate of the incoming signal, or NL_uR_{ss}, where N is the number of input or output switch ports, L_u is the number of CDMA traffic channels per port and R_{ss} is the symbol rate per traffic channel. (In equivalent application, the rate in a TMS is $NL_u(log_2M)R_{ss}$ assuming M-PSK modulated signal.)
7. The CDS provides a capacity of $NL_u \times NL_u$ simultaneous connections. (TMS provides only $N \times N$ connections per time slot, which is an inefficient way of routing CDMA signals, since they must time multiplexed before switching.)
8. The distribution of the signal amplitude in the CDB is asymptotically Gaussian (for large N) with zero mean and variance proportional to N, while its dynamic range is $[-NA_i,\ NA_i]$ (exclusive of the noise component, discussed in Section 4.3.2), where A_i is the amplitude of the input signal.
9. The switch control assignments are made on demand. The switch throughput can be maximized by an optimum assignment algorithm, while a computationally simple random assignment algorithm achieves results that are near optimum. (This is not the case in TMS where a random assignment algorithm achieves 15% less throughput than the optimum one.)

4.5 Conclusions

In this chapter we have presented architectures of a switching system based on code division technology. This system, called a Code Division Switch (CDS) may be applied in switched CDMA networks such as SS/CDMA, BS/CDMA and CS/CDMA for routing CDMA user-channels. We have analyzed the CDS performance and characterized its complexity. As shown, the proposed CDS architecture is nonblocking and has a hardware complexity and speed which is proportional to the size of the switch.

We have also shown that the CDS routes the CDMA user channels without introducing interference. However, any existing interference or noise in the input channel will be transferred to the output, assuming that no demodulation or channel decoding takes place at the switch. Although the CDS does not require demodulation/remodulation and channel decoding/re-encoding in order to perform the switching function, in certain applications demodulation/remodulation and decoding/re-encoding is needed in order to provide satisfactory end-to-end link performance. An example of such an application is the BS/CDMA for wireless terrestrial networks. On the other hand, SS/CDMA and CS/CDMA applications do not require demodulation/remodulation in the CDS. Further, the CDS does not require source decoding/re-encoding and data buffering at the switch node in any of the proposed applications.

The amplitude distribution of the combined signal in the code division bus of the CDS has been evaluated, and is found to be asymptotically Gaussian with zero mean and variance proportional to the number switch ports. In addition, the CDS throughput under the control of a random algorithm is near optimum, which is not the case in time multiplexed switching.

References

[1] D. Gerakoulis and E. Geraniotis 'A Code Division Switch Architecture for Satellite Applications' *IEEE Journal on Selected Areas in Commun.*, Vol. 18, No. 3, March 2000, pp. 481–495.
[2] D. Gerakoulis and R.H. Erving 'Method and Apparatus for Switching Code Division Multiple Access Modulated Beams' U.S. Patent No. 5,815,527, September 29 1998.
[3] R.H. Erving, D. Gerakoulis and R.R. Miller 'Symbol Switching of CDMA Channels' U.S. Patent No. 5,805,579, September 8, 1998.
[4] D. Gerakoulis, E. Drakopoulos 'A Demand Assignment System for Mobile Users in a Community of Interest' *IEEE Trans. on Vehic. Tech.*, Vol. 44, No. 3, August 1995, pp. 430–442.
[5] A.S. Acampora, T-S. Chu, C. Dragone and M.J. Gans 'A Metropolitan Area Radio System Using Scanning Pencil Beams' *IEEE Trans. on Commun.*, Vol. 39, No. 1, January 1991, pp. 141–151.
[6] M. Grimwood and P. Richardon *Terayon Communications Systems.* 'S-CDMA as a High-Capacity Upstream Physical Layer' *IEEE802.14a/98-016*, June 15 1998.

Appendix 4A: A Switch Design Example

In this appendix we present an implementation example of the code division switch. In this example we assume that the uplink spreading and modulation process is based on the SE-CDMA shown in Figures 3.27 and 3.28. In particular, we consider the Fully Orthogonal (FO) SE-CDMA having $L_u = 60$ orthogonal traffic channels per beam within a CDMA bandwidth of $\approx$10 MHz. Beams are separated by orthogonal and PN-codes. There are $L_b = 4$ orthogonal codes for separating four adjacent beams. The modulation scheme in 8-PSK ($M = 2^3$), the symbol rate and spreading rates are as shown in Figure 4.6.

The number of satellite beams per frequency band is equal to the number of input or output ports in each switch module, which is $N = 32$. Each frequency band is reused in every beam (frequency reuse one). The are as many $N \times N$ switch modules as the number of available (uplink, downlink) pairs of frequency bands. The bit, symbol and chip rates and the size of switch modules considered above are based on AT&T's VoiceSpan satellite project.

On board the satellite the received signal is down-converted from the RF carrier to baseband (RF/BB). Figure 4.12 shows an implementation example of the RF/BB signal processing in which the output is a digitized modulated signal at the sampled waveform level. After the A/D converter the signal is digitized at a 4$\times$ oversampling rate to produce an 8-bit resolution digital sample stream. The 8-bit digital domain sampled signal is applied to a root raised cosine matched filter which minimizes intersymbol interference. The I and Q signal components are separated by the quadrature modulator. After multiplying with 1 bit sine and cosine and accumulating over 4 bits, the 4$\times$ oversampling rate is converted to 1$\times$ the sampling rate.

Figure 4.13 shows the implementation of the CDS. Since the I and Q componets use the same orthogonal and PN-codes, the I and Q signal combiners and the I and Q CDB-must be kept separate within the switch. The Traffic Channel Despreading

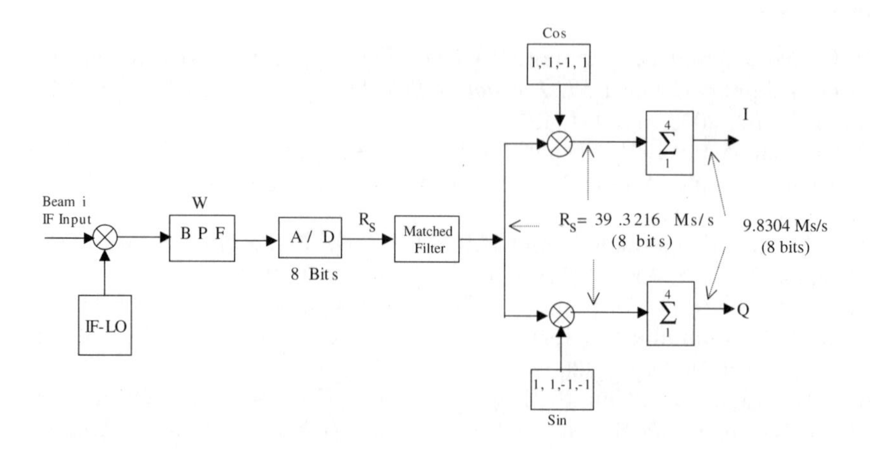

Sampling Rate: R_S=39.3216 10^6 , 8 bit samples/sec

Bandwidth: W~10 MHz
BPF : Band Pass Filter
Matched Filter: Inter Chip Interference Filter

Figure 4.12 The RF to baseband converter (RF/BB).

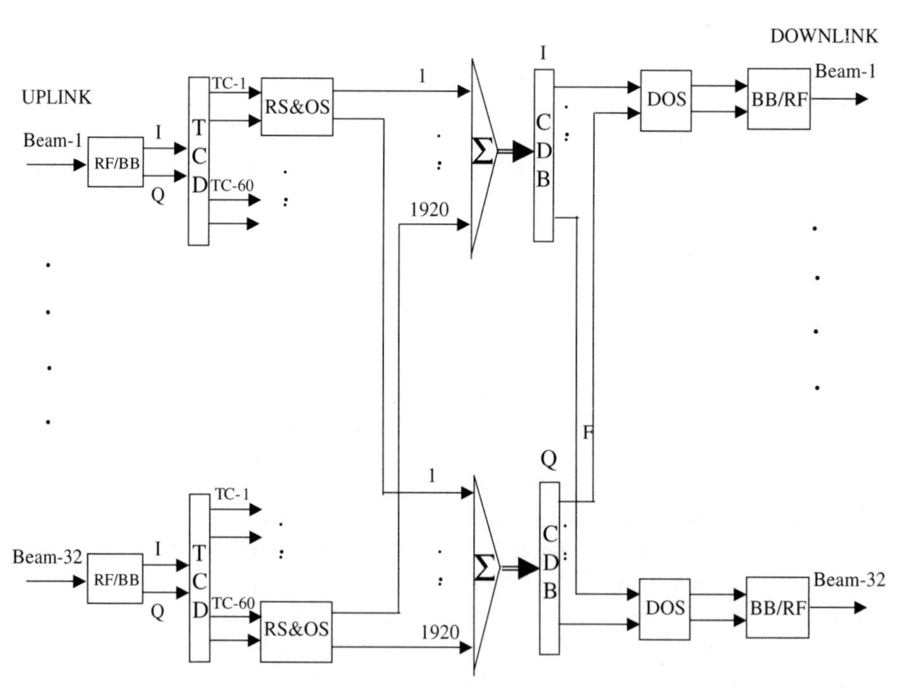

RF/BB : RF to baseband converter.

TCD : Traffic Channel Despreader.

RS&OS : Respreading & Overspreading.

CDB : Code Division Bus.

DOS : De-overspreading.

BB/RF : Baseband to Rconverter.

Figure 4.13 A CDS implementation example.

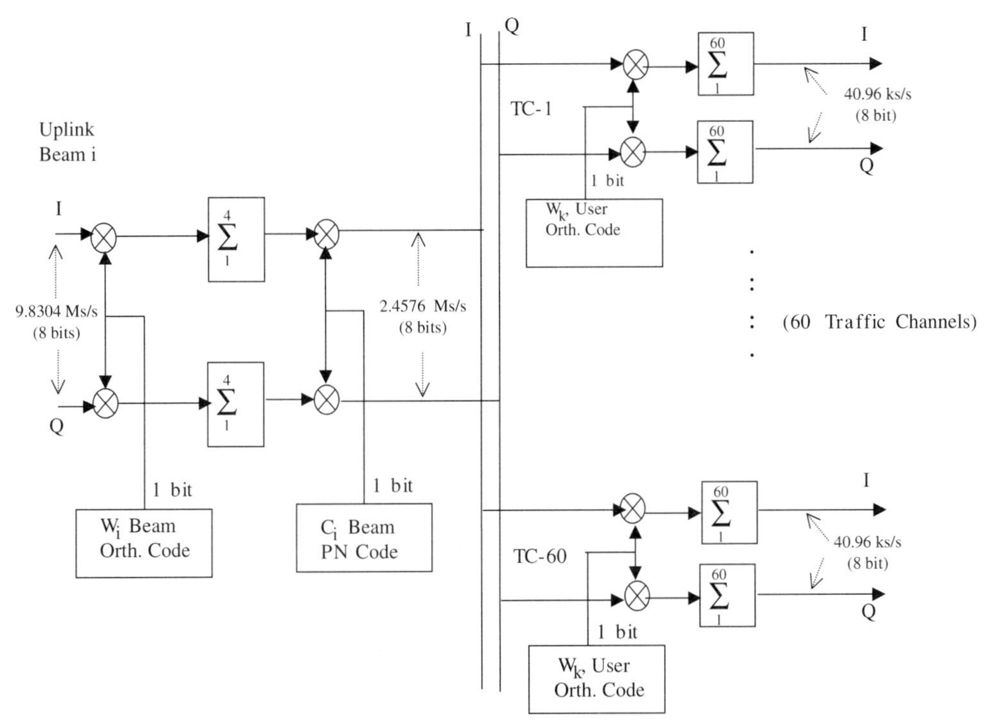

Figure 4.14 The Traffic Channel Despreading (TCD) circuit.

(TCD) circuit for beam and user (Traffic channel) recovery is shown in Figure 4.14. The output of TCD provides the information symbols having a rate of 40.96 ks/s. Each symbol is represented as an 8-bit sample which is a real number (value of the quantized level). This value is carried via the switch to the output beam. Thus the input signal noise will also be carried to the downlink beam. The respreading and overspreading (RS&OS) circuit is the same as shown in Figure 4.5, and the de-overspreading (DOS) circuit in Figure 4.4. As shown, there are 60 RS&OSs per beam, or a total of 1920 RS&OS, for the 32 beams, equal to the number of traffic channels in the switch. The number of DOS circuits is 32. The CDB chip rate is 78.6432 Ms/s. If we also had demodulation (i.e. phase detection and symbol recovery), then the rate would be increased by $\log_2 8 = 3$. (i.e. 3×40.96 ks/s).

5

The Satellite Switched CDMA Throughput

5.1 Overview

As we have discussed in Chapters 3 and 4, the satellite switched CDMA system provides on-board switching which operates with demand assignment control. This approach resolves both the multiple access and switching problems while it allows efficient utilization of the system resources. This is achieved with Traffic channel assignment algorithms which can maximize throughput and integrate the traffic of circuit calls and data packets. A large population of end users may then access the geostationary satellite network, which routes the circuit calls and data packets directly between them. A related method based on Time Division Multiple Access (TDMA), called Satellite Switched TDMA (SS/TDMA), has been proposed for packet switched data (see reference [1]).

In this chapter we provide channel assignment algorithms for optimum, sub-optimum and random switch operation. In each case, the system throughput has been evaluated by simulation and the performance results are compared. Performance analysis has been carried out for the case of optimum switch scheduling. The analysis is based on a discrete time Markovian model, and provides the call blocking probabilities and data packet delays. In Section 5.2 we describe the demand assignment system, and present the Traffic channel assignment control algorithms. In Section 5.3 we provide the throughput analysis, and in Section 5.4 we present the performance results. This work originally appeared in reference [2].

5.2 The Demand Assignment System

The SS/CDMA demand assignment network is illustrated in Figure 5.1. The satellite has an $N \times N$ Code Division Switch (CDS) and a Control Unit (CU). The interface between the satellite and the SUs, called the Common Air Interface (CAI), consists of control and traffic channels. The control channels deliver signaling messages to and from the satellite while the traffic channels carry information data directly between the end SUs. The multiple access scheme of the uplink control channel is based on a Spread Spectrum Random Access (SSRA) protocol, while the traffic channel access is an orthogonal CDMA scheme (in both uplink and downlink) called Spectrally Efficient Code Division Multiple Access (SE-CDMA) (see Chapters 3 and 6). Each SE-CDMA

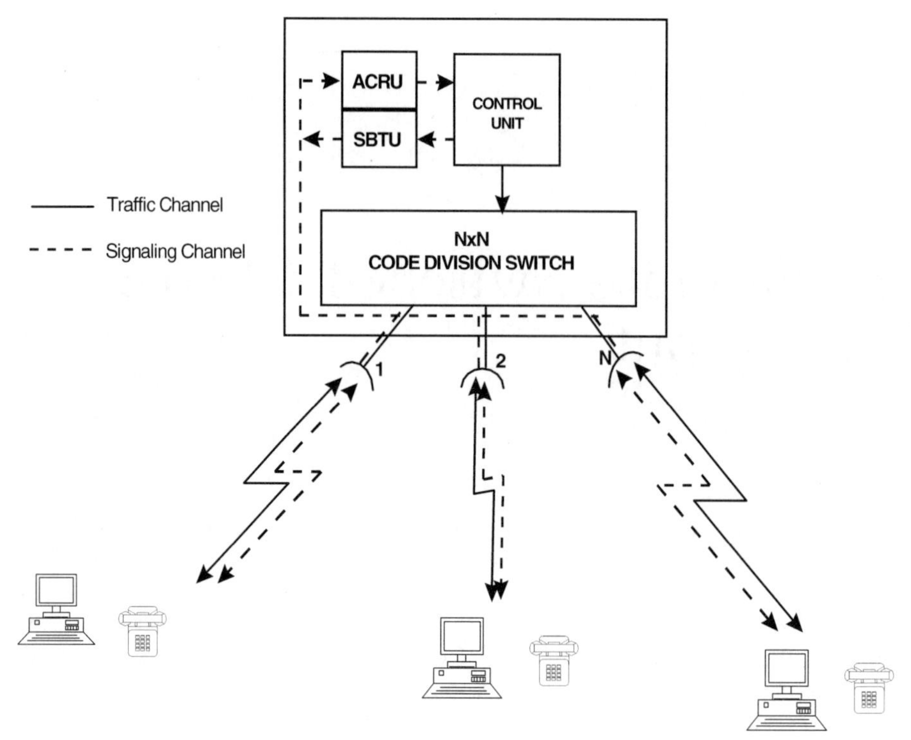

Figure 5.1 The SS/CDMA demand assignment network.

frequency band (W) can support up to L traffic channels. The services assumed here are both *circuit switched* and *packet switched*. The circuit switched services are for voice, video and data, while the packet switched service is only for data.

The on-board switching and control architecture is illustrated in Figure 4.2 of Chapter 4. Traffic channels are routed via the CDS module while signaling control messages are transmitted via the control channels, and are processed at the CU. The switching system consists of Code Multiplexed Switch (CMS) modules. Each CDS module routes calls between N uplink and N downlink beams within the *same* frequency band, where each frequency band in each beam provides L traffic channels. The size of the CDS module then is $NL \times NL$. The CDS module system design has been described in Chapter 4.

The frequency band and the traffic channel allocations for circuit or packet switched services are made upon user request. Message requests and assignments are sent via the control channels, while the information data are transmitted via the traffic channels. This allows dynamic sharing of the available switching resources between different types of traffic. While the intra-band switching is performed by a single CDS mode, the intermodule or interband switching of traffic will be handled by the demand assignment method. That is, given that the available spectrum consists of q pairs of uplink and downlink bands, the module assignments are made upon call arrival. This means that while the control channel has a pre-assigned frequency band, the traffic channel band is assigned to each SU upon call arrival, which tunes to it for transmitting and use the corresponding module for routing its call.

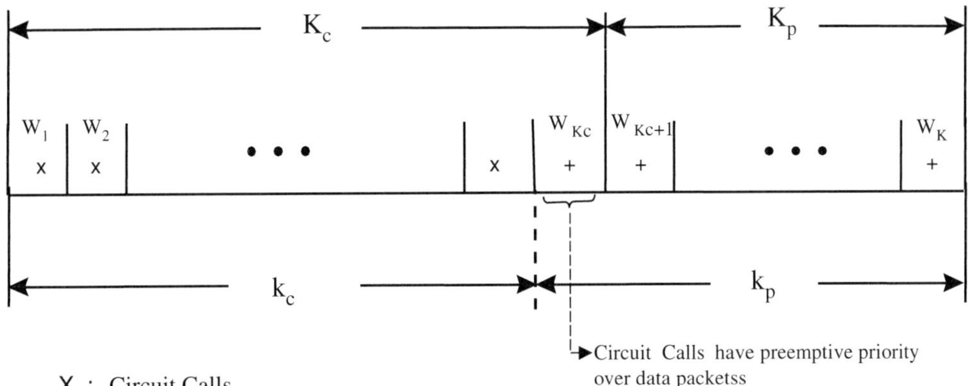

X : Circuit Calls

+ : Data Packets

K_c : Number of orthogonal Traffic Channels for circuit calls

K_p : Number of orthogonal Traffic Channels for data packets.

W_k : Orthogonal Traffic Channel codes $k = 1, 2, \ldots, K$

$K = K_c + K_p$

$K = k_c + k_p$

Figure 5.2 The movable boundary method for integrating circuit and packets.

The demand assignment system also provides an efficient method for integrating packet and circuit switched services. The proposed method is based on the *movable boundary* approach and is described as follows: Given a pool of K orthogonal traffic channels, K_c out of K will be allocated for circuit switched calls and K_p for packet switched data ($K = K_c + K_p$) (see Figure 5.2). Any unused circuit traffic channel may be assigned momentarily for packets. Traffic channels allocated for packet services are not assigned for circuits. Let k_c be the number of active circuit calls and k_p the number of packets in transmission at a given time frame. The traffic channel assignment rules will be based on the conditions:

$$\text{(a)} \quad k_c \leq K_c \quad \text{and} \quad \text{(b)} \quad k_c + k_p \leq K$$

Furthermore, the integration of circuit and packet switched services is extended in both the satellite links and the code division switch modules (see Figure 5.3). Based on the movable boundary method, the assignment conditions (a) and (b) should hold true in each uplink beam i as well as each downlink beam j. That is, no more than K_c circuit calls may be routed from any uplink beam i to any downlink beam j. Moreover, the total number of circuits and packets admitted in to the uplink beam i and downlink beam j cannot exceed the beam capacity K. If condition (a) does not hold true after the arrival of any new circuit call, the call will be blocked. Similarly, if condition (b) does not hold true after the arrival of a new data packet, the packet will be buffered.

The assignments of circuit and packet switched services will be made out of a pool of traffic channels. In an uplink beam i, a traffic channel $y_k^{(i)}$ will be identified by the

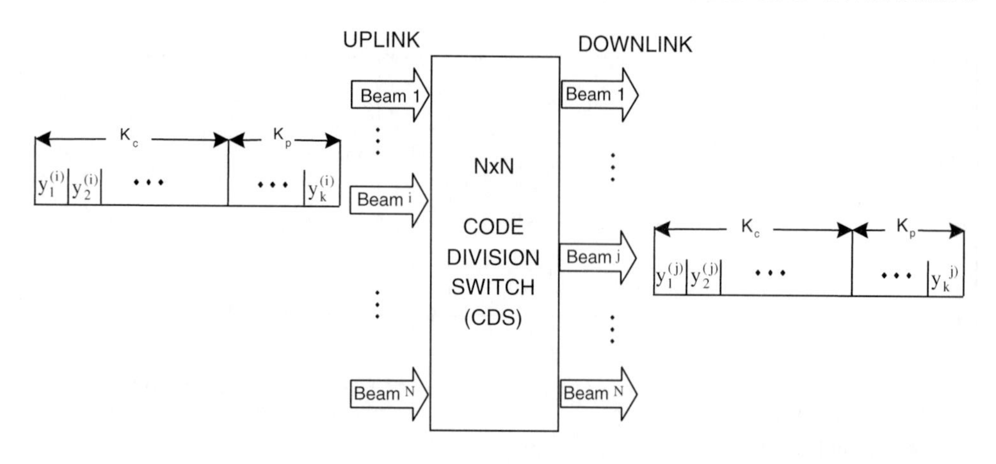

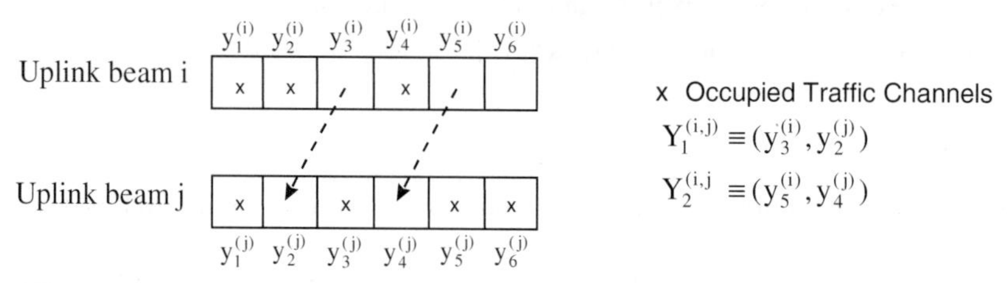

Figure 5.3 The integration of circuit and packets within the switch module.

frequency band f_n, the orthogonal Traffic channel code W_u, and the beam code w_i:

$$y_k^{(i)} \equiv (f_n, W_u, w_i) \text{ for } n = 1, \ldots, q; u = 1, \ldots, L; i = 1, \ldots, N \text{ and } k = 1, \ldots, K$$

Similarly, in a downlink beam j, a traffic channel $y_k^{(j)}$ is identified by

$$y_k^{(j)} \equiv \ (f_m, W_v, w_j) \text{ for } m = 1, \ldots, q; v = 1, \ldots, L; j = 1, .., N \text{ and } k = 1, \ldots, K$$

where $K = q \times L$ is the total number of traffic channels in all the frequency bands, q is the number of allocated CDMA frequency bands (W), and L is the number of traffic channels in each frequency band.

An end-to-end traffic channel (from uplink beam i to downlink beam j), $Y_k^{(i,j)}$, is then defined as an ordered pair of traffic channels. Thus,

$$Y_k^{(i,j)} \equiv (y_k^{(i)}, y_k^{(j)}) \quad \text{ for } k = 1, \ldots, K$$

The timing delay of the traffic channel assignment process is $T_A = w + 2t_p$, where $2t_p$ is the round trip propagation delay and w is the waiting time for the assignment to be made. (All requests are kept on-board until a decision is made.) In the above equation, the assumption is that transmissions over the access channel (i.e. the uplink control channel) is always successful. Although the access channel transmissions have a high probability of success (0.9 or better), they are not always successful. (The access channel will be designed to operate at a point of low-throughput and low-delay in order to meet such a requirement.) Therefore, the actual delay is

$$T_A = w + 2t_p + \alpha(w + 2t_p)$$

where α is the average number of retransmissions required over the access channel.

The system will also provide full duplex communication based on Frequency Division Duplexing (FDD). Each interbeam call requires the assignment of an uplink and a downlink band in each direction (a total of four bands). For intrabeam calls, the two uplink (or downlink) traffic channels may be separated either by frequency or by code.

5.2.1 System Control Algorithm

The system model described here is based on a single CDS module (and thus one pair of frequency bands) of size $N \times N$ for switching traffic between N uplink and N downlink beams. Each beam has a capacity of L traffic channels in each frequency band. L_c and L_p traffic channels in each beam are used by circuit services and packet services, respectively ($L_c + L_p = L$). Any unused circuit channels can also be assigned momentarily for packets. However, traffic channels for data packets cannot be used for circuit calls.

During each frame, SUs send reservation requests for new circuit calls and data packets to the CU via the uplink control channels. The CU collects all requests in matrix form, with rows and columns representing the uplink and downlink beams, respectively. Let $\mathbf{T}_a(k-1)$ and $\mathbf{D}_a(k-1)$ be the circuit and data requests, respectively, in frame $(k-1)$. The CU also maintains traffic matrix $\mathbf{T}_o(k-1)$ of active (ongoing) call connections and matrix $\mathbf{D}_b(k-1)$ of unassigned (buffered) packet requests from previous frames, which are waiting in the SUs' buffers to be assigned in subsequent frames. In addition, at the end of the circuit call, the SU sends an indication to the CU that its traffic channel becomes available, which is represented by $\mathbf{T}_e(k-1)$.

Based on the given matrices in frame $(k-1)$, $\mathbf{T}_a(k-1)$, $\mathbf{D}_a(k-1)$, $\mathbf{T}_e(k-1)$, $\mathbf{T}_o(k-1)$ and $\mathbf{D}_b(k-1)$, the CU applies an algorithm that makes assignment decisions to determine matrices $\mathbf{T}_o(k)$, $\mathbf{T}_b(k)$, $\mathbf{D}_o(k)$ and $\mathbf{D}_b(k)$ for the next frame k. $\mathbf{T}_o(k)$ represents the updated ongoing circuit calls, $\mathbf{T}_b(k)$ represents the blocked circuit calls, $\mathbf{D}_o(k)$ represents the assigned data packets, and $\mathbf{D}_b(k)$ is the updated buffered data requests. The objective of the algorithm is to maximize the matrices $\mathbf{T}_o(k)$ and $\mathbf{D}_o(k)$ for the given set of inputs. The CU then passes the assignment decisions to the SUs via the downlink control channels. Each SU then transmits circuit calls and data packets on the assigned traffic channels, while the CU provides the appropriate connections to the CDS module. Note that while the traffic channel is reserved for the entire duration of a circuit call, each data packet transmission lasts only one frame.

Figures 5.4-A and -B show the traffic flow for circuits and packets, respectively. In steady-state operation the flow equations for circuit calls and data packets can be written as:

$$\mathbf{T}(k-1) = \mathbf{T}_o(k-1) - \mathbf{T}_e(k-1) + \mathbf{T}_a(k-1)$$
$$\mathbf{T}(k-1) = \mathbf{T}_o(k) + \mathbf{T}_b(k)$$
$$\mathbf{D}(k-1) = \mathbf{D}_b(k-1) + \mathbf{D}_a(k-1)$$
$$\mathbf{D}(k-1) = \mathbf{D}_o(k) + \mathbf{D}_b(k)$$

A.

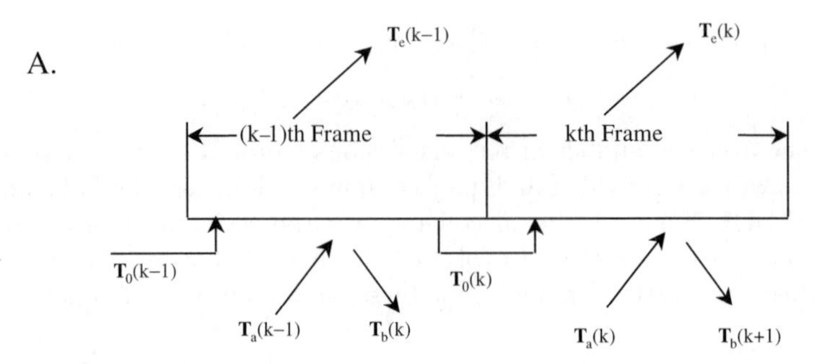

$\mathbf{T}_0$: Active calls
$\mathbf{T}_a$: Newly arrived calls
$\mathbf{T}_r$: Ended calls
$\mathbf{T}_b$: Blocked calls

B.

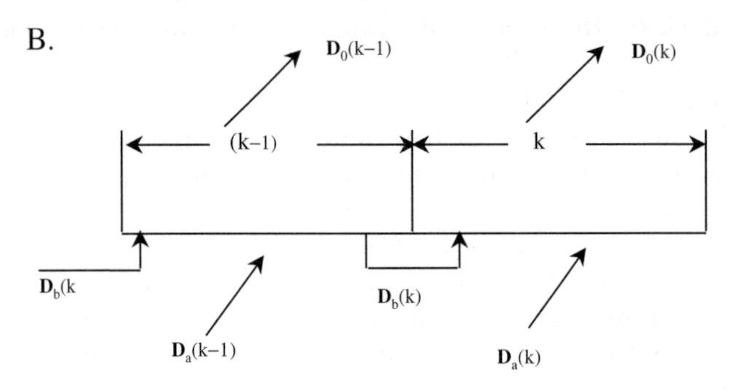

$\mathbf{D}_a$:New packet arrivals
$\mathbf{D}_r$: Packets in the buffer to be scheduled in subsequent frames
$\mathbf{D}_0$:Packets scheduled fo transmission

Figure 5.4 The traffic flow between frames *k-1* and *k* for circuits-1 and packets-2.

Matrices $\mathbf{T}_o$ and $\mathbf{D}_o$ (the number of ongoing circuit calls and assigned data packets, respectively) must satisfy the assignment conditions at any time:

$$\sum_{i=1}^{N} t_{ij}(\mathbf{T}_o) \leq L_c \qquad \text{for } j = 1, \ldots, N \qquad (a-1)$$

$$\sum_{j=1}^{N} t_{ij}(\mathbf{T}_o) \leq L_c \qquad \text{for } i = 1, \ldots, N \qquad (a-2)$$

$$\sum_{i=1}^{N} t_{ij}(\mathbf{T}_o + \mathbf{D}_o) \leq L \qquad \text{for } j = 1, \ldots, N \qquad (b-1)$$

$$\sum_{j=1}^{N} t_{ij}(\mathbf{T}_o + \mathbf{D}_o) \leq L \qquad \text{for } i = 1, \ldots, N \qquad (b-2)$$

where the notation $t_{ij}(\mathbf{X})$ stands for the (i, j) entry of matrix $\mathbf{X}$. Condition (a-1) says that the total number of traffic channels used for circuit calls destinated for downlink beam j cannot exceed L_c, and similarly, (a-2) restricts those originated from uplink beam i. (b-1) and (b-2) restrict the total number of traffic channels used by both circuit calls and data packets to L.

Traffic Channel Assignment Algorithms

The Traffic Channel Assignment Algorithms (TCAAs) allow the CU to assign circuit calls and data packets in order to achieve a high degree of channel utilization and meet capacity constraints. Three such algorithms are proposed here, TCAA-1 (Optimum), TCAA-2 (Fast/Sub-Optimum) and the Random Traffic Channel Assignment (RTCA) algorithm. These algorithms are applied on matrices $\mathbf{T}_o(k-1)$, $\mathbf{T}_e(k-1)$ and $\mathbf{T}_a(k-1)$ of circuit calls and on matrices $\mathbf{D}_b(k-1)$ and $\mathbf{D}_a(k-1)$ of data packets of the current frame, and provide the matrices $\mathbf{T}_o(k)$ and $\mathbf{T}_b(k)$ of circuit calls and the matrices $\mathbf{D}_o(k)$ and $D_b(k)$ of data packets for the next frame.

In the description of the algorithms below, let

$$\mathbf{T}_r(k-1) \triangleq \mathbf{T}_o(k-1) - \mathbf{T}_e(k-1) \qquad \mathbf{T}(k-1) \triangleq \mathbf{T}_r(k-1) + \mathbf{T}_a(k-1)$$

$$\mathbf{D}(k-1) \triangleq \mathbf{D}_b(k-1) + \mathbf{D}_a(k-1)$$

Also, let $r_i(\mathbf{X})$ and $c_j(\mathbf{X})$ denote the ith-row and j^{th}-column sums of matrix $\mathbf{X}$, respectively.

Traffic Channel Assignment Algorithm-1 (Optimum)

TCAA-1 utilizes a maximum flow algorithm to maximize the number of accepted calls. A bipartite graph is set up based on the number of traffic channels available and the number of new requests in each uplink and downlink beam. The 'maximum flow' of that graph is computed which represents the requests that are accepted.

Step 1a:

Initialize matrix $\mathbf{A} = 0$. Consider all (i, j) with $t_{ij}(\mathbf{T}_a(k-1)) > 0$, $r_i(\mathbf{T}_r(k-1)) < L_c$ and $c_j(\mathbf{T}_r(k-1)) < L_c$. Construct matrix $\mathbf{A}$ with $t_{ij}(\mathbf{A}) = \min\{L_c - r_i(\mathbf{T}_r(k-1)), L_c - c_j(\mathbf{T}_r(k-1)), t_{ij}(\mathbf{T}_a(k-1))\}$. If no such (i, j) exists, set $A_r = 0$ and goto Step 1c.

Step 1b:

Set up a network associated with $\mathbf{A}$ (see Remark 1). Find the maximum flow in the network and the corresponding matrix $\mathbf{A}_r$ (see Remark 2).

Step 1c:

Set $\mathbf{T}_o(k) = \mathbf{T}_r(k-1) + \mathbf{A}_r$ and $\mathbf{T}_b(k) = \mathbf{T}_a(k-1) - \mathbf{A}_r$.

Step 2a:

Initialize matrix $\mathbf{B} = 0$. Consider all (i,j) with $t_{ij}(\mathbf{D}(k-1)) > 0$, $r_i(\mathbf{T}_o(k)) < L$ and $c_j(\mathbf{T}_o(k)) < L$. Construct matrix $\mathbf{B}$ with
$t_{ij}(\mathbf{B}) = \min\{L - r_i(\mathbf{T}_o(k)), L - c_j(\mathbf{T}_o(k)), t_{ij}(\mathbf{D}(k-1))\}$.
If no such (i,j) exists, set $\mathbf{B}_r = 0$ and goto Step 2c.

Step 2b:

Set up a network associated with $\mathbf{B}$. Find the maximum flow in the network and the corresponding matrix $\mathbf{B}_r$.

Step 2c:

Set $\mathbf{D}_o(k) = \mathbf{B}_r$ and $\mathbf{D}_b(k) = \mathbf{D}(k-1) - \mathbf{B}_r$.

Remarks on TCAA-1

1. Let us represent matrix $\mathbf{A}$ with the bipartite graph $G_A(I, J, E)$, where the nodes $i \in I$ correspond to the rows of $\mathbf{A}$ and $j \in J$ correspond to the columns of $\mathbf{A}$. The edges $e \in E$ joining the nodes i and j have capacity $C_{ij} = t_{ij}(A)$. Let us add to $G_A(I, J, E)$ a source node S_p and a sink node S_q. S_p is connected to any node $i \in I$ by an edge with capacity $C_{pi} = L_c - r_i(T_r(k-1))$. Similarly, any node $j \in J$ is connected to sink S_q by an edge with capacity $C_{jq} = L_c - c_j(T_r(k-1))$. The resulting graph is then called 'the network associated with A.' For data packets, i.e. matrix $\mathbf{B}$, replace L_c with L and $\mathbf{T}_r(k-1)$ with $\mathbf{T}_o(k)$.
2. A maximal flow algorithm is given in reference [3], using the labeling method. This algorithm gives a maximum flow network which is presented by matrix $\mathbf{A}_r$. For an $N \times N$ matrix, the complexity of the algorithm is bounded by $O(N^3)$. Therefore, TCAA-1 has complexity of about the same order.
3. TCAA-1 is optimal in the sense of maximizing the number of accepted calls or minimizing the number of blocked calls. This follows from the fact that TCAA-1 is based on a maximal flow algorithm. Matrix $\mathbf{T}_o(k)$ is maximized (i.e. the sum of all the entries in the matrix is maximized) for given $\mathbf{T}_a(k-1)$ and $\mathbf{T}_r(k-1)$. Similarly, $\mathbf{D}_o(k)$ is maximized for given $\mathbf{T}_o(k)$ and $\mathbf{D}_a(k-1)$. Note, however, that the maximum flow, and hence $\mathbf{T}_o(k)$, provided in Step 1b is not unique. Therefore, further maximizing of $\mathbf{D}_o(k)$ is possible if a different maximum flow or optimal matrix $\mathbf{T}_o(k)$ is used. An example of TCAA-1 (for circuit calls only) is given in Figure 5.5.

Traffic Channel Assignment Algorithm-2 (Fast/Sub-Optimum)

TCAA-2 attempts to maximize the accepted calls in a forward blind manner by blocking new calls that violate the scheduling conditions. The matrix $\mathbf{T}(k-1) = \mathbf{T}_r(k-1) + \mathbf{T}_a(k-1)$ is reduced in each iteration of the algorithm until the capacity constraints are satisfied. After that, the iterations are repeated with matrix $\mathbf{D}(k-1)$.

Given Matrices : $T_r^{(k-1)} = T_o^{(k-1)} - T_e^{(k-1)}, T_a^{(k-1)}$ *and* $L_c = 8, \quad L = 10$

$$T_r = \begin{bmatrix} 2 & 0 & 2 & 1 \\ 0 & 3 & 2 & 1 \\ 1 & 2 & 2 & 3 \\ 2 & 2 & 0 & 3 \end{bmatrix} \begin{matrix} - & 5 \\ - & 6 \\ - & 8 \\ - & 7 \end{matrix} \qquad T_a = \begin{bmatrix} 1 & 2 & 1 & 0 \\ 2 & 1 & 1 & 0 \\ 0 & 0 & 1 & 0 \\ 2 & 2 & 1 & 1 \end{bmatrix}$$

$$\begin{matrix} | & | & | & | & \uparrow r_i \\ 5 & 7 & 6 & 8 & \leftarrow c_j \end{matrix}$$

Find Matrix $\mathbf{A}=[a_{ij}]$ *such that*

$$a_j = \min\{[L_c - r_i(T_r)],[L_c - c_j(T_r)],t_{ij}(T_a)\} \qquad A=[a_{ij}]=\begin{bmatrix} 1 & 1 & 1 & 0 \\ 2 & 1 & 1 & 0 \\ 0 & 0 & 0 & 0 \\ 1 & 1 & 1 & 0 \end{bmatrix}$$

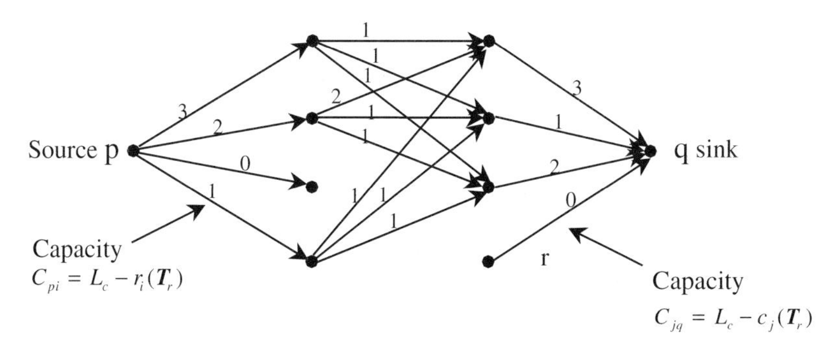

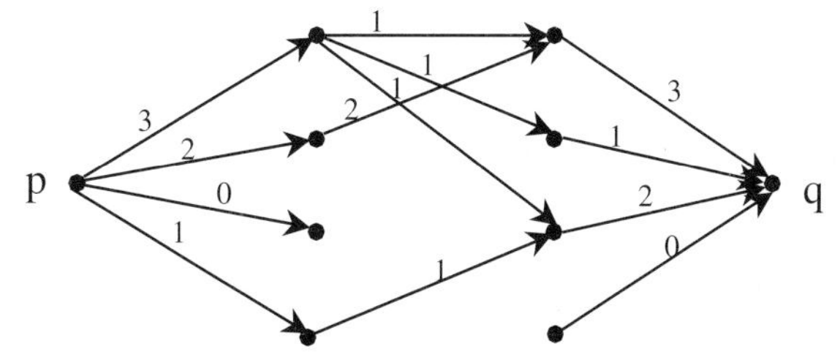

Using the network associated with $\mathbf{A}$, *we find its maximum flow matrix* $\mathbf{A}_r$

$$A_r = \begin{bmatrix} 1 & 1 & 1 & 1 \\ 2 & 0 & 0 & 0 \\ 0 & 0 & 0 & 0 \\ 0 & 0 & 1 & 0 \end{bmatrix} \qquad T_o = T_r + A_r = \begin{bmatrix} 3 & 1 & 3 & 1 \\ 2 & 3 & 2 & 1 \\ 1 & 2 & 2 & 3 \\ 2 & 2 & 1 & 3 \end{bmatrix} \begin{matrix} - & 8 \\ - & 8 \\ - & 8 \\ - & 8 \end{matrix} \qquad \text{and} \qquad T_b = T_a - T_r$$

$$\begin{matrix} | & | & | & | \\ 8 & 8 & 8 & 8 \end{matrix}$$

Step 2 of TCAA- operates on matrices T_0 and $\mathbf{D}$ in a similar manner.

Figure 5.5 An example of TCAA-1.

Step 0:

$m = 0$. Set matrices $\mathbf{T}_a^m$ and $\mathbf{T}^m$ such that $t_{ij}(\mathbf{T}_a^m) = t_{ij}(\mathbf{T}_a(k-1))$, if $r_i(\mathbf{T}_r(k-1)) < L_c$ and $c_j(\mathbf{T}_r(k-1)) < L_c$. $t_{ij}(\mathbf{T}_a^m) = 0$, otherwise. $\mathbf{T}^m \leftarrow \mathbf{T}_r(k-1) + \mathbf{T}_a^m$.

Step 1:

Choose any (i, j) with $t_{ij}(\mathbf{T}_a^m) > 0$, $r_i(\mathbf{T}^m) > L_c$ and $c_j(\mathbf{T}^m) > L_c$; Set
$t_{ij}(\mathbf{T}_a^{m+1}) = t_{ij}(\mathbf{T}_a^m) - \min\{r_i(\mathbf{T}^m) - L_c, c_j(\mathbf{T}^m) - L_c, t_{ij}(\mathbf{T}_a^m)\}$;
$\mathbf{T}^{m+1} = \mathbf{T}_r(k-1) + \mathbf{T}_a^{m+1}$, $m \leftarrow m + 1$;
Goto Step 1.
If no such (i, j) exists, goto Step 2.

Step 2:

Choose any row i with $r_i(\mathbf{T}^m) > L_c$; Goto Step 2a.
If no such row exists, choose any column j with $c_j(\mathbf{T}^m) > L_c$; Goto Step 2b.
If no such column exists, goto Step 3.

Step 2a:

Choose any column j with $c_j(\mathbf{D}(k-1) + \mathbf{T}^m) > L$ and $t_{ij}(\mathbf{T}_a^m) > 0$; Set $t_{ij}(\mathbf{T}_a^{m+1}) = t_{ij}(\mathbf{T}_a^m) - \min\{r_i(\mathbf{T}^m) - L_c, c_j(\mathbf{D}(k-1) + \mathbf{T}^m) - L, t_{ij}(\mathbf{T}_a^m)\}$.
If no such column exists, then choose any column j with $t_{ij}(\mathbf{T}_a^m) > 0$;
Set $t_{ij}(\mathbf{T}_a^{m+1}) = t_{ij}(\mathbf{T}_a^m) - \min\{r_i(\mathbf{T}^m) - L_c, t_{ij}(\mathbf{T}_a^m)\}$.
$\mathbf{T}^{m+1} = \mathbf{T}_r(k-1) + \mathbf{T}_a^{m+1}$, $m \leftarrow m + 1$;
Goto Step 2.

Step 2b:

Choose any row i with $r_i(\mathbf{D}(k-1) + T^m) > L$ and $t_{ij}(\mathbf{T}_a^m) > 0$; Set $t_{ij}(\mathbf{T}_a^{m+1}) = t_{ij}(\mathbf{T}_a^m) - \min\{c_j(\mathbf{T}^m) - L_c, r_i(\mathbf{D}(k-1) + \mathbf{T}^m) - L, t_{ij}(\mathbf{T}_a^m)\}$.
If no such row exists, then choose any row i with $t_{ij}(\mathbf{T}_a^m) > 0$;
Set $t_{ij}(\mathbf{T}_a^{m+1}) = t_{ij}(\mathbf{T}_a^m) - \min\{c_j(\mathbf{T}^m) - L_c, t_{ij}(\mathbf{T}_a^m)\}$.
$\mathbf{T}^{m+1} = \mathbf{T}_r(k-1) + \mathbf{T}_a^{m+1}$, $m \leftarrow m + 1$;
Goto Step 2.

Step 3:

Set $\mathbf{T}_o(k) = \mathbf{T}_r(k-1) + \mathbf{T}_a^m$, $\mathbf{T}_b(k) = \mathbf{T}_a(k-1) - \mathbf{T}_a^m$.
$n = 0$, $\mathbf{D}^n = \mathbf{D}(k-1)$; Goto Step 4.

Step 4:

Choose any (i, j) with $t_{ij}(\mathbf{D}^n) > 0$, $r_i(\mathbf{D}^n + \mathbf{T}_o(k)) > L$ and $c_j(\mathbf{D}^n + \mathbf{T}_o(k)) > L$;
Set $t_{ij}(\mathbf{D}^{n+1}) = t_{ij}(\mathbf{D}^n) - \min\{r_i(\mathbf{D}^n + \mathbf{T}_o(k)) - \mathbf{L}, c_j(\mathbf{D}^n + \mathbf{T}_o(k)) - L, t_{ij}(\mathbf{D}^n)\}$;
$n \leftarrow n + 1$; Goto Step 4.
If no such (i, j) exists, goto Step 5.

Step 5:

> Choose any row i with $r_i(\mathbf{D}^n + \mathbf{T}_o(k)) > L$; Goto Step 5a.
> If no such row exists, choose any column j with $c_j(\mathbf{D}^n + \mathbf{T}_o(k)) > L$; Goto Step 5b.
> If no such column exists, goto Step 6.

Step 5a:

> Choose any column j with $t_{ij}(\mathbf{D}^n) > 0$; Set $t_{ij}(\mathbf{D}^{n+1}) = t_{ij}(\mathbf{D}^n) - \min\{r_i(\mathbf{D}^n + T_o(k)) - L, t_{ij}(\mathbf{D}^n)\}$;
> $n \leftarrow n+1$; Goto Step 5.

Step 5b:

> Choose any row i with $t_{ij}(\mathbf{D}^n) > 0$; Set $t_{ij}(\mathbf{D}^{n+1}) = t_{ij}(\mathbf{D}^n) - \min\{c_j(\mathbf{D}^n + \mathbf{T}_o(k)) - L, t_{ij}(\mathbf{D}^n)\}$;
> $n \leftarrow n+1$; Goto Step 5.

Step 6:

> $\mathbf{D}_o(k) = \mathbf{D}^n$, $\mathbf{D}_b(k) = \mathbf{D}(k-1) - \mathbf{D}_o(k)$.

Remarks on TCAA-2

TCAA-2 is a sub-optimum algorithm for providing traffic channel assignments. TCAA-2 has reduced computational complexity as compared to TCAA-1. (Its complexity in the worst case is $O(N^2)$.) The algorithm attempts to maximize the circuit calls in $\mathbf{T}_o$ also considers maximizing the traffic of data packets in $\mathbf{D}_o$ (see Steps 2a and 2b).

Random Traffic Channel Assignments (RTCA)

The Random Traffic Channel Assignment algorithm given below chooses the entries in matrix $\mathbf{T}_a(k-1)$ one by one randomly, and determines if that entry can be accommodated. The process is then repeated with $\mathbf{D}_a(k-1)$. This algorithm has the least possible computation complexity.

Step 0:

> $m = 0$, $\mathbf{T}_o^m = \mathbf{T}_r(k-1)$, $\mathbf{T}_x^m = 0$.

Step 1:

> For each $t_{ij}(\mathbf{T}_a(k-1)) > 0$, set
> $t_{ij}(\mathbf{T}_x^m) = \min\{t_{ij}(\mathbf{T}_a(k-1)),$
> $L_c - r_i(\mathbf{T}_o^m), L_c - c_j(\mathbf{T}_o^m)\}$,
> $\mathbf{T}_o^{m+1} = \mathbf{T}_o^m + \mathbf{T}_x^m$, $m \leftarrow m+1$;
> Goto Step 1.

Step 2:

> $\mathbf{T}_o(k) = \mathbf{T}_o^m$, $\mathbf{T}_b(k) = \mathbf{T}_a(k-1) - \mathbf{T}_x^m$.
> $n = 0$, $\mathbf{D}_x^n = 0$.

Step 3:

> For each $t_{ij}(\mathbf{D}(k-1)) > 0$, set
> $t_{ij}(\mathbf{D}_x^{n+1}) = \min\{t_{ij}(\mathbf{D}(k-1)),$
> $L - r_i(\mathbf{T}_o(k) + \mathbf{D}_x^n), L - c_j(\mathbf{T}_o(k) + \mathbf{D}_x^n)\}$,
> $n \leftarrow n+1$. Goto Step 3.

Step 4:

$$\mathbf{D}_o(k) = \mathbf{D}_x^n\,,\ \mathbf{D}_b(k) = \mathbf{D}(k-1) - \mathbf{D}_o(k)\,.$$

Remarks on RTCA
RTCA treats all the entries in the arrival matrix equally, and does not have a systematic way to determine which entry should be accepted or blocked ahead of others (c.f. TCAA-2 where the entries are 'ordered' according to the row or column sums of the matrix).

Time Versus Code Multiplexed Switching

Comparing the time slot assignment problem in a Time Multipled Switch (TMS) with the equivalent problem in a CDS, we can make the following observations. In each time slot of a TMS, in order to avoid conflict we must make no more than one assignment to any input port i or output port j. Now, given the assignments of ongoing calls over a frame, a new call request is said to be blocked if it results in conflict in every time slot of the frame. An optimum time slot assignment algorithm may then have to rearrange the ongoing calls in order to accommodate the newly arrived calls. One such example is shown in Figure 5.6. One of the newly arrived calls in the example would have been blocked if rearrangement of the ongoing calls was not allowed. It has been shown [4] that for TMS, the throughput of an optimum assignment algorithm is 10% to 15% higher than that of a random algorithm which does not allow rearrangement of ongoing calls.

In code switching, on the other hand, a new call request can be assigned to the requested input-output port (i.e. uplink-downlink beam) as long as there are codes in these beams available. In other words, the requested route can be formed by pairing up any of the available codes of the input-output beams without rearranging the ongoing calls. Hence in code switching, optimization means collecting requests for a time period (a frame), and then assigning them so that the input-output flow is maximized, as described by the above algorithms. When this time frame becomes very small, the optimum, sub-optimum and random assignments will converge. If, for instance, only one new call request has arrived in the time frame, then there would be no difference between the three assignments. These observations are verified by the results given in Section 5.4 under light traffic load.

5.3 System Throughput Analysis

In this section we present the system performance analysis when the traffic channel assignment algorithm is *optimum*. The analysis provides the blocking probability for circuit calls and the delay for data packets. The performance of the optimum algorithm may also be considered as the limiting case of the *sub-optimum* or *random* algorithm, since the performance of the two systems will be the same when the number of new call requests per frame is at most one. This is often the case for circuit switched calls. When there is more than one request per frame, the performance analysis given here is a tight bound of actual performance as obtained by simulation.

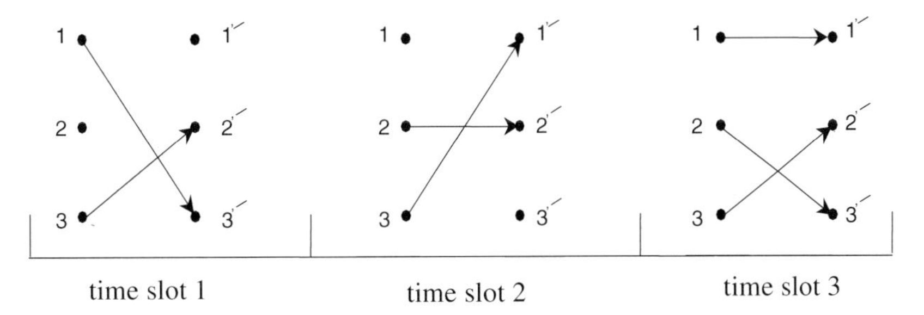

Newly arrived circuit calls $1 \rightarrow 1'$ and $2 \rightarrow 3'$

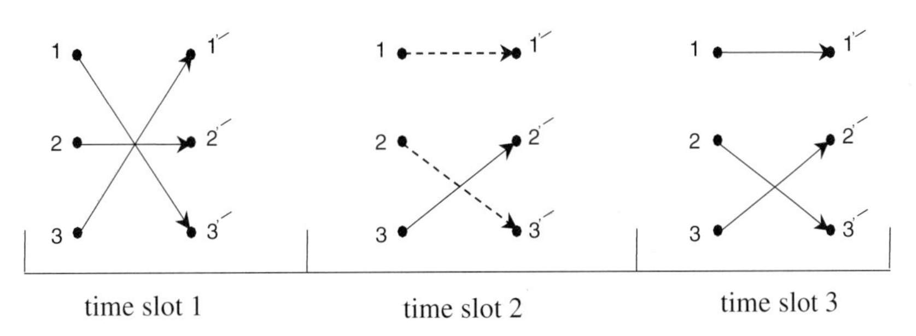

Figure 5.6 TMS call rearrangement for accommodating new calls.

5.3.1 Distribution of Circuit Calls

Each active circuit user is modeled as a two-state ON/OFF Markov process. A transition from the OFF state to the ON state takes place with probability σ_c, which is the probability of a call arrival in a frame by a single user. The transition from ON to OFF state, representing the call termination in a frame, takes place with probability ρ. Let the total number of circuit call users in each uplink beam be M_c. We combine all the users in the N uplink beams into one population of size NM_c. The call arrival process is described by the following conditional probability:

$$\Pr\left[r_i(\mathbf{T}_a(k-1)) = m | r_i(\mathbf{T}_o(k-1)) = l\right] = b(m, NM_c - l, \sigma_c) \quad 0 \leq m \leq NM_c - l$$

where $b(l, n, p) = \binom{n}{l} p^l (1-p)^{n-l}$, the random variable $r_i(\mathbf{T}_a(k-1))$ is the row sum of matrix $\mathbf{T}_a(k-1)$ and represents the number of new call arrivals in uplink beam i in frame $(k-1)$ for any destination. The condition $r_i(\mathbf{T}_o(k-1)) = l$ represents the row sum of matrix $\mathbf{T}_o(k-1)$ of ongoing calls in frame $(k-1)$, with $0 \leq l \leq L_c$. The capacity for circuit calls is denoted by L_c. The probability distribution of random variable $r_i(\mathbf{T}_e(k))$, of calls ending transmission in uplink beam i, is the following:

$$\Pr\left[r_i(\mathbf{T}_e^{(k)}) = l | r_i(\mathbf{T}_o^{(k)}) = h\right] = b(l, h, \rho) \qquad 0 \leq h \leq L_c$$

The call duration in number of frames is geometrically distributed:

$$\Pr[t_c = l] = \rho(1-\rho)^{l-1}$$

The average call duration is then $\bar{t}_c = 1/\rho$.

The random variables $r_i(\mathbf{T}_o(k))$ and $r_i(\mathbf{T}_b(k))$ represent the row sum of matrices $\mathbf{T}_o(k)$ of ongoing calls and $\mathbf{T}_b(k)$ of blocked requests in frame k, respectively, for uplink beam i. The dependence relations between these random variables (given that $\mathbf{T}(k-1) = \mathbf{T}_o(k-1) - \mathbf{T}_e(k-1) + \mathbf{T}_a(k-1)$) are the following:

$$r_i(\mathbf{T}_o(k)) = \min\{r_i(\mathbf{T}(k-1)), L_c\}$$
$$r_i(\mathbf{T}_b(k)) = \max\{0, r_i(\mathbf{T}(k-1)) - L_c\}$$

The above equations are based on the assumption that no more than L_c traffic channels can be assigned to circuit calls in each uplink beam. In addition, we assume that the algorithm always assigns exactly L_c calls if $r_i(T) \geq L_c$, i.e. then we assume that the traffic channel assignment algorithm is *optimum.*

Now, let the total number of calls $r_i(T_o)$ handled at input beam i, be the state of a Markov chain. The transition probabilities from frame $(k-1)$ to frame k will then be the following:

$$\Pr[r_i(\mathbf{T}_o(k)) = m | r_i(\mathbf{T}_o(k-1) = n)]$$
$$= \begin{cases} \sum_{l=0}^{n} b(l,n,\rho) b(m+l-n, NM_c - n, \sigma_c) & \text{if } 0 \leq m < L_c,\ 0 \leq n \leq L_c \\ \sum_{l=0}^{n} b(l,n,\rho) \sum_{k=L_c+l-n}^{NM_c - n} b(k, NM_c - n, \sigma_c) & \text{if } m = L_c,\ 0 \leq n \leq L_c \end{cases}$$

The steady state distribution, $\Pr(r_i(T_o))$ will then be found from the following iteration:

$$\Pr[r_i(\mathbf{T}_o^{(k)}) = m] = \sum_{n=0}^{L_c} \Pr[r_i(\mathbf{T}_o^{(k)}) = m | r_i(\mathbf{T}_o^{(k-1)} = n)] \Pr[r_i(\mathbf{T}_o^{(k-1)}) = n]$$

given that $\sum_{n=0}^{L_c} \Pr[r_i(\mathbf{T}_o(k-1) = n] = 1$.

Now, given N, the number of input or output beams, we assume that each call has equal probability of being addressed to any input i or output j. As N becomes large, the number of calls routed from a given input to a given output becomes negligible relative to the total number of calls handled. This implies that the number of calls currently being handled by input i becomes independent of the number of calls handled by output j. In addition, as N becomes large the input and output Markov chains approach dynamic independence, because the settling time of the Markov chain becomes negligible compared to the time between connection requests for a given input-output pair. The assumption of independence is shown to be a reasonable approximation (see reference [4]).

Therefore, since the input and output Markov chains are independent and identical, the steady-state distribution of the total number of calls destined for output j, $c_j(\mathbf{T}_o)$ is the same as $r_i(\mathbf{T}_o)$.

For a call from input i to output j, the call will be blocked if input i and/or output j is currently operating at its peak capacity. Because of this blocking, the traffic load is reduced. This is accounted for by replacing σ_c with $\sigma_c(1 - \Pr[r_i(\mathbf{T}_o(k-1)) = L_c])$. The Markov chain is then solved recursively to obtain the steady-state distribution of $r_i(\mathbf{T}_o)$. (see reference [5]).

Call Blocking Probability

The call blocking probability P_B is the probability that the call will be blocked at either the input and/or the output. That is,

$$P_B = \Pr[\text{call blocked at the input} \cup \text{call blocked at the output}]$$

P_B can be expressed as a function of the distributions of the random variables $r_i(\mathbf{T}_o)$ and $c_j(\mathbf{T}_o)$, as follows:

$$P_B = \Pr[r_i(T_o) = L_c] + \Pr[c_j(T_o) = L_c] - \Pr[r_i(\mathbf{T}_o) = L_c]\Pr[c_j(\mathbf{T}_o) = L_c]$$

Now, since $\Pr[r_i(\mathbf{T}_o) = L_c] = \Pr[c_j(\mathbf{T}_o) = L_c]$, we have

$$P_B = \big(2 - \Pr[r_i(\mathbf{T}_o) = L_c]\big)\Pr[r_i(\mathbf{T}_o) = L_c]$$

5.3.2 Distribution of Data Packets

Data users are assumed to have no memory. Each data user can generate up to one packet per frame with probability σ_p and each packet has the length of one frame. Let M_p be the number of data users in beam. We combine all the data users in the N beams into one population of size NM_p. The packet arrival process then is given by:

$$\Pr[r_i(\mathbf{D}_a(k-1)) = m] = b(m, NM_p, \sigma_p) \qquad 0 \leq m \leq NM_p$$

where random variable $r_i(\mathbf{D}_a(k-1))$ is the row sum i of the packet arrival matrix in frame $(k-1)$.

Let us also consider the random variables $r_i(\mathbf{D}_b(k-1))$, $r_i(\mathbf{D}_b(k))$ and $r_i(\mathbf{D}_o(k))$, representing the row sum of the matrices of the number of buffered packets in frame $(k-1)$, the buffered packets and the packets transmitted in frame k, respectively. The dependence relations between the above random variables are:

$$\begin{aligned} r_i(\mathbf{D}_o(k)) &= \min\{r_i(\mathbf{D}_b(k-1)) + r_i(\mathbf{D}_a(k-1)), c(k)\} \\ r_i(\mathbf{D}_b(k)) &= \max\{0, r_i(\mathbf{D}_b(k-1)) + r_i(\mathbf{D}_a(k-1)) - c(k)\} \end{aligned}$$

where, $c(k) = L - r_i(\mathbf{T}_o(k))$ is the total number of traffic channels available for packets in frame k $(c(k) \geq L_p)$.

The above relations are based on the assumption that unused circuit traffic channels are assigned for packets. It is also assumed that the assignment algorithm has the capability of assigning all available channels (l_{p_i}) to packets if $r_i(\mathbf{D}) \geq l_{p_i}$.

Let us assume the duration of a circuit call is much longer than one frame. Hence, we can assume that the number of ongoing circuit calls remains unchanged while steady-state is reached by the packet calls. Let the number of ongoing circuit calls be $r_i(\mathbf{T}_o) = m$. Let us now consider a Markov chain with state $[r_i(\mathbf{D}_b)]$. Based on this assumption, the transition probabilities for data packets are the following:

$$\begin{aligned} &\Pr[r_i(\mathbf{D}_h^{(k)}) = p | r_i(\mathbf{D}_h^{(k-1)}) = q] \\ &= \begin{cases} \sum_{l=0}^{c-q} b(l, NM_p, \sigma_p) & \text{if } p = 0,\ 0 \leq q \leq NB_p \\ b(c-q+p, NM_p, \sigma_p) & \text{if } 1 \leq p \leq B_p - 1,\ 0 \leq q \leq NB_p \\ \sum_{l=c-q+NB_p}^{NM_p} b(l, NM_p, \sigma_p) & \text{if } p = NB_p,\ 0 \leq q \leq NB_p \end{cases} \end{aligned}$$

where B_p is the buffer size for each beam, and $c = L - m$ is the number of traffic channels available for packet calls given that there are m ongoing circuit calls (packet arrivals when the buffer is full are discarded).

The steady-state distribution of the data packet queue size $r_i(\mathbf{D}_b)$ can be obtained by solving the above Markov chain, and then averaging with respect to the steady-state distribution of the number of ongoing circuit calls.

Average Packet Delay

The probability $q_{ij}(l)$ of having a queue length of l for an input-output pair (i, j) is given by:

$$q_{ij}(l) = \Pr[r_i(\mathbf{D}_b) = l] \cdot \Pr[r_i(\mathbf{D}_b) = l] + 2 \cdot \Pr[r_i(\mathbf{D}_b) = l] \cdot \Pr[r_i(\mathbf{D}_b) < l]$$

where we have used the same argument of dynamic independence between input and output, and the distribution of the output queue length is the same as the input queue length.

The average waiting time (in number of frames) for scheduling the packet, W_s, is given by

$$W_s = \frac{\sum_1^{B_p} l q_{ij}(l)}{N M_p \sigma_p}$$

The total average waiting time until packet transmission is then

$$W = W_s + 2t_p + \alpha(W_s + 2t_p) + (2t_p + t_f)$$

where $2t_p$ is the round trip delay, t_f is the frame length and α is the average number of retransmissions of packet requests over the Access channel ($\alpha = 0.01$ by design of the Access channel). The length of a packet request message is assumed to be much smaller than the frame length, and is assumed to arrive in the middle of the frame, (see Figure 5.7).

5.4 Performance Results

In this section, we present the performance results for three cases of the movable boundary (L_c, L): (I) $L_c = 0 < L$, (II) $0 < L_c < L$ and (III) $L_c = L$. An 8×8 switch ($N = 8$) is used in all three cases.

(I) $L_c = 0 < L$. In this case, all traffic channels are used for data packet assignments only. We present the data packet delay versus the offered packet load for the three TCAAs presented in Section 5.3. The number of traffic channels is $L = 8$ with $L_c = 0$. We assume the user population for each uplink beam is $M = 50,000$ and the buffer size for each user is $B_p = 10$. (New requests are considered lost if the buffer is full). The probability that a user will transmit a packet in a frame is σ_p. The packet load (for each input beam) then is defined as $M\sigma_p$.

The average packet queuing delays are plotted in Figure 5.8. The results are obtained by Markovian analysis for the TCAA-1 and by simulations for the TCAA-1 (Optimum), TCAA-2 (Fast/Sub-Optimum) and RTCA (Random). The average delays are expressed in terms of the number if frames. From the results, we observe that when

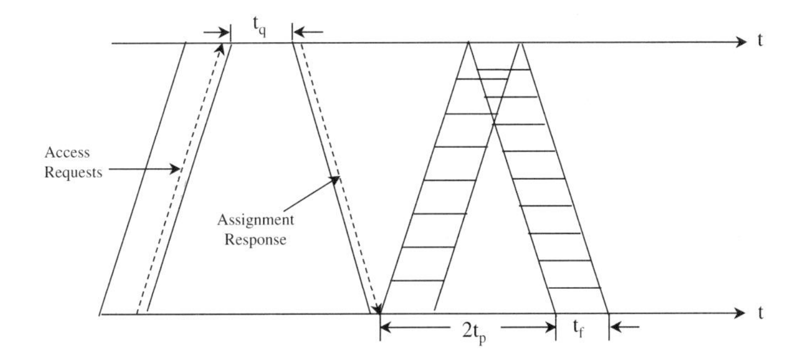

$2t_p$: Round trip propagation delay
t_f: Frame length or packet length
t_q: Queing delay of the packet request
α: Average number of retransmissions of the packet request over the access channel ($\alpha = 0.01$)

$$T_D = t_q + 2\,t_p + \alpha(t_q + 2t_p) + (2t_p + t_f)$$

Figure 5.7 The total data packet delay, T_D.

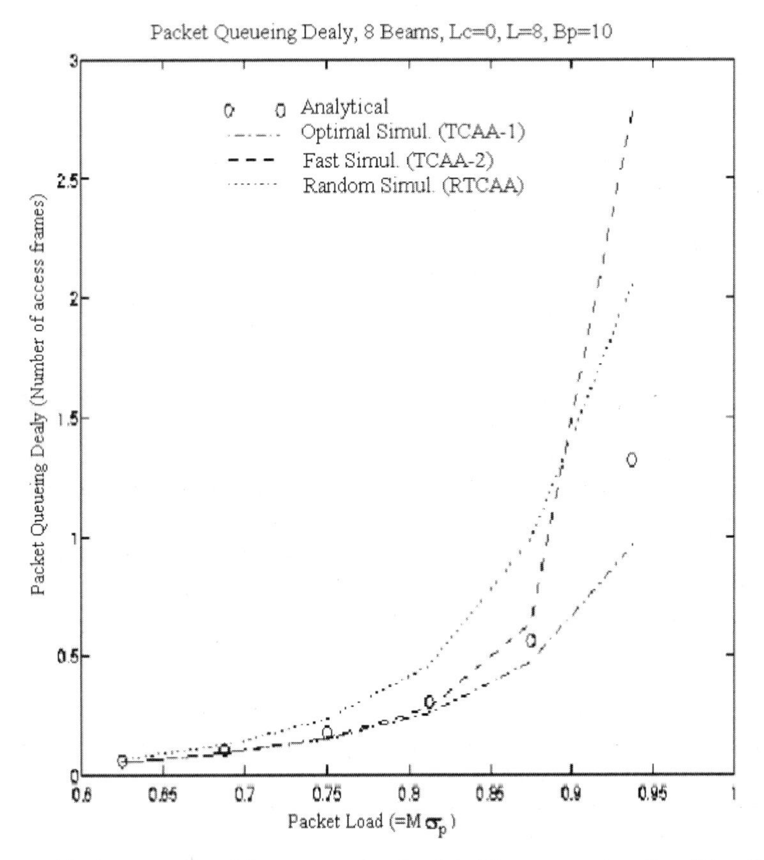

Figure 5.8 Packet queuing delay vs. packet load (without any circuit-call load). System parameters: Lc = 0, L = 8, Bp = 10. The queuing delays are evaluated analytically for TCAA-1 and by simulation for TCAA-1, -2 and RTCAA.

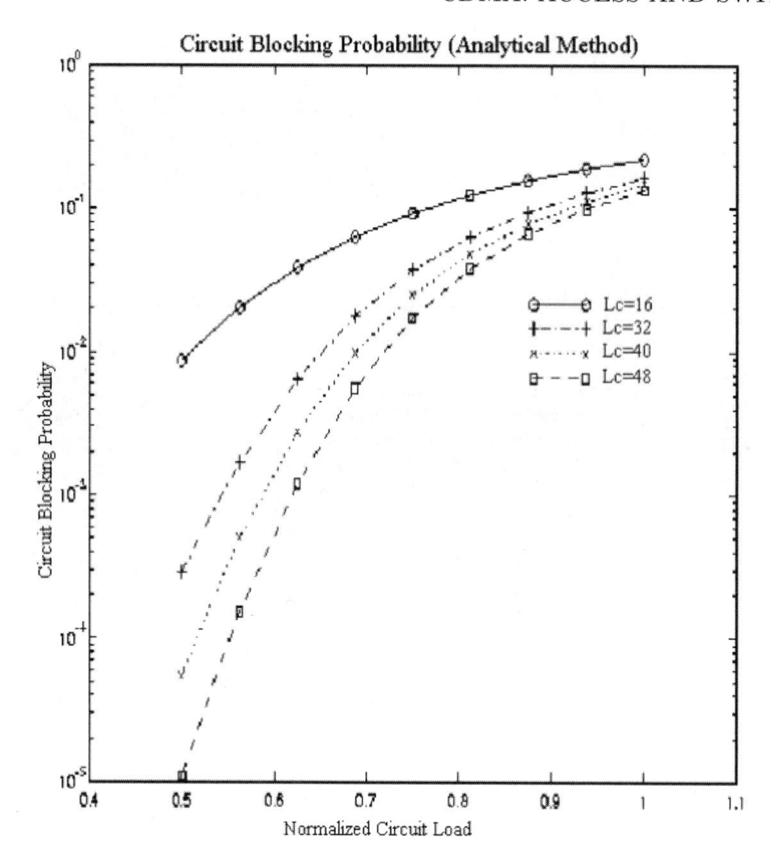

Figure 5.9 Circuit blocking probability vs. normalized circuit load with circuit call only. System parameters: Lc = 16, 32, 40, 48. Circuit blocking probabilities are evaluated analitically.

the traffic load is heavy, the optimum assignment algorithm (TCAA-1) outperforms the other two algorithms. When the traffic load is light, the fast assignment algorithm (TCAA-2) gives a performance close to the optimum algorithm. Also, the TCAA-2 packet delay appears to increase rapidly for packet loads above 0.88, due to the greedy nature of the algorithm when the number of arrivals per frame is large. In addition, we find that the results obtained from the Markovian analysis and the simulation of the TCAA-1 are very close, except at a higher packet load (above 0.85). The reason is that in the simulation, each beam has one buffer of size B_p, while the analysis combines the buffer of all beams into one of size NB_p, which gives a slightly larger delay than simulation.

(II) $0 < L_c < L$. In this case we consider both circuit and packet calls, and compare the performances of the three TCAAs. In Figure 5.9, we present the circuit blocking probabilities versus the normalized circuit load for different circuit call capacities, $L_c = 16, 32, 40, 48$. Figure 5.10 shows the blocking probabilities of circuit calls versus the normalized load of circuit calls. The packet traffic load is fixed at $M\sigma_p = 0.75$. The probability that an ongoing circuit call will terminate in the next frame, ρ_c, is

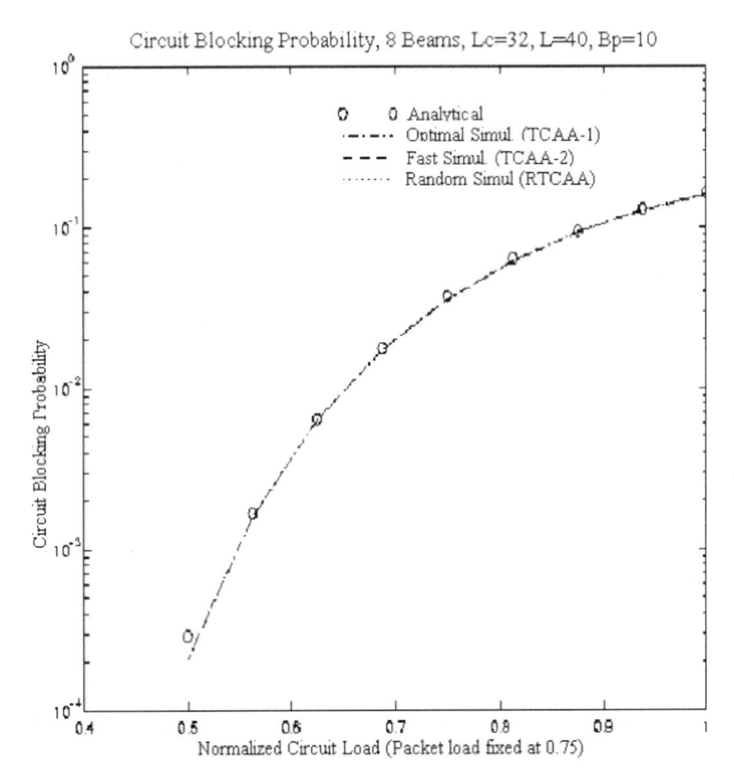

Figure 5.10 Circuit blocking probability vs. normalized circuit load when the packet load is fixed at 0.75. System parameters: Lc = 32, L = 40, Bp = 10.

fixed at 3.333×10^{-4}, which corresponds to an average circuit call duration of 3000 frames (that is, for frame length 40 ms, the call duration is 2 min). and the probability that a circuit user will become active, σ_c, is varied. The circuit traffic load is defined as $M\sigma_c/\rho_c$, where $M = 50000$ is the number of circuit users in each uplink beam. The normalized circuit load is defined as $\frac{M\sigma_c/\rho_c}{L_c/N}$. The system parameters used are $L_c = 32$, $L = 40$ and $B_p = 10$. Here we observe that the blocking probabilities for TCAA-1, TCAA-2 and RTCA obtained from both analysis and simulation are almost identical. This is due to the fact that circuit calls are much longer than the frame length, and thus the average number of call arrival or terminations in a frame is very small. In this case, the optimum and the random TCAA give the same result for code division switching systems. Figure 5.11 shows the average queueing delays for data packets versus the normalized circuit load for a fixed value of packet load $M\sigma_p = 0.75$ packets/frame. Here we observe the the packet delay for the RTCA is higher than the other TCAAs for higher circuit loads.

(III) $L_c = L$. This is when circuit calls have pre-emptive priority over packets for all traffic channels in the pool. In this case, we present the average queueing delays for data packets versus the normalized circuit load for a fixed value of packet load $M\sigma_p = 0.5$ packets/frame in Figure 5.12. The system parameters are $L_c = 32$, $L = 32$ and $B_p = 10$. We observe that packet traffic can be routed through the switch even

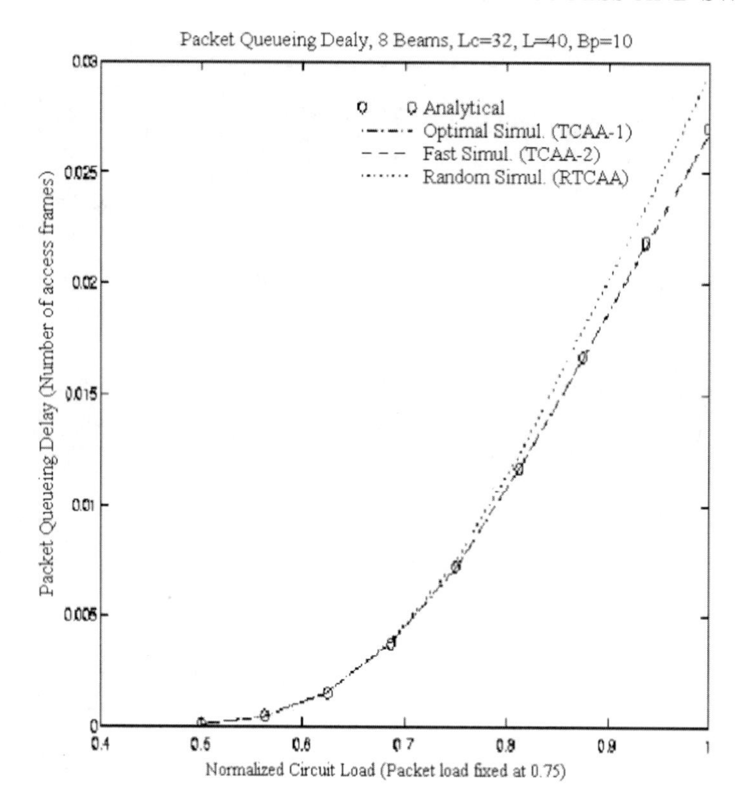

Figure 5.11 Packet queuing delay vs. normalized circuit load when packet load is fixed at 0.75. System parameters: Lc = 32, L = 40, Bp = 10. Queuing delays are evaluated analytically for TCAA-1 and by simulation for TCAA-1, -2 and RTCAA.

when all its capacity is allocated for circuit calls. We also observe that TCAA-1 and TCAA-2 perform better than the Random TCAA (RTCA).

5.5 Conclusions

In this chapter, we have proposed a Satellite Switched CDMA (SS/CDMA) demand assignment system which provides multiple access and switching for both voice calls and data packets. The system operates under the control of channel assignment algorithms. Three such algorithms have been presented here, an optimum, a sub-optimum and a random one. Performance analysis and simulation has been carried out in order to evaluate the switch throughput under the control of the proposed algorithms.

Performance results indicate that the blocking probabilities for circuit calls under optimal and random control are almost the same. This is an advantage of the code division switching method since a computational simple random assignment algorithm can offer almost the same results as the optimum one. In time multiplexed switching, on the other hand, a random algorithm has 15% lower throughput of circuit calls than the optimum one. The data packet delay via the switch has also been evaluated for

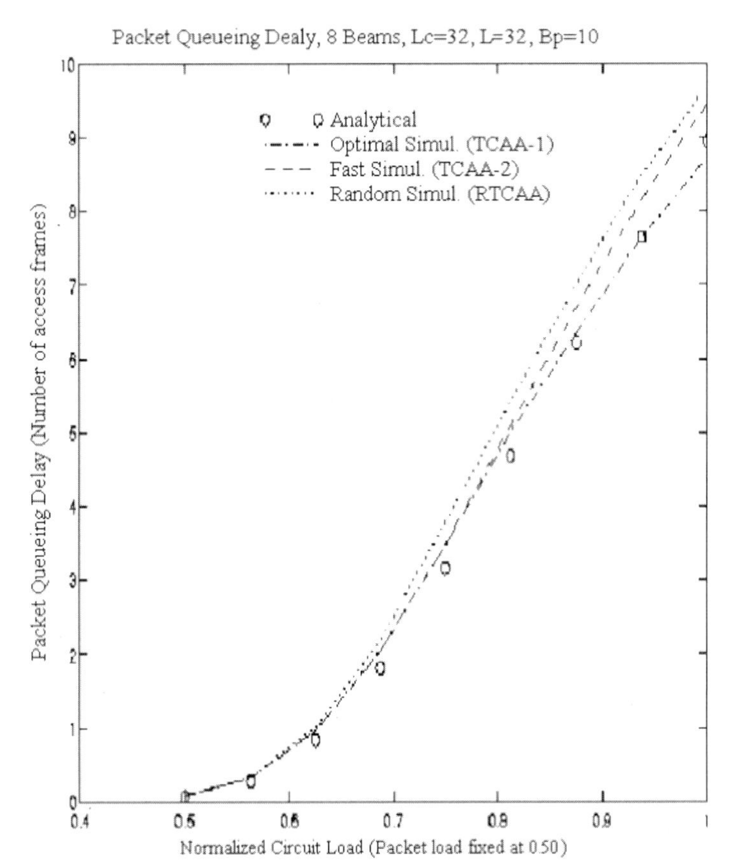

Figure 5.12 Packet queuing delay vs. normalized circuit load when packet load is fixed at 0.5. System parameters: L = Lc = 32, Bp = 10. Packet delays are evaluated analytically for TCAA-1 and by simulation for TCAA-1, -2 and RTCAA.

the cases: (I) no traffic channels for circuit calls, (II) traffic channels for both circuit calls and data packets, and (III) no traffic channels for data packets. In all cases we observed that the sub-optimum algorithm packet delay is near the optimum one, and the random one has slightly higher delays, for packet loads of 0.75 or less and for any circuit load. We also observed that in case (III) where there are no traffic channels for packets, a packet load of 0.5 can be routed via the switch with limited delays.

References

[1] T. Inukai 'An efficient SS/TDMA time-slot assignment algorithm' *IEEE Trans. Commun.*, Vol. 27, No. 10, October 1979 pp. 1449–1455.

[2] D. Gerakoulis, W-C. Chan and E. Geraniotis 'Throughput Evaluation of a Satellite-Switched CDMA (SS/CDMA) Demand Assignment System' *IEEE J. Select. Areas Commun.* Vol. 17, No. 2, February 1999 pp. 286–302.

[3] T.C. Hu *Integer Programming and Network Flows.* Addison-Wesley, Massachusetts, 1970.

[4] C. Rose and M.G. Hluchyj 'The performance of random and optimal scheduling in a time-multiplex switch' *IEEE Trans. Commun.*, Vol. 35, No. 8, August 1987 pp. 813–817.

[5] F. Kelly 'Blocking probabilities in large circuit-switched networks' *Adv. in App. Prob.*, Vol. 18, 1986 pp. 473–505.

6

The Spectrally Efficient CDMA Performance

6.1 Overview

As we have discussed in Chapter 3, the SS/CDMA Traffic channels are based on a spectrally efficient CDMA (SE-CDMA). The SE-CDMA is designed to reuse each frequency channel in every satellite beam (frequency reuse one), and also achieve a very low bit error rate (10^{-6} to 10^{-10}) at a very low signal-to-noise ratio (E_b/N_o). The low E_b/N_o value will allow the use of an Ultra Small Aperture Terminal (USAT) (antenna dish 26" in diameter), and provide a sufficient margin to mitigate the Output Back-Off (OBO) at the on-board downlink power amplifier (TWT).

In this chapter we first present the system description and the signal and channel models (Section 6.2). Then, in Section 6.3, we provide the intra- and inter-beam interference analysis. In Section 6.4, we examine the on-board signal processing and the impact of the uplink-downlink coupling. In Section 6.5, we evaluate the Bit Error Rate (BER) using a concatenated channel encoder and M-ary PSK modulation. In Section 6.6, we present the performance results, and in Section 6.7 a discussion the conclusions. This work was originally presented in reference [1].

6.2 System Description and Modeling

The Traffic channels in the SS/CDMA system carry voice and data directly between the end subscriber units. The multiple access and the modulation of the traffic channel is based on the Spectrally Efficient Code Division Multiple Access (SE-CDMA) scheme, which is analyzed in this chapter. Each SE-CDMA channel is comprised of three segments: the uplink and downlink channels and the on-board routing circuit. Both the uplink and downlink are orthogonal CDMA channels. A generalized block diagram of the SE-CDMA is shown in Figure 3.27 of Chapter 3. The concatenated channel encoder consists of an outer Reed–Solomon RS(x,y) code (rate y/x) and an inner Turbo-code with rate k/n. The *Turbo-Code* is a parallel concatenation of recursive systematic convolutional codes linked by an interleaver. The Turbo encoder output generates n (parallel) symbols which are mapped into the M-ary PSK signal set ($M = 2^n$). The signal phases Φ_i are then mapped into the inphase and quadrature components (a, b), $\Phi_i \rightarrow (a, b)$.

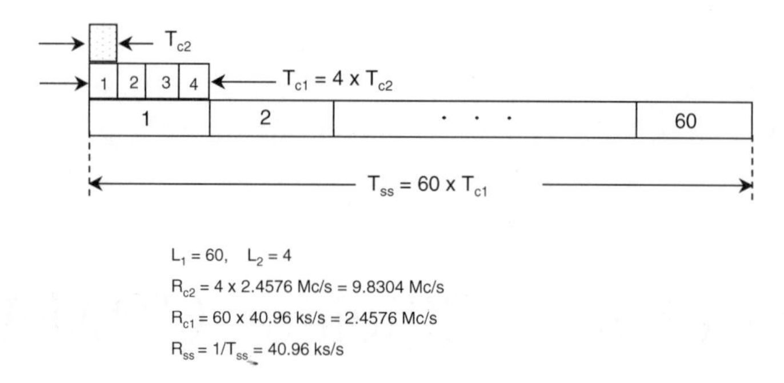

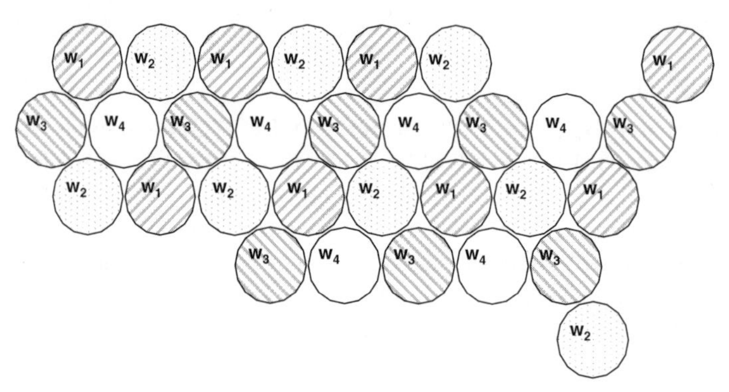

Figure 6.1 The spreading and overspreading symbols for FO/SE-CDMA and the beam-code re-use over continental USA.

The SE-CDMA spreading operation takes place in two steps. The first step provides orthogonal separation of all users within the CDMA channel of bandwidth W, and the second one orthogonal and/or PN code separation between the satellite beams. Depending on the particular implementation of the spreading process, the SE-CDMA can be Fully Orthogonal (FO), Mostly Orthogonal (MO) or Semi-Orthogonal (SO). In all implementations there is orthogonal separation of the users within each beam. In addition, the FO/SE-CDMA provides orthogonal separation of the first tier of the satellite beams (four beams). The MO/SE-CDMA has two orthogonal beams in the first tier, while the SO/SE-CDMA has all beams separated by PN-codes. The spreading operations for the FO and MO/SE-CDMA are shown in Figure 3.12-A and for the SO/SE-CDMA in Figure 3.12-B in Chapter 3. The inphase and quadrature components are spread by the same orthogonal and PN-codes. The FO and MO SE-CDMA require code generators L_1 and L_2 for the user and the beam separation, respectively. The first spreading step generates the chip rate R_{c1}, and the second generates the chip rate R_{c2} (overspreading). The FO/SE-CDMA has $R_{c2} = 4 \times R_{c1}$ and the MO/SE-CDMA has $R_{c2} = 2 \times R_{c1}$. The (I,Q) PN code generator has a rate of R_{c2}, and is used to isolate the interference from the second tier of beams. The SO/SE-CDMA spreading consists of L-orthogonal user codes and a PN beam code. The satellite beams in this case are separated only by the PN code, having a rate of $R_c = R_{c2}$.

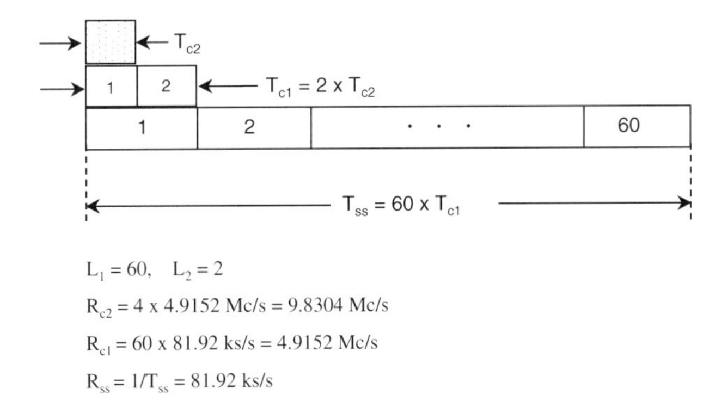

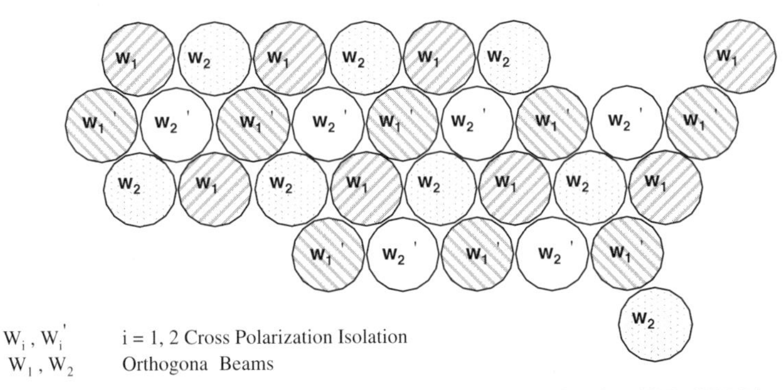

Figure 6.2 The spreading and overspreading symbols for SM/SE-CDMA and the beam-code re-use over continental USA.

The spreading orthogonal code of length L chips will span over the entire length of a symbol. Also, in order to maintain the code orthogonality, the SE-CDMA requires synchronization for the uplink channel. That is, the chips of all orthogonal codes of the uplink SE-CDMA channel must be perfectly aligned at the satellite despreaders.

The specific SE-CDMA implementations are described in Table 6.1. These are the Fully Orthogonal (FO-1), the Mostly Orthogonal (MO-1) and the Semi-Orthogonal (SO-1). In all implementations the outer Reed–Solomon code has a rate of 15/16, the inner Turbo encoder rate is 2/3 for FO-1, 1/2 for MO-1 and 1/3 for SO-1. The modulation scheme is 8-PSK for FO-1 and QPSK for MO-1 and SO-1. FO-1 has a beam code reuse of 1/4, MO-1 1/2 and SO-1 of 1. The above set of parameters has

Table 6.1 SE-CDMA selected implementations.

SE-CDMA IMPLEMENT.	OUTER CODE	INNER CODE	MPSK SCHEME	CODE REUSE
FO-1	$RS(16\lambda, 15\lambda)$	Turbo, 2/3	8-PSK	1/4
MO-1	$RS(16\lambda, 15\lambda)$	Turbo, 1/2	QPSK	1/2
SO-1	$RS(16\lambda, 15\lambda)$	Turbo, 1/3	QPSK	1

Table 6.2 Bit, symbol and chip rates for each SE-CDMA implementation.

RATE	FO-1	MO-1	SO-1
R (kb/s)	64	64	64
R_b(kb/s)	76.8	76.8	76.8
R_s(ks/s)	81.92	81.92	81.92
R_{ss}(ks/s)	40.69	81.92	122.88
R_{c1}(Mc/s)	2.4576	4.9152	$R_{c1} = R_{c2}$
$R_c = R_{c2}$(Mc/s)	9.8304	9.8304	9.8304
R_{c1}/R_{ss}	60	60	80
R_{c2}/R_{c1}	4	2	1

been selected so that the system capacity is maximized while the BER and E_b/N_o are the lowest possible for the particular implementation.

Table 6.2 shows examples of specific values of the bit, symbol and chip rates for the FO-1, MO-1 and SO-1 when the source rate is $R = 64$ kb/s. Figure 6.1 illustrates the *overspreading* operation for the beam reuse pattern FO implementation. The FO/SE-CDMA provides 60 orthogonal codes for user channels, having a chip rate of $R_{c1} = 2.4576$ Mc/s. Overspreading by a factor of 4 will raise the chip rate to $R_{c2} = 9.8304$ Mc/s, and will provide four orthogonal codes for separating the satellite beams. The resulting pattern has all beams orthogonal in the first tier, while in the second tier, beams are separated by PN-codes.

[Rates R, R_b, R_s, R_{ss}, and $R_c = R_{c2}$ are measured at points shown in Figure 3.7]

The MO/SE-CDMA has a similar implementation. The spreading rate on the first step is $R_{c1} = 4.9152$ Mc/s. The overspreading rate is $R_{c2} = 2 \times R_{c1}$, and provides, two orthogonal codes for beam isolation. In the resulting pattern, four out of six beams in the first tier are orthogonally isolated, and two by cross-polarization and PN-codes. (Cross-polarization will be used for further reduction of the other beam interference in this case.) Figure 6.2 illustrates the overspreading and the beam reuse pattern for MO/SE-CDMA. In the SO/SE-CDMA the spreading operation has the user orthogonal code and the beam I and Q PN codes. All codes have the same rate $R_c = 9.8304$ Mc/s. Beams are only separated by PN codes.

Following the spreading operation, the resulting I and Q waveforms will be band-limited by a digital FIR filter. The FIR filter is a Raised Cosine filter with a roll-off factor of 0.15 or more. After the digital filter the signal will be converted into analog form and modulated by a quadrature modulator, as shown in Figure 3.11. The resulting IF signal bandwidth will be W ($W \approx 10MHz$).

The SE-CDMA receiver is illustrated in Figure 6.3. The chip synchronization and tracking for despreading the orthogonal and PN codes is provided by a mechanism specifically developed for this system which is presented in Chapter 7. This analysis,

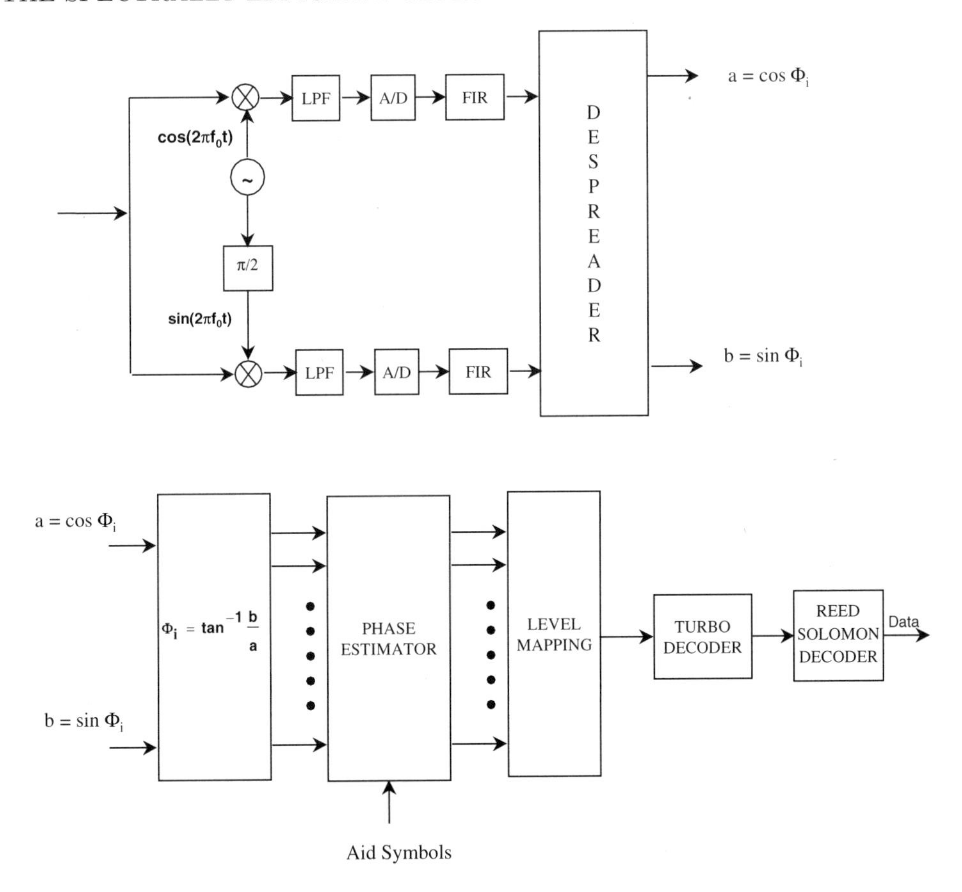

Figure 6.3 The SE-CDMA receiver.

however, will assume perfect chip synchronization at the despreader. Coherent detection will also be provided using reference or aid symbols. The aid symbols that have a known phase are inserted at the transmitter at a low rate and extracted at the receiver in order to provide the phase estimates for the information symbols. The analysis in this paper, however, will consider ideal coherent detection. The channel decoding for the Reed–Solomon and Turbo codes will only take place at the receiver of the end user. On board the satellite we consider three possible options: (a) baseband despreading-respreading without demodulation or channel decoding; (b) baseband despreading-respreading with demodulation but not channel decoding; and (c) Intermediate Frequency (IF) despreading-respreading without demodulation or channel decoding. However, the analysis and numerical results presented in this paper are limited only to case (a).

6.2.1 Signal and Channel Models

In this subsection we provide a brief description of the signal and channel model. The signal model includes the data and spreading modulation, while the channel is described by a 'Rician' flat fading model.

a)

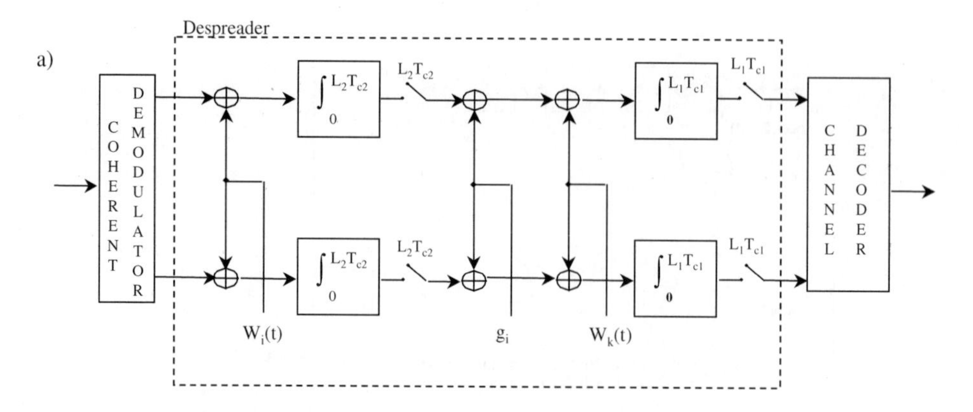

b)

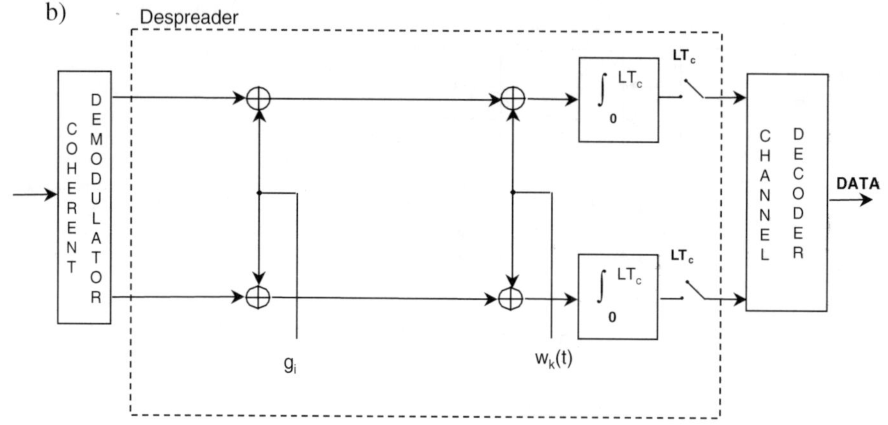

Figure 6.4 The despreading operations for (a) FO and MO/SE-CDMA, and (b) SO/SE-CDMA.

Data Modulation

The transmitted signal from the kth user is

$$s_k(t) = \sqrt{2P_k}\left\{b_I^{(k)}(t)c_I^{(k)}(t)\cos[\omega_c^u t + \theta^{(k)}] + b_Q^{(k)}(t)c_Q^{(k)}(t)\sin[\omega_c^u t + \theta^{(k)}]\right\}$$

where P_k is the transmitted power of the kth signal (this includes transmitter antenna gains and power control); $\omega_c^u = 2\pi f_c^u$ is the frequency carrier of the system for the uplink (actually it corresponds to the center frequency of one of the 10 MHz channels) $\theta^{(k)}$ is the phase angle of the kth signal (user) local oscillator. It is modeled as a slowly changing random variable uniformly distributed in $[0, 2\pi]$.

The data waveforms $b_I^{(k)}(t)$ and $b_Q^{(k)}(t)$ are given by

$$b_I^{(k)}(t) = \sum_{n=-\infty}^{\infty} b_I^{(k)}[n]p_{T_s}(t - nT_s)$$

$$b_Q^{(k)}(t) = \sum_{n=-\infty}^{\infty} b_Q^{(k)}[n]p_{T_s}(t - nT_s)$$

and represent the inphase and quadrature components of the data waveform (sequence of M-ary symbols) of the kth user. In this notation $p_{T_s}(t)$ is a rectangular pulse of duration T_s, and the symbol duration; $T_s = (\log_2 M)T_b$, where T_b is the bit duration (this relationship is modified later in the paper due to the Turbo inner coding and the RS outer coding used). $\left(b_I^{(k)}[n], b_Q^{(k)}[n]\right)$ are defined to be the inphase and quadrature components of the nth M-ary symbol of the kth user. They are defined as $b_I^{(k)}[n] = \cos\phi^{(k)}[n]$ and $b_Q^{(k)}[n] = \sin\phi^{(k)}[n]$, where $\phi^{(k)}[n]$ denotes the phase angle of the nth M-ary symbol of the kth signal (user); they take values in the sets

$$b_I^{(k)}[n] \in \left\{\cos\left[\frac{(2m-1)\pi}{M}\right], m = 1, 2, \ldots, M\right\}$$

$$b_Q^{(k)}[n] \in \left\{\sin\left[\frac{(2m-1)\pi}{M}\right], m = 1, 2, \ldots, M\right\}$$

It is assumed that the sequences of phase angles (symbols) $\phi^{(k)}[n]$ of the $k = 1, 2, \ldots, K$ signals are i.i.d, i.e. independent for different n (symbols) and for different k (signals/users), and are identically distributed. With respect to the latter, it is assumed that the phase angle $\phi^{(k)}[n]$ of the nth symbol of the kth signal is uniformly distributed in the set $\{\pi/M, 3\pi/M, \ldots, (2M-1)/M\}$, and subsequently the inphase and quadrature components $b_I^{(k)}[n]$ and $b_Q^{(k)}[n]$ are i.i.d (for different k and n) and uniformly distributed (take each value with equal probability $1/M$) in the above sets. For the same k and n, $b_I^{(k)}[n]$ and $b_Q^{(k)}[n]$ are not independent of each other, but are uncorrelated; thus we can easily show that the expected value over the above sets results in

$$E\left\{b_I^{(k)}[n]b_Q^{(k)}[n]\right\} = 0 \text{ and } E\left\{\left[b_I^{(k)}[n]\right]^2\right\} = E\left\{\left[b_Q^{(k)}[n]\right]^2\right\} = \frac{1}{2}$$

CDMA Spreading Modulation

For a CDMA system using inphase and quadrature codes $c_I^{(k)}[l]$ and $c_Q^{(k)}[l]$ we have

$$c_I^{(k)}(t) = \sum_{l=-\infty}^{\infty} c_I^{(k)}[l]g_{T_c}(t - lT_c)c_Q^{(k)}(t) = \sum_{l=-\infty}^{\infty} c_Q^{(k)}[l]g_{T_c}(t - lT_c)$$

The *chip shaping waveform*, which takes values

$$g_{T_c}(t) = \sin c(W_{ss}t)\frac{\cos(\pi\rho W_{ss}t)}{1 - 4\rho^2 W_{ss}^2 t^2} \quad \text{for all } t$$

where $\sin c(x) = \sin(\pi x)/(\pi x)$, $W_{ss} = 1/T_c$ (for SO/SE-CDMA) or $W_{ss} = 1/T_{cc}$ (for FO/SE-CDMA and MO/SE-CDMA) is the total spread signal bandwidth, and $g(t)$ has as a Fourier Transform the raised cosine pulse (in the frequency domain)

$$G(f) = \frac{1}{W_{ss}} \text{ for } |f| < f_1 \frac{1}{2W_{ss}}\left\{1 + \cos\left[\frac{\pi(|f| - f_1)}{W_{ss} - 2f_1}\right]\right\} \quad \text{for } f_1 < |f| < W_{ss} - f_1$$

$G(f) = 0$ for $|f| > W_{ss} - f_1$.

This represents the transfer function of the *chip filter* used at the transmitter to band-limit the spread-spectrum signal. The parameter f_1 is related to ρ, the *roll-off factor*, and the *total one-sided bandwidth* for the chip filter as

$$\rho = 1 - \frac{2f_1}{W_{ss}}, \quad W_c^F = W_{ss} - f_1 = (1+\rho)\frac{W_{ss}}{2}$$

For example, for a roll-off factor of $\rho = 0.15$ (15%), the two-sided bandwidth of the chip filter is given by $2W_c^F = 1.15W_{ss}$.

For the SO/SE-CDMA system there is one chip duration T_c and one processing gain (due to spreading) L (chips per symbol) such that $T_s = LT_c$. In this system $c^{(b_k)}[l]$ is the unique PN code (the beam address) characterizing beam b_k, where $b_k \in \{1, 2, \ldots, N\}$ is the index of the beam at which the kth user resides, and $\left(w_I^{(k)}[l], w_Q^{(k)}[l]\right)$ is the pair of orthogonal codes (Quadrature Residue) assigned to user k; these codes are reused in each of the N beams. In this case we can write

$$c_I^{(k)}(t) = \sum_{l=-\infty}^{\infty} w_I^{(k)}[l]c^{(b_k)}[l]g_{T_c}(t - lT_c) c_Q^{(k)}(t) = \sum_{l=-\infty}^{\infty} w_Q^{(k)}[l]c^{(b_k)}[l]g_{T_c}(t - lT_c)$$

If only one orthogonal code per user is used, then $w_I^{(k)}[l] = w_Q^{(k)}[l]$ for all l, and thus $c_I^{(k)}(t) = c_Q^{(k)}(t)$ for all t.

For the FO/SE-CDMA and MO/SE-CDMA systems there are two chip durations T_c and T_{cc} corresponding to the two stages of spreading. Besides the PN beam code $c^{(b_k)}[l]$ and the pair of orthogonal user codes $\left(w_I^{(k)}[l], w_Q^{(k)}[l]\right)$, there is a Walsh orthogonal code $w^{(b_k)}[m]$ assigned to beam b_k. The following relationships are now true:

$$T_s = L_u T_c \quad \text{and} \quad T_c = L_b T_{cc}$$

where L_u (user chips per symbol) and L_b (beam chips per user chip) are the two processing (spreading) gains, and

$$L = L_u L_b$$

is the total spreading gain. There are two possible choices for $L_b = 2$ corresponding to MO/SE-CDMA and $L_b = 4$ corresponding to FO/SE-CDMA. There are L_b orthogonal beam codes, and these codes are re-used between the N beams, as shown in Figure 6.1 for FO/SE-CDMA and in Figure 6.2 for MO/SE-CDMA. In this case, we can write

$$c_I^{(k)}(t) = \sum_{l=-\infty}^{\infty} w_I^{(k)}[l]c^{(b_k)}[l] \sum_{m=lL_b}^{(l+1)L_b-1} w^{(b_k)}[n]g_{T_{cc}}(t - mT_{cc})$$

where $g_{T_{cc}}(t - mT_{cc})$ is the same as $g_{T_c}(t - mT_c)$ above, with T_{cc} replacing T_c. Similarly,

$$c_Q^{(k)}(t) = \sum_{l=-\infty}^{\infty} w_Q^{(k)}[l]c^{(b_k)}[l] \sum_{m=lL_b}^{(l+1)L_b-1} w^{(b_k)}[m]g_{T_{cc}}(t - mT_{cc})$$

We may have $c_Q^{(k)}(t) = c_I^{(k)}(t)$ if each user uses only one orthogonal code.

The Channel Model

The K_a band SATCOM channel is well approximated by a flat fading channel having a standard Rician pdf $p(x) = \frac{x}{\sigma^2} \exp\left(-\frac{x^2+\mu^2}{2\sigma^2}\right) I_0\left(\frac{\mu x}{\sigma^2}\right)$ or equivalently the pdf

$$p(r) = \frac{2(K_f+1)r}{S^2} \exp\left[-(K_f+1)\frac{r^2}{S^2} - K_f\right] I_0\left[2\frac{r}{S}\sqrt{K_f(K_f+1)}\right]$$

where

$$K_f = \frac{\mu^2}{2\sigma^2} S^2 = E\{X^2\} = \mu^2 + 2\sigma^2 = \frac{K_f+1}{K_f}\mu^2 = (K_f+1)(2\sigma^2)$$

is the 'Rician factor' equal to the ratio of the power in the LOS (line of sight) path (μ^2) and the power in the reflected paths ($2\sigma^2$), and S^2 is the total received power in the LOS and reflected paths.

Rain fade statistics determine the values of the parameters. Under severe rain fades the channel model will be better approximated by a Raleigh pdf (special case of the above for $\mu = 0 = K_f$). There is no delay spread in this flat fading channel model. We assume that all signals are fading independently and according to the above Rician distribution (with the same parameters for all signals).

In our analysis and numerical results we assumed that the SATCOM channel is equivalent to an AWGN channel. This approximation is only good for *clear-sky conditions*, but allows us to focus on the effects of other-user interference (intra-beam and other-beam) of the system under full-load (high capacity conditions).

The analysis of this chapter can be easily modified to account for the Rician fading model above. Specifically, the variance of all other-user interference terms should be multiplied by the factor $1 + \frac{1}{K_f}$ (or its square for cross-terms of interference, see Section 6.5), and the final expression for the Bit Error Rate (BER) of the user of interest should be obtained by first conditioning on the Rician amplitude and then integrating with respect to the Rician distribution. However, this was not included in the chapter due to space limitations, and because of the selected emphasis of the paper on other-user interference issues.

6.3 Interference Analysis

In this section we first evaluate the cross-correlation functions of the CDMA codes of the interfering users from the various beams. Then we compute the power of other-user interference, assuming that perfect power control is employed to calibrate for the different received signal strengths of the user signals.

6.3.1 Cross-correlation of Synchronous CDMA Codes

Under fully synchronous system operation (time-jitter = 0) the normalized (integrated over the period of one symbol) cross-correlation between different users takes the form

$$C_{k,i} = \frac{1}{L}\sum_{l=0}^{L-1} w^{(k)}[l]w^{(i)}[l]c^{(b_k)}[l]c^{(b_i)}[l]$$

for SO/SE-CDMA and

$$C_{k,i} = \frac{1}{L_u L_b} \sum_{l=0}^{L_u-1} w^{(k)}[l] w^{(i)}[l] c^{(b_k)}[l] c^{(b_i)}[l] \sum_{m=lL_b}^{(l+1)L_b-1} w^{(b_k)}[m] w^{(b_i)}[m]$$

for FO/SE-CDMA and MO/SE-CDMA.

Code Cross-correlation for SO/SE-CDMA

Recall that for the SO/SE-CDMA system $T_s = LT_c$. Let b_k and b_i be the beams that users k and i reside in. If $b_k = b_i, k \neq i$ (users in the same beam), then

$$C_{k,i} = \frac{1}{L} \sum_{l=0}^{L-1} w^{(k)}[l] w^{(i)}[l] = 0$$

since the codes are orthogonal (Quadrature Residue); see Chapter 2. If instead, Quasi-Orthogonal (QO) preferred phase Gold codes (see Chapter 2) are used, we have

$$C_{k,i} = \frac{1}{L} \sum_{l=0}^{L-1} w^{(k)}[l] w^{(i)}[l] = \frac{1}{L} \cdot 1 = \frac{1}{L}$$

If $b_k \neq b_i, k \neq i$ (users in different beams), the concatenation of orthogonal (or quasi-orthogonal) user codes and PN beam codes results in codes that have (approximately) PN properties, and thus

$$E\{C_{k,i}\} = 0 \text{ and } Var\{C_{k,i}\} = \frac{1}{L}$$

where the averages are taken with respect to the PN sequence taking values +1 and −1 with equal probability and independently from chip to chip, and from user to user (different users). This is the random sequence model of PN sequences that has been widely used in the literature; it is very accurate when L is large (larger than 30).

In conclusion, for the SO/SE-CDMA system and two users k and i we have

$$E\{C_{k,i}^2\} = \begin{cases} 0 & k \text{ and } i \text{ in same beam, orthogonal codes used} \\ \frac{1}{L^2} & k \text{ and } i \text{ in same beam, quasi-orthogonal codes used} \\ \frac{1}{L} & k \text{ and } i \text{ in different beams, all codes used} \end{cases}$$

Here we assumed that the same polarization is used over all of the beams.

Code Cross-correlation for FO/SE-CDMA and MO/SE-CDMA

Recall that for the FO/SE-CDMA (and MO/SE-CDMA) system $T_s = L_u T_c$ and $T_c = L_b T_{cc}$. Again let b_k and b_i be the beams that users k and i reside in. If $b_k = b_i, k \neq i$ (k and i in the same beam), we have

$$C_{k,i} = \frac{1}{L_u} \sum_{l=0}^{L_u-1} w^{(k)}[l] w^{(i)}[l] = \begin{cases} 0 & k \text{ and } i \text{ in same beam, orthogonal codes used} \\ \frac{1}{L_u} & k \text{ and } i \text{ in same beam, quasi-orthog. codes used} \end{cases}$$

If $b_k \neq b_i, k \neq i$ (k and i in different beams) we must distinguish between the first tier of beams b_k surrounding beam b_i and the second tier of beams b_k surrounding beam b_i.

For FO/SE-CDMA all first-tier beams use distinct orthogonal codes (for overspreading), and thus $\sum_{n=0}^{L_b-1} w^{(b_k)}[n]w^{(b_i)}[n] = 0$ because of the code reuse factor of 4. Thus

$$C_{k,i} = 0 \ , \text{ if } b_k \text{ is a first-tier beam surrounding beam } b_i$$

For MO/SE-CDMA, half of them use a distinct orthogonal code, the other half use the same orthogonal code but different polarization (than the beam of interest), so interference from them is not 0, but rather it is the same as PN-type interference (since the concatenation of the (orthogonal or quasi-orthogonal) user code and the PN (address) beam code is (approximately) equivalent to a PN code) but lower by 6 db (=1/4 of the power of the beam of interest) for a polarization isolation of 6 dB. Thus, for MO/SE-CDMA, considering b_k to be a second-tier beam with respect to beam b_i, we have

$$E\{C_{k,i}^2\} = \begin{cases} \frac{1}{L_u} & \text{using the same beam code and same polarization} \\ \frac{1}{4L_u} & \text{using the same beam code and different polarization} \\ 0 & \text{uses a different beam code} \end{cases}$$

Both of the above results are independent of the user codes $w^{(k)}$ and $w^{(i)}$ being orthogonal or quasi-orthogonal.

Similarly, for the second-tier beams we have for either FO/SE-CDMA or MO/SE-CDMA (no distinction now)

$$E\{C_{k,i}^2\} = \begin{cases} \frac{1}{L_u} & \text{using the same beam code and same polarization} \\ \frac{1}{4L_u} & \text{that using the same beam code and different polarization} \\ 0 & \text{using a different beam code} \end{cases}$$

6.3.2 Normalized Power of Other-User Interference

Analysis of Interference at the Matched Filter Output

In computing the variance of the interference caused to user i by any other user k, we note that the input to a correlation receiver (coherent demodulator) for M-ary PSK modulation takes the form

$$\begin{aligned} r^u(t) &= s_i^u(t) + \sum_{k \neq i} s_k^u(t) + n^u(t) \\ &= \sqrt{2P_i^u} b_I^{(i)}(t) c_I^{(i)}(t) \cos[\omega_c^u t + \theta^{(i)}] + \sum_{k \neq i} \sqrt{2P_k^u} b_I^{(k)}(t) c_I^{(k)}(t) \cos[\omega_c^u t + \theta^{(k)}] \\ &\quad + \sqrt{2P_i^u} b_Q^{(i)}(t) c_Q^{(i)}(t) \sin[\omega_c^u t + \theta^{(i)}] + \sum_{k \neq i} \sqrt{2P_k^u} b_Q^{(k)}(t) c_Q^{(k)}(t) \sin[\omega_c^u t + \theta^{(k)}] \\ &\quad + n^u(t) \end{aligned}$$

where $n^u(t)$ is a Gaussian noise process (AWGN) with zero mean and two-sided spectral density $\frac{1}{2}N_0$. P_k^u is the received uplink power at the satellite receiver; it includes the propagation and flat fading losses (no multipath or delay spread), the satellite receiver antenna gain, and of course, the transmitted power P_k.

At this point it is assumed that the correlation receiver performs baseband processing of the signal (despreading and coherent demodulation). In this sense this analysis is valid for the uplink of the GEO SATCOM system (despreading and demodulation take place onboard the satellite), provided that the transmitted powers are appropriately adjusted. However, for the subsequent analysis of this section to be valid for the downlink (despreading and demodulation take place at the user receiver), we must assume that full regeneration of the signal takes place on board the satellite, and thus the uplink and downlink performances are decoupled (except for the cascade AWGN noise). These issues are revisited and discussed in full detail in Section 6.4, where one of the on-board processing options is modeled and analyzed.

Under these conditions, the inphase and quadrature components of the output of the ith correlation receiver take the form

$$
\begin{aligned}
Z_I^{(i),u} &= \int_0^{T_s} r^u(t) c_I^{(i)}(t) 2\cos[\omega_c^u t + \theta^{(i)}] dt Z_Q^{(i),u} \\
&= \int_0^{T_s} r^u(t) c_Q^{(i)}(t) 2\sin[\omega_c^u t + \theta^{(i)}] dt
\end{aligned}
$$

respectively. It is assumed that the receiver is in perfect time, frequency and phase synchronization with the ith transmitter, and that the demodulation of the 0th symbol of duration $T_s = LT_c$ for SO/SE-CDMA (or $T_s = L_u L_b T_{cc}$ for FO/SE-CDMA and MO/SE-CDMA) is performed. Thus, without loss of generality, we can assume that $\theta^{(i)} = 0$. Let us define N_I^u as the noise component at the output of the inphase branch of the correlator as $N_I^u = \int_0^{T_s} n^u(t) 2\cos(\omega_c^u t) dt$. Using the above decomposition of the received signal $r(t)$ in inphase and quadrature components, assuming (as is typical in DS/CDMA systems) that $\omega_c^u T_c >> 1$ so that $\frac{\sin(\omega_c^u T_c)}{\omega_c^u T_c} << 1$, and performing the integrations, we obtain

$$
\begin{aligned}
Z_I^{(i),u} &= \sqrt{2P_i^u} T_s b_I^{(i)}[0] \\
&+ \sum_{k \neq i} \sqrt{2P_k^u} T_s \Bigg(b_I^{(k)}[0] \cos[\theta^{(k)}] \frac{1}{L} \sum_{l=0}^{L-1} w_I^{(k)}[l] w_I^{(i)}[l] c^{(b_k)}[l] c^{(b_i)}[l] \\
&+ b_Q^{(k)}[0] \sin[\theta^{(k)}] \frac{1}{L} \sum_{l=0}^{L-1} w_Q^{(k)}[l] w_I^{(i)}[l] c^{(b_k)}[l] c^{(b_i)}[l] \Bigg) + N_I^u
\end{aligned}
$$

A similar development provides the quadrature component $Z_Q^{(i),u}$, which we omit here.

Next we evaluate the first and second moments of $Z_I^{(i),u}$ and $Z_Q^{(i),u}$ with respect to the phase angle θ_k of the local oscillator of user k ($1 \leq k \leq K$, $k \neq i$) and its M-ary PSK symbols $b_I^{(k)}[0]$ and $b_Q^{(k)}[0]$. First, we use the fact that averaging with respect to $\theta^{(k)}$ gives $E\{\cos\theta_k \sin\theta_k\} = 0$ and $E\{\cos^2\theta_k\} = E\{\sin^2\theta_k\} = \frac{1}{2}$ to obtain for the squares of desired (inphase and quadrature) signal terms

$$\left[D_I^{(i),u}\right]^2 = \left[E\left\{Z_I^{(i),u}\right\}\right]^2 = (2P_i^u)T_s^2[b_I^{(i)}[0]]^2 \left[D_Q^{(i),u}\right]^2$$
$$= \left[E\left\{Z_Q^{(i),u}\right\}\right]^2 = (2P_i^u)T_s^2[b_Q^{(i)}[0]]^2$$

and for the variance of other-user interference

$$Var\left\{Z_I^{(i),u}\right\} = \frac{1}{4}\sum_{k\neq i}(2P_k^u)T_s^2\left\{\left[\frac{1}{L}\sum_{l=0}^{L-1} w_I^{(k)}[l]w_I^{(i)}[l]c^{(b_k)}[l]c^{(b_i)}[l]\right]^2 + \left[\frac{1}{L}\sum_{l=0}^{L-1} w_Q^{(k)}[l]w_I^{(i)}[l]c^{(b_k)}[l]c^{(b_i)}[l]\right]^2\right\} + N_0T_s$$

A similar expression is valid for $Var\left\{Z_Q^{(i),u}\right\}$ (omitted). If every user is assigned a single orthogonal code, i.e. $w_Q^{(k)}[l] = w_I^{(k)}[l]$ for all $k = 1, 2, \ldots, K$, the above expressions simplify further to

$$Var\left\{Z_I^{(i),u}\right\} = Var\left\{Z_Q^{(i),u}\right\} = \frac{1}{2}\sum_{k\neq i}(2P_k^u)T_s^2E\{C_{k,i}^2\} + N_0T_s$$

where

$$C_{k,i} = \frac{1}{L}\sum_{l=0}^{L-1} w_I^{(k)}[l]w_I^{(i)}[l]c^{(b_k)}[l]c^{(b_i)}[l]$$

and $E\{C_{k,i}^2\}$ was as evaluated in Section 6.3. Notice that the above calculations were conducted for the SO/SE-CDMA system. They are all also valid for the FO/CDMA system if we replace terms like $\frac{1}{L}\sum_{l=0}^{L-1} w_I^{(k)}[l]w_I^{(i)}[l]c^{(b_k)}[l]c^{(b_i)}[l]$ by the corresponding terms $\frac{1}{L_u}\sum_{l=0}^{L_u-1} w_I^{(k)}[l]w_I^{(i)}[l]c^{(b_k)}[l]c^{(b_i)}[l]\frac{1}{L_b}\sum_{n=0}^{L_b-1} w^{(b_k)}[n]w^{(b_i)}[n]$ which involve the orthogonal beam codes $w^{(b_k)}[n]$ and $w^{(b_i)}[n]$ of the beams b_k and b_i, respectively, where the users k and i reside.

Thus if we normalize the variance (power) of the other-user interference by the power of the desired signal, we obtain the normalized interference power

$$Var\left\{Z_I^{(i),u}\right\} = Var\left\{Z_Q^{(i),u}\right\} = \frac{1}{2}\sum_{k\neq i}\frac{P_k^u}{P_i^u}C_{k,i}^2 + \frac{N_0T_s}{2P_i^u} = \frac{1}{2}\sum_{k\neq i}\frac{P_k^u}{P_i^u}C_{k,i}^2 + \left(\frac{2E_s^u}{N_0}\right)^{-1}$$

for the I and Q branches of the correlation receiver, with E_s^u and E_b^u

$$E_s^u = P_i^uT_s = P_i^u(\log_2 M)T_b = (\log_2 M)E_b^u$$

as the (received at uplink) energy per symbol and energy per bit of the signal, respectively.

Therefore, observing that $E\{C_{k,i}^2\}$ (Section 6.3.5) is basically independent (for large user populations) of the specific pair (k, i) of user codes, we obtain

$$\hat{I}^u = \left[\sum_{k \neq i} \frac{P_k^u}{P_i^u}\right] \bar{I}_s + \left(\frac{2E_s^u}{N_0}\right)^{-1}$$

We represent the (power of the) interference caused by one user to another by $\bar{I}_s$ as if they had the same power. The factor $\sum_{k \neq i} \frac{P_k^u}{P_i^u}$ which depends on power control is evaluated in detail in Section 6.3.3. Using these results, we next derive the values of $\bar{I}_s$.

Power of Interference from Single User for SO, FO and MO/SE-CDMA

The power of interference from a single user in a SO/SE-CDMA system is given by

$$\bar{I}_s = \begin{cases} 0 & \text{if interfering user in same beam, orthogonal user codes} \\ \frac{1}{2L^2} & \text{if interfering user in same beam, preferred Gold codes} \\ \frac{1}{2L} & \text{if interfering user in different beam, any type of user code} \end{cases}$$

where L is the processing gain due to spreading (number of chips per user symbol), and only one type of polarization is used over all beams.

Similarly, the power of interference from a single interfering user in a FO/SE-CDMA system is given by

$$\bar{I}_s = \begin{cases} 0 & \text{interfering user in same beam, orthogonal user codes} \\ \frac{1}{2L_u^2} & \text{interfering user in same beam, preferred Gold codes} \\ 0 & \text{interfering user in first tier surrounding beam} \\ \frac{1}{2L_u} & \text{user in second tier beam with same beam code and same polarization} \\ \frac{1}{8L_u} & \text{user in second tier beam with same beam code but different polarization} \\ 0 & \text{user in second tier beam with different beam code} \end{cases}$$

where L_u is the processing gain due to user spreading (number of user chips per user symbol).

Finally, the power of interference from a single user in a MO/SE-CDMA system is given by

$$\bar{I}_s = \begin{cases} 0 & \text{interfering user in same beam, orthogonal user codes} \\ \frac{1}{2L_u^2} & \text{if interfering user in same beam, preferred Gold codes} \\ 0 & \text{interfering user in any tier surrounding beam with different beam code} \\ \frac{1}{2L_u} & \text{user in any tier beam with same beam code and same polarization} \\ \frac{1}{8L_u} & \text{user in any tier beam with same beam code but different polarization} \end{cases}$$

6.3.3 Effects of Power Control on Interference

In this section we incorporate the effects of power control and location of users on other-user interference. Center-of-beam (boresight) antenna gains, propagation losses, and attenuation due to flat fading (with no frequency or time selectivity) have not

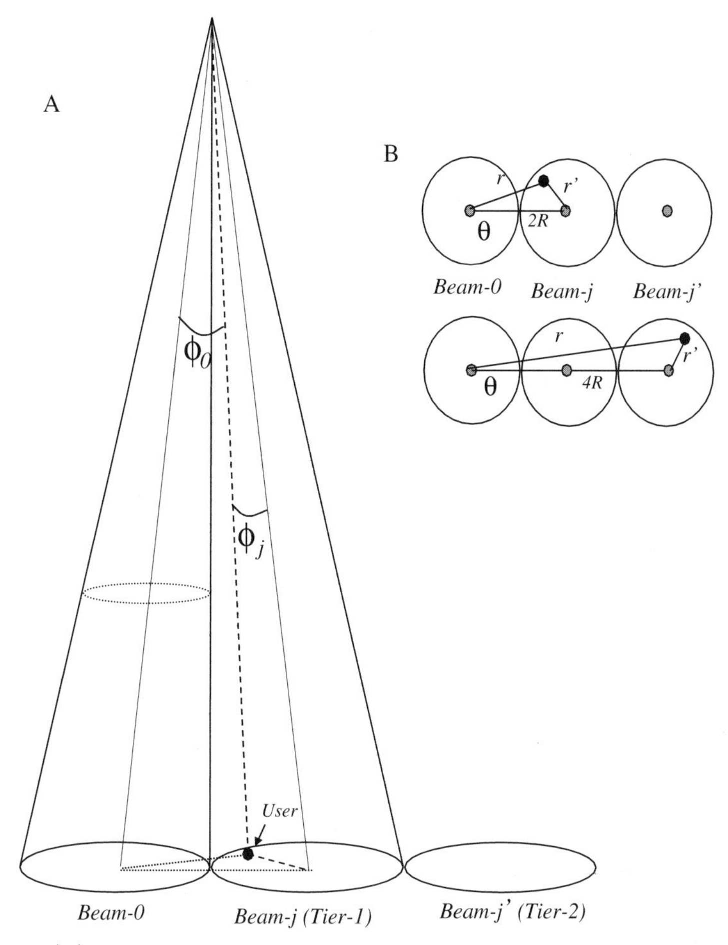

Figure 6.5 (A) The antenna pointing angle to a user in beam j from beam 0. (B) The geometry of the antenna pattern in beams 0, j and j'.

been taken into consideration in this analysis, which is performed under clear-sky conditions, since they only add or subtract dBs to or from the E_b/N_0 (in dB) required to achieve a particular BER value. However, we do actually evaluate the average $E\{P_k^u/P_i^u\}$ (refer to section 6.3.2) of the ratio of the received powers of the kth interfering user and i^{th} (desirable) user with respect to the user location in the beams, and after power control and satellite antenna beam patterns (transmit and receive) have been taken into account.

Notation

First we introduce the transmitter and receiver antenna power gain functions of the satellite, and of the user receiver (CPE, or Customers Premises Equipment):

$$G_t(\phi) = \text{satellite transmit antenna power gain function}$$
$$G_r(\phi) = \text{satellite receive antenna power gain function}$$

$$\hat{G}_t(\phi) = \text{CPE transmit antenna gain power function}$$
$$\hat{G}_r(\phi) = \text{CPE receive antenna power gain function}$$

These functions are provided in Appendix 6A.

Next we introduce all of the necessary satellite pointing angles (called *stereo angles*), distances and planar angles that are involved in modeling and evaluating the interference caused to the user of interest (assumed to be in the footprint of beam 0) from another user located in the footprint of beam j (see Figure 6.5):

$$\phi_0^{(0)} = \text{stereo angle to user of interest in beam 0 with respect to center of beam 0}$$
$$\phi_0^{(j)} = \text{stereo angle to interfering user in beam } j \text{ with respect to center of beam 0}$$
$$\phi_j^{(0)} = \text{stereo angle to user of interest in beam 0 with respect to center of beam } j$$
$$\phi_j^{(j)} = \text{stereo angle to interfering user in beam } j \text{ with respect to center of beam } j$$
$$\theta_0^{(0)} = \text{planar angle of user of interest in beam 0 with respect to center of beam 0}$$
$$\theta_0^{(j)} = \text{planar angle of interfering user in beam } j \text{ with respect to center of beam 0}$$
$$r_0^{(0)} = \text{distance of user of interest in beam 0 from the center of beam 0}$$
$$r_0^{(j)} = \text{distance of interfering user in beam } j \text{ from the center of beam 0}$$
$$r_j^{(0)} = \text{distance of user of interest in beam 0 from the center of beam } j$$
$$r_j^{(j)} = \text{distance of interfering user in beam } j \text{ from the center of beam } j$$

Refer to Figure 6.5-A for the implied beam architecture. The pointing angles and distances shown in Figure 6.5-B pertain to the interfering user (residing in beam j), that is $\phi_0 = \phi_0^{(j)}$, $\phi_j = \phi_j^{(j)}$, $r = r_0^{(j)}$ and $r' = r_j^{(j)}$.

It is assumed that all beam footprints are approximated by circles of radius $R = 256$ km, and the distance of the GEO satellite from the center of all beams is $D_0 = 35,786$ km equal to the distance from the earth's surface; that is, the curvature of the earth is not taken into account.

We further assume that both the user of interest (in beam 0) and the interfering user (in beam j) are drawn from populations that can be modeled with the simple location model of Appendix 6A.

Uplink Power Control and Other-User Interference

Let us assume that the normalized transmitted power from any CPE is P_t, together with the transmit antennae gain $P_t\hat{G}_t$. Then at the satellite receive antennae of beam 0, the received power for the signal of the user of interest is $P_t\hat{G}_t \cdot G_r^{(0)}\left(\phi_0^{(0)}\right) \cdot (\lambda/4\pi)^2$, while the power of an interfering user in beam j received at the antenna of beam 0 is $P_t\hat{G}_t \cdot G_r^{(0)}\left(\phi_0^{(j)}\right) \cdot (\lambda/4\pi)^2$, where we have used the definitions of pointing angles above. Notice that the received power of the useful signal varies with its location within beam 0.

This undesirable variation can be eliminated (or at least mitigated) if we assume that a *power control scheme* (open-loop) is used, according to which the transmitted power from each user's CPE is inversely proportional to the satellite receive antenna gain for the satellite pointing angle to that user. In this case the user of interest transmits $\frac{P_t}{G_r^{(0)}(\phi_0^{(0)})}\hat{G}_t$ and the interfering user transmits $\frac{P_t}{G_r^{(j)}(\phi_j^{(j)})}\hat{G}_t$. Therefore, these two signals are received at the receive antenna of beam 0 as $P_t\hat{G}_t(\lambda/4\pi)^2$ and $P_t\hat{G}_t(\lambda/4\pi)^2\frac{G_r^{(0)}(\phi_0^{(j)})}{G_r^{(j)}(\phi_j^{(j)})}$, respectively.

Clearly, the received power of the useful signal is now independent of the location of the user within beam 0, and the ratio of received interference power to received useful signal power (= normalized interference) is $\frac{G_r^{(0)}(\phi_0^{(j)})}{G_r^{(j)}(\phi_j^{(j)})}$, which depends upon the pointing angles $\phi_0^{(j)}$ and $\phi_j^{(j)}$ on the distances $r_0^{(j)}$ and $r_j^{(j)}$ (see Appendix 6A) of the interfering user from the centers of the footprints of beam 0 and beam j, respectively. From these two distances, $r_0^{(j)}$ varies in the range $R \leq r_0^{(j)} \leq 3R$ for a first-tier beam j and in the range $3R \leq r_0^{(j)} \leq 5R$ for a second-tier beam j, while $r_j^{(j)}$ is evaluated from $r_0^{(j)}$ and from $\theta_0^{(j)}$ according to the formulas of Appendix 6A.

Notice that the transmitter and receive antennae gain functions $\hat{G}_t(\phi)$ and $\hat{G}_r(\phi)$ of the CPE do not come into the interference (and BER) evaluation, because the CPE transmitter and receiver are viewed as a single point (dimensionless or point sources) from the satellite due to the enormous distance between the GEO satellite and the CPE. However, the transmitter and receive antennae gains in the centers of the beams $\hat{G}_t(0) = \hat{G}_t$ and $\hat{G}_r(0) = \hat{G}_r$ do come into the calculation.

Taking the average with respect to all possible locations of the (single) interfering user in a first-tier beam gives

$$\bar{I}_1^u = Var\{I_1^{u,(j)}\} = E\left\{\frac{P_k^u}{P_i^u}\right\} = \frac{1}{\pi R^2}\int_{-\pi/6}^{\pi/6}\int_R^{3R}\frac{G_r^{(0)}\left(\phi_0^{(j)}\left[r_0^{(j)},\theta_0^{(j)}\right]\right)}{G_r^{(j)}\left(\phi_j^{(j)}\left[r_0^{(j)},\theta_0^{(j)}\right]\right)}r_0^{(j)}dr_0^{(j)}d\theta_0^{(j)}$$

for a user k belonging to first-tier beam j interfering with user i of beam 0 and

$$\bar{I}_2^u = Var\{I_2^{u,(j)}\} = E\left\{\frac{P_k^u}{P_i^u}\right\} = \frac{1}{\pi R^2}\int_{-\pi/12}^{\pi/12}\int_{3R}^{5R}\frac{G_r^{(0)}\left(\phi_0^{(j)}\left[r_0^{(j)},\theta_0^{(j)}\right]\right)}{G_r^{(j)}\left(\phi_j^{(j)}\left[r_0^{(j)},\theta_0^{(j)}\right]\right)}r_0^{(j)}dr_0^{(j)}d\theta_0^{(j)}$$

for a user k belonging to the second-tier beam j interfering with user i of beam 0. The functions $G(\cdot)$ are obtained from Appendix 6A. Owing to symmetry, the above integrals are the same for all beams $j = 1, 2, \ldots, 6$ of the first tier or all beams $j = 1, 2, \ldots, 12$ of the second tier.

To the above equations we must now add the power of interference from a single user in the same beam 0 as the user of interest. Although the fully synchronous SO, FO and MO-SE/CDMA systems avoid intra-beam (same beam) interference through the use of orthogonal (quadrature-residue) codes, such interference from users in the same beam is present due to time-jitter resulting from inaccuracies in the synchronization algorithm. Arguments such as those used above for evaluating the interference power from beams j of the first or second tiers can be used to show that $\bar{I}_0^u = Var\{I_0^{u,(0)}\} = 1$, since the application of the power control law will now give $\frac{G_r^{(0)}(\phi_0^{(*)})}{G_r^{(0)}(\phi_0^{(*)})} = 1$, where $\phi_0^{(*)}$ is the pointing angle from the satellite to a user $*$ in beam 0. Notice that, although the distance of the user of interest and of user $*$ from the center of the beam footprint are different, their distance from the satellite is practically the same, so that distance-based (rather than pointing angle-based) power control has no effect on the received power at the satellite.

To obtain the total normalized interference power from all K users of a beam j while taking the properties of the CDMA codes into account, we must make use of the expression for $\bar{I}_s$ cited at the end of Section 6.3.2. The total interference from any single first-tier beam or any single second-tier beam is then obtained by multiplying $\bar{I}_1^u$ and $\bar{I}_2^u$, respectively, by $K\bar{I}_s$. Similarly, the total interference from beam 0 (if at all present) is obtained as $(K-1)\bar{I}_s$ (since $\bar{I}_0^u = 1$); only if PN-codes are used is this non-zero; it is zero for the SO, FO and MO/SE-CDMA systems of this paper.

Total Uplink Interference Power for SO, FO and MO/SE-CDMA

The final expression of the total normalized variance (power) of the other-user interference from all adjacent beams now depends on the CDMA system in question.

For the SO/SE-CDMA system under fully synchronous conditions we get no interference from the same beam and PN-type interference from the adjacent beams of the first and second tiers. Thus, if the *same polarization* is used for the signal of all beams, the total other-user interference power $\bar{I}_{0,t}^u$ becomes

$$\begin{aligned} \bar{I}_{0,t}^u &= K\bar{I}_s \left[\sum_{j=1}^{6} Var\{I_1^{u,(j)}\} + \sum_{j=1}^{12} Var\{I_2^{u,(j)}\} \right] \\ &= \frac{K}{2L} \left(6\bar{I}_1^u + 12\bar{I}_2^u \right) \end{aligned}$$

provided all first and second tier beams are involved in the coverage of a particular area; otherwise, only those beams involved in coverage must be accounted for in the factors 6 and 12 of the above equation. For example, for coverage of CONUS by 30–32 beams, a maximum of 8 (rather than 12) second-tier beams are surrounding one beam (worst case), resulting in

$$\bar{I}_{0,t}^u = \frac{K}{2L} \left(6\bar{I}_1^u + 8\bar{I}_2^u \right)$$

If half of the beams use *different polarization* from the other half, a 6 dB (= of 1/4) isolation factor is introduced, and the power of interference is

$$\bar{I}_{0,t}^u = \frac{K}{2L} \left(3.75\bar{I}_1^u + 5\bar{I}_2^u \right)$$

where the factors $3.75 = 3\cdot 1 + 3\cdot \frac{1}{4}$ and $5 = 4\cdot 1 + 4\cdot \frac{1}{4}$ account for the use of different polarization by half of the beams.

Similarly, for FO/SE-CDMA under fully synchronous conditions, we get no interference from the same beam, no interference from the adjacent first tier beams,

Table 6.3 Normalized power of total other-user interference for uplink.

	$\bar{I}_{0,t}^u$	
System	One Polarization	Two Polarizations
SO/SE-CDMA	$\frac{K}{2L}\left(6\bar{I}_1^u + 8\bar{I}_2^u\right)$	$\frac{K}{2L}\left(3.75\bar{I}_1^u + 5\bar{I}_2^u\right)$
FO/SE-CDMA	$\frac{K}{2L_u}\left(4\bar{I}_2^u\right)$	$\frac{K}{2L_u}\left(2\bar{I}_2^u\right)$
MO/SE-CDMA	N/A	$\frac{K}{2L_u}\left(0.5\bar{I}_1^u + 4.5\bar{I}_2^u\right)$

and PN-type interference from four adjacent beams (out of the 12) of the second tier. Thus, if the *same polarization* is used for the signal of all beams,

$$\bar{I}_{0,t}^u = \frac{K}{2L_u}\left(4\bar{I}_2^u\right)$$

If different polarizations are used then only two second-tier beams contribute PN-type interference, and thus

$$\bar{I}_{0,t}^u = \frac{K}{2L_u}\left(2\bar{I}_2^u\right)$$

Finally, for the MO/SE-CDMA system and fully synchronous conditions, we get no interference from the same beam, no interference from four of the adjacent first-tier beams, and PN-type interference from two of the adjacent first-tier beams (with different polarizations and from six adjacent beams of the second-tier (four with the same polarization and two with different polarization). Thus,

$$\bar{I}_{0,t}^u = \frac{K}{2L_u}\left(0.5\bar{I}_1^u + 4.5\bar{I}_2^u\right)$$

where $0.5 = 2 \cdot \frac{1}{4}$ and $4.5 = 4 \cdot 1 + 2 \cdot \frac{1}{4}$ account for the different polarizations of beams. The results of this section are summarized in Table 6.3, where $\bar{I}_1^u$ and $\bar{I}_2^u$ are obtained from the double integrals derived in the previous subsections.

Power Control and Other-User Interference for Downlink

Let us assume that the normalized transmitted power of any signal from the satellite is P_t^s. Then because of the satellite transmit antenna gain of beam 0 and the receive antenna gain of the CPE $\hat{G}_r$, the received power of the signal of the user of interest at the CPE is $P_t^s G_t^{(0)}(\phi_0^{(0)}) \cdot \hat{G}_r(\lambda/4\pi)^2$. Similarly, the transmitted power of an interfering user in beam j from the antenna of beam 0 is $P_t^s G_t^{(0)}(\phi_j^{(0)})$, and thus the corresponding received power at the CPE is $P_t^s G_t^{(0)}(\phi_j^{(0)}) \cdot \hat{G}_r(\lambda/4\pi)^2$. Again, the definitions of the pointing angles in Section 6.3.3 are used.

Notice that the received power of the useful signal varies with its location within beam 0. This undesirable variation can be eliminated (or at least mitigated) if we assume that a *power control scheme* is used according to which the transmitted power from the satellite of each user's signal is inversely proportional to the satellite transmit antenna gain for the satellite pointing angle to that CPE. In this case the signal generated at the satellite (before the satellite transmit antennae of beam 0) is $\frac{P_t^s}{G_t^{(0)}(\phi_0^{(0)})}$ for the the user of interest and $\frac{P_t^s}{G_t^{(j)}(\phi_j^{(j)})}$ for the signal of the interfering user. Therefore, these two signals are received at the receive antenna of the CPE of the user of interest in beam 0 as $P_t^s\hat{G}_r(\lambda/4\pi)^2$ and $P_t^s\hat{G}_r(\lambda/4\pi)^2\frac{G_t^{(0)}(\phi_j^{(0)})}{G_t^{(j)}(\phi_j^{(j)})}$, respectively. Clearly, the received power of the useful signal is now independent of the location of the user within beam 0, and the ratio of received interference power to received useful signal power (= normalized interference) is $\frac{G_t^{(0)}(\phi_j^{(0)})}{G_t^{(j)}(\phi_j^{(j)})}$, which depends (through the pointing angles $\phi_j^{(0)}$ and $\phi_j^{(j)}$) on the distances $r_j^{(0)}$ and $r_j^{(j)}$ of the user of interest from the center of the footprint of beam j and of the interfering user from the center of the footprint of beam j, respectively; $r_j^{(0)}$ and $r_j^{(j)}$ are evaluated for first and second tier beams as in Appendix 6A.

The difference to the corresponding calculation at the beginning of the section is that the pointing angle $\phi_j^{(0)}$ depends not only on the location of the interfering user of beam j, but also on the location of the user of interest in beam 0; this is reflected in an additional integration with respect to $r_0^{(0)}$ and $\theta_0^{(0)}$ (from which $r_j^{(0)}$ is evaluated) when averages of interference with respect to user locations are calculated.

Taking the average with respect to all possible locations of the (single) interfering user in a first-tier beam and the user of interest in beam 0 gives

$$\bar{I}_1^d = Var\{I_1^{d,(j)}\} = \frac{1}{(\pi R^2)^2}\int_{-\pi}^{\pi}\int_0^R [\, X \,]\, r_0^{(0)} dr_0^{(0)} d\theta_0^{(0)}$$

where X is given by

$$X = \int_{-\pi/6}^{\pi/6}\int_R^{3R} \frac{G_t^{(0)}\left(\phi_j^{(0)}\left[r_0^{(j)}, \theta_0^{(j)}; r_0^{(0)}, \theta_0^{(0)}\right]\right)}{G_t^{(j)}\left(\phi_j^{(j)}\left[r_0^{(j)}, \theta_0^{(j)}\right)\right]} r_0^{(j)} dr_0^{(j)} d\theta_0^{(j)}$$

For the interference from the first-tier beam j we have

$$\bar{I}_2^d = Var\{I_2^{d,(j)}\} = \frac{1}{(\pi R^2)^2}\int_{-\pi}^{\pi}\int_0^R [\, Y \,]\, r_0^{(0)} dr_0^{(0)} d\theta_0^{(0)}$$

where Y is given by

$$Y = \int_{-\pi/12}^{\pi/12}\int_{3R}^{5R} \frac{G_t^{(0)}\left(\phi_j^{(0)}\left[r_0^{(j)}, \theta_0^{(j)}; r_0^{(0)}, \theta_0^{(0)},\right]\right)}{G_t^{(j)}\left(\phi_j^{(j)}\left[r_0^{(j)}, \theta_0^{(j)}\right]\right)} r_0^{(j)} dr_0^{(j)} d\theta_0^{(j)}$$

for interference from the second-tier beam j. Owing to the symmetry, the above integrals are the same for all beams $j = 1, 2, \ldots, 6$ of the first tier or all beams $j = 1, 2, \ldots, 12$ of the second tier.

Table 6.4 Normalized power of total other-user interference for downlink.

	$\bar{I}^d_{0,t}$	
System	One polarization	Two polarizations
SO/SE-CDMA	$\frac{K}{2L}\left(6\bar{I}^d_1 + 8\bar{I}^d_2\right)$	$\frac{K}{2L}\left(3.75\bar{I}^d_1 + 5\bar{I}^d_2\right)$
FO/SE-CDMA	$\frac{K}{2L_u}\left(4\bar{I}^d_2\right)$	$\frac{K}{2L_u}\left(2\bar{I}^d_2\right)$
MO/SE-CDMA	N/A	$\frac{K}{2L_u}\left(0.5\bar{I}^d_1 + 4.5\bar{I}^d_2\right)$

The final results for the total interference power in downlink $\bar{I}^d_{0,t}$ for the SO, FO and MO/SE-CDMA systems of interest are then shown for quick reference in Table 6-4. where $\bar{I}^d_1$ and $\bar{I}^d_2$ are obtained from the triple integrals derived in this section.

Power Control Implementation Issues

In deriving the expressions for the power of the total other-user interference in the uplink and downlink $\bar{I}^u_{0,t}$ and $\bar{I}^d_{0,t}$, respectively, we assume that power control is used both at the CPE (for the uplink) and on-board the satellite (for the downlink). Specifically, we assume that a CPE residing in beam j transmits its signal with power $\frac{P_t}{G_r^{(j)}(\phi_j^{(j)})}\hat{G}_t$ over the uplink, where P_t is the normalized CPE transmitted power (for all signals), $\hat{G}_t$ is the CPE transmitter antenna gain, and $G_r^{(j)}(\phi_j^{(j)})$ is the satellite receiver antenna gain at the (stereo) angle $\phi_j^{(j)}$ pointing to the CPE in question (refer to Section 6.3.3 for notation).

In order to implement such an *uplink power control scheme*, the SU must have knowledge of

- the satellite receive antenna gain pattern $G_r^{(j)}(\cdot)$ for beam j, and
- its position with respect to the center of the beam it resides in.

$\phi_j^{(j)}$ can be approximated as $\phi_j^{(j)} = \tan^{-1}\left(r_j^{(j)}/D_0\right)$, where $r_j^{(j)}$ and D_0 are defined in Section 6.3.3 and Appendix 6A. The above required information is easy to obtain, and the SU can effectively use this power control scheme.

Similarly, for a signal in beam j the satellite generates power (before it is sent over the satellite transmit antenna) $\frac{P_t^s}{G_t^{(j)}(\phi_j^{(j)})}$, where P_t^s is the normalized satellite transmitted power (for all signals) and $G_t^{(j)}\left(\phi_j^{(j)}\right)$ is the satellite transmit antenna gain at the (stereo) angle $\phi_j^{(j)}$ pointing to the recipient SU in question.

In order to implement such a *downlink power control scheme*, the satellite must have knowledge of

- the satellite transmit antenna gain pattern $G_t^{(j)}(\cdot)$ for the beam j that contains the recipient SU, and
- the position of the recipient SU with respect to the center of its beam j.

For the latter, we can again use the approximation $\phi_j^{(j)} = \tan^{-1}(r_j^{(j)}/D_0)$. Although the above required information is easy to obtain, it may be more difficult for the satellite to execute this power control scheme than it is for the corresponding uplink power control scheme executed by the CPE. This is because the satellite must have knowledge of the pertinent information for all CPEs, and this list should be updated as more subscribers are served by the system. However, it should be possible to program the on-board controller so that different signals are amplified by different amounts before being transmitted over the downlink.

Finally, we should emphasize that the power control schemes discussed in this section pertain to the mitigation of the near-far problem of CDMA and the reduction (actually balancing) of the other-user interference from the same beam. They are open-loop and control the power of the signals within each beam only, without taking into account the adjacent beams. It is possible to consider more elaborate power control schemes that coordinate the transmissions of CPEs and satellite signals over more than one beam at a time (for example, over the first-tier beams surrounding each beam). Also, dynamic closed-loop power control that mitigates the effects of signal fluctuation due to fading (e.g. rain fades) is of interest, but has not been considered in this paper.

6.4 On-board Processing and Uplink/Downlink Coupling

Three options were considered for processing of the spread-spectrum signals on board the GEO satellite: (a) baseband despreading and respreading; (b) despreading/demodulation followed by respreading/remodulation; and (c) IF despreading and respreading.

Baseband despreading/respreading implies that the CDMA signals are first despread with their uplink (transmit) codes, and then respread with their downlink (destination) codes; all processing is performed in the baseband; no demodulation/remodulation or decoding/encoding is taking place on board the satellite.

On-board demodulation/remodulation means that the CDMA signals are fully despread and demodulated at the uplink (satellite receiver) before being modulated and spread a new for the downlink transmission; no decoding/encoding is taking place on board the satellite.

IF despreading/respreading is the same with baseband despreading/respreading, except that the signals are processed at an IF frequency (different for each 10 MHz channel) and are not downconverted all the way to basedband.

During our detailed study of the SS/CDMA system, it turned out that IF despreading significantly degrades the overall system performance so that the options were narrowed down to (a) and (b). From these two we selected to present only (b) (baseband despreading/respreading) in this book, since it requires a more interesting analysis (involving coupling of uplink and downlink other-user interference), and because it has inferior performance to (a) (despreading/demodulation followed by remodulation/respreading) it can serve as a lower bound on the achievable performance of the SS/CDMA system (i.e. as an upper bound on the BER).

6.4.1 Baseband Despreading/Respreading: Interference Model

Under baseband despreading/respreading the CDMA signals are first downconverted to basedband and then despread with their uplink (transmit) codes. The inphase and quadrature 'coordinates' (projections) $b_I^{(k)}[n]$ and $b_Q^{(k)}[n]$ of an M-ary data symbol with phase $\phi^{(k)}[n]$ do not need to be extracted. This corresponds to not deciding which symbol is sent during the nth data symbol interval (that is, we do not determine in which decision region the two-dimensional vector $(Z_I^{(k)}[n], Z_I^{(k)}[n])$ of the outputs of the inphase and quadrature branches of the correlation receiver lies). However, two branches (inphase and quadrature) for the correlation receiver are still necessary to ensure that $b_I^{(k)}[n]$ and $b_Q^{(k)}[n]$ are transferred from the uplink to the downlink. The above scheme has a lower implementation complexity compared to scheme (b) above, but at the expense of increased interference. The details of the model are complex; only the key parts are shown here.

The uplink (received at the satellite signal) is

$$Z_I^{(i),u}[n] = D_I^{(i),u}[n] + I_I^{(k,i),u}[n] + N_I^{(i),u}[n]$$

with its desired signal, interference and AWGN components.

The transmitted signal from the satellite over the downlink is

$$\hat{s}_i(t) = \sqrt{2\hat{P}_i}\hat{b}_I^{(i)}(t)c_I^{(i)}(t)\cos[\omega_c^d t + \hat{\theta}^{(i)}] + \sqrt{2\hat{P}_i}\hat{b}_Q^{(i)}(t)c_Q^{(i)}(t)\cos[\omega_c^d t + \hat{\theta}^{(i)}]$$

where $\hat{b}_I^{(i)}(t) = \sum_{n=-\infty}^{\infty} Z_I^{(i),u}[n]p_{T_s}(t - nT_s)$ and $\hat{b}_Q^{(i)}(t) = \sum_{n=-\infty}^{\infty} Z_Q^{(i),u}[n]p_{T_s}(t - nT_s)$

The output of the inphase branch of the correlation receiver of the CPE matched to the ith signal becomes

$$\begin{aligned} Z_I^{(i),d}[n] &= D_I^{(i),d}[n] + \sum_{k' \in \mathcal{I}^d(i)} I_I^{(k',i),d}[n] + N_I^d[n] \\ &= \sqrt{2P_i^d T_s}\Big\{(\ X_I\ )\ \cos[\theta^{(b_i)} - \hat{\theta}_i^{(b_i)}] \\ &\quad +(\ X_Q\ )\ \sin[\theta^{(b_i)} - \hat{\theta}_i^{(b_i)}]\frac{1}{L}\sum_{l=0}^{L-1} w_Q^{(i)}[l]w_I^{(i)}[l]\Big\} \\ &\quad + \sum_{k' \in \mathcal{I}^d(i)} \sqrt{2P_{k'}^d T_s}\Big\{(\ Y_I\ )\ \cos[\theta^{(b_{k'})} - \hat{\theta}_i^{(b_i)}]\frac{1}{L}\sum_{l=0}^{L-1} w_I^{(k')}[l]w_I^{(i)}[l]c^{(b_{k'})}[l]c^{(b_i)}[l] \\ &\quad +(\ Y_Q\ )\ \sin[\theta^{(b_{k'})} - \hat{\theta}_i^{(b_i)}]\frac{1}{L}\sum_{l=0}^{L-1} w_Q^{(k')}[l]w_I^{(i)}[l]c^{(b_{k'})}[l]c^{(b_i)}[l]\Big\} + N_I^d[n] \end{aligned}$$

where X_I, X_Q, Y_I, Y_Q are given below:

$$X_j = D_j^{(i),u}[n] + \sum_{k \in \mathcal{I}^u(i)} I_j^{(k,i),u}[n] + N_j^{(i),u}[n] \quad for\ j = I, Q$$

$$Y_j = D_j^{(k'),u}[n] + \sum_{k \in \mathcal{I}^u(k')} I_j^{(k,k'),u}[n] + N_j^{(k'),u}[n] \quad for\ j = I, Q$$

The set $\mathcal{I}^d(i)$ contains all the users interfering with user i during its downlink transmission, and the sets $\mathcal{I}^u(i)$ and $\mathcal{I}^u(k')$ contain all the users interfering with users i and k', respectively, during the uplink transmission.

6.4.2 Power of End-to-End Other-User Interference

The evaluation here parallels that in Section 6.3.2, and is conducted in two stages. During the first stage the power of interference in the uplink must be evaluated. To this end, we use the expression obtained at the end of the analysis of Section 6.3.1. During the second stage, we evaluate the power of signal and interference for the end-to-end system.

With all of the terms defined in Section 6.3.2, recall that

$$\hat{I}^u = \left[\sum_{k \in \mathcal{I}^u(i)} \frac{P_k^u}{P_i^u} \right] \bar{I}_s + \left(\frac{2E_s^u}{N_0} \right)^{-1}$$

Next we evaluate the power for the end-to-end interference. In the absence of a phase estimation error, that is, $\hat{\theta}^{(i)} \approx \theta^{(i)}$ or equivalently $\hat{\theta}_i^{(b_i)} \approx \theta^{(b_i)}$, we obtain for the desired signal and the other-user interference

$$Z_I^{(i),d}[n] = D_I^{(i),d}[n] + \sum_{k' \in \mathcal{I}^d(i)} I_I^{(k',i),d}[n] + N_I^d[n]$$

where the desired i^{th} signal is contained in the term

$$D_I^{(i),d}[n] = \sqrt{2P_i^d T_s} \left[D_I^{(i),u}[n] + \sum_{k \in \mathcal{I}^u(i)} I_I^{(k,i),u}[n] + N_I^{(i),u}[n] \right]$$

$N_I^{(i),u}[n]$ denotes the downlink AWGN term, and the other-user interference term $I_I^{(k',i),d}[n]$ due to the k'^{th} user $[k' \in \mathcal{I}^d(i)]$ is given by

$$I_I^{(k',i),d}[n] = \sqrt{2P_{k'}^d T_s} \left[D_I^{(k'),u}[n] \right.$$
$$\left. + \sum_{k \in \mathcal{I}^u(k')} I_I^{(k,k'),u}[n] + N_I^{(k'),u}[n] \right] \frac{1}{L} \sum_{l=0}^{L-1} w_I^{(k')}[l] w_I^{(i)}[l] c^{(b_{k'})}[l] c^{(b_i)}[l]$$

when k' and i are in the same uplink beam and by a similar (more complicated expression) when when users k' and i are in different downlink beams.

Computing the power in the desired signal as

$$\left[D_I^{(i),d} \right]^2 = \left[E\left\{ Z_I^{(i),d} \right\} \right]^2 = [(2P_i^u)T_s^2][b_I^{(i)}[0]]^2[(2P_i^d)T_s^2]$$

and the variance of the other-user interference and the AWGN, and normalizing by the power of the desired signal, we obtain

$$\hat{I}^e = \frac{1}{2}\sum_{k\in\mathcal{I}^u(i)} \frac{P_k^u}{P_i^u} C_{k,i}^2 + \left(\frac{2E_s^u}{N_0}\right)^{-1} + \\ + \sum_{k'\in\mathcal{I}^d(i)} \left[\frac{1}{2} + \frac{1}{2}\sum_{k\in\mathcal{I}^u(k')} \frac{P_k^u}{P_{k'}^u} C_{k,k'}^2 + \left(\frac{2E_s^u}{N_0}\right)^{-1}\right] \frac{P_{k'}^d}{P_i^d} C_{k'i}^2 + \left(\frac{2E_s^d}{N_0}\right)^{-1}$$

where the cross-correlation functions $C_{k,i}$ $C_{k',i}$ and $C_{k',k}$ were defined in Section 6.3.1, the uplink symbol energy E_s^u was defined in Section 6.3.2, and the downlink symbol energy E_s^d is similarly defined.

If the functions of the form $C_{k,i}$ are independent of the pair (k, i) (which the case with high accuracy as argued in Section 6.3.1) we can simplify as

$$\hat{I}^e = \left[\sum_{k\in\mathcal{I}^u(i)} \frac{P_k^u}{P_i^u}\right] \bar{I}_s + \left(\frac{2E_s^u}{N_0}\right)^{-1} + \\ + \sum_{k'\in\mathcal{I}^d(i)} \left\{\frac{1}{2} + \left[\sum_{k\in\mathcal{I}^u(k')} \frac{P_k^u}{P_{k'}^u}\right] \bar{I}_s + \left(\frac{2E_s^u}{N_0}\right)^{-1}\right\} \frac{P_{k'}^d}{P_i^d}(2\bar{I}_s) + \left(\frac{2E_s^d}{N_0}\right)^{-1}$$

where $\bar{I}_s$ was defined in Section 6.3.2 to denote the interference caused by a single user. Moreover, if the term $\sum_{k\in\mathcal{I}^u(k')} \frac{P_k^u}{P_{k'}^u}$ is nearly independent of k', which will be the case (on average) for a large size CDMA system, then we can further simplify $\hat{I}^e$ as

$$\hat{I}^e \approx \left[\bar{I}_s \sum_{k\in\mathcal{I}^u(i)} \frac{P_k^u}{P_i^u}\right] + \left[\bar{I}_s \sum_{k'\in\mathcal{I}^d(i)} \frac{P_{k'}^d}{P_i^d}\right] + \left[(2\bar{I}_s) \sum_{k'\in\mathcal{I}^d(i)} \frac{P_{k'}^d}{P_i^d}\right] \cdot \left[\bar{I}_s \sum_{k\in\mathcal{I}^u(k')} \frac{P_k^u}{P_{k'}^u}\right] \\ + \left(\frac{2E_s^u}{N_0}\right)^{-1} \cdot \left[(2\bar{I}_s) \sum_{k'\in\mathcal{I}^d(i)} \frac{P_{k'}^d}{P_i^d}\right] + \left(\frac{2E_s^u}{N_0}\right)^{-1} + \left(\frac{2E_s^d}{N_0}\right)^{-1}$$

This last expression includes (in order of appearance) uplink MAI (Multiple-Access Interference), downlink MAI, the cross-product of downlink MAI and uplink MAI, the cross-product of downlink MAI and uplink AWGN, uplink AWGN and downlink AWGN; that is, six terms instead of the typical two terms (MAI and AWGN) involved in single-hop transmission (e.g. for bent-pipe satellite or for on-board demodulation/remodulation).

6.4.3 Total End-to-End Other-User Interference Power

Applying the methodology of Section 6.3.3 for computing the power of other-user interference when power control is used and the location of the users is taken into account, we can obtain the results summarized in Table 6.5.

Table 6.5 Normalized power of end-to-end interference for baseband despreading/respreading.

System	End-to-End Interference $\bar{I}_0^e$, (One Polarization).
SO/SE-CDMA	$\frac{K}{2L}\left(6\bar{I}_1^u+8\bar{I}_2^u\right)+\frac{K}{2L}\left(6\bar{I}_1^d+8\bar{I}_2^d\right)$ $+\frac{K}{2L}\left(6\bar{I}_1^u+8\bar{I}_2^u\right)\cdot\frac{K}{L}\left(6\bar{I}_1^d+8\bar{I}_2^d\right)$ $+\left(\frac{2E_s^u}{N_0}\right)^{-1}\cdot\frac{K}{L}\left(6\bar{I}_1^d+8\bar{I}_2^d\right)+\left(\frac{2E_s^u}{N_0}\right)^{-1}+\left(\frac{2E_s^d}{N_0}\right)^{-1}$
FO/SE-CDMA	$\frac{K}{2L_u}\left(4\bar{I}_2^u\right)+\frac{K}{2L_u}\left(4\bar{I}_2^d\right)+\frac{K}{2L_u}\left(4\bar{I}_2^u\right)\cdot\frac{K}{L_u}\left(4\bar{I}_2^d\right)$ $+\left(\frac{2E_s^u}{N_0}\right)^{-1}\cdot\frac{K}{L_u}\left(4\bar{I}_2^d\right)+\left(\frac{2E_s^u}{N_0}\right)^{-1}+\left(\frac{2E_s^d}{N_0}\right)^{-1}$

System	End-to-End Interference $\bar{I}_0^e$, (Two Polarizations)
SO/SE-CDMA	$\frac{K}{2L}\left(3.75\bar{I}_1^u+5\bar{I}_2^u\right)+\frac{K}{2L}\left(3.75\bar{I}_1^d+5\bar{I}_2^d\right)$ $+\frac{K}{2L}\left(3.75\bar{I}_1^u+5\bar{I}_2^u\right)\cdot\frac{K}{L}\left(3.75\bar{I}_1^d+5\bar{I}_2^d\right)$ $+\left(\frac{2E_s^u}{N_0}\right)^{-1}\cdot\frac{K}{L}\left(3.75\bar{I}_1^d+5\bar{I}_2^d\right)+\left(\frac{2E_s^u}{N_0}\right)^{-1}+\left(\frac{2E_s^d}{N_0}\right)^{-1}$
FO/SE-CDMA	$\frac{K}{2L_u}\left(2\bar{I}_2^u\right)+\frac{K}{2L_u}\left(2\bar{I}_2^d\right)+\frac{K}{2L_u}\left(2\bar{I}_2^u\right)\cdot\frac{K}{L_u}\left(2\bar{I}_2^d\right)$ $+\left(\frac{2E_s^u}{N_0}\right)^{-1}\cdot\frac{K}{L_u}\left(2\bar{I}_2^d\right)+\left(\frac{2E_s^u}{N_0}\right)^{-1}+\left(\frac{2E_s^d}{N_0}\right)^{-1}$
MO/SE-CDMA	$\frac{K}{2L_u}\left(0.5\bar{I}_1^u+4.5\bar{I}_2^u\right)+\frac{K}{2L_u}\left(0.5\bar{I}_1^d+4.5\bar{I}_2^d\right)$ $+\frac{K}{2L_u}\left(0.5\bar{I}_1^u+4.5\bar{I}_2^u\right)\cdot\frac{K}{L_u}\left(0.5\bar{I}_1^d+4.5\bar{I}_2^d\right)$ $+\left(\frac{2E_s^u}{N_0}\right)^{-1}\cdot\frac{K}{L}\left(0.5\bar{I}_1^d+4.5\bar{I}_2^d\right)+\left(\frac{2E_s^u}{N_0}\right)^{-1}+\left(\frac{2E_s^d}{N_0}\right)^{-1}$

6.5 Bit Error Rate (BER) Evaluation

6.5.1 BER Evaluation for M-ary PSK SE-CDMA

In this section we provide the BER evaluation for an M-ary PSK CDMA system. Critical to this evaluation are the statistics of the interference at the output of the two branches of the matched filter receiver. We show that two separate calculations are necessary, depending on the absence or presence of transponder nonlinearities.

The starting point for our analysis are the expressions for the outputs of the two branches of the correlation receiver (matched filter) of the ith user. The calculation of this section is valid for both the uplink (on-board demodulation) and the downlink (demodulation at the CPE); therefore we suppress the superscripts u or d in our notation. We saw in Sections 6.3.3 and 6.4.1 that the outputs of the two (I and Q) branches of the correlation receiver always consist of the sum of the desired signal, (possibly) crosstalk interference, other-user interference and AWGN.

A key assumption is that the sum of desired signal, other-user interference and AWGN at the two branches of the output of the correlation receiver $Z_I^{(i)}$ and $Z_Q^{(i)}$ are jointly Gaussian random variables. For a moderately large number of users in the CDMA system, this approximation is justified by the Central Limit Theorem (CLT), since the outputs of the I and Q branches each consist of a large number of terms. The individual Gaussianness is justified by applying the CLT to each of the two variables. The joint Gaussianness is justified by applying the CLT to any linear combination of these variables $aZ_I^{(i)} + bZ_Q^{(i)}$ for arbitrary real numbers a and b.

Moreover, it is not difficult to show that in the *absence of transponder nonlinearities*, the random variables representing the outputs of the I and Q branches of the correlation receiver are uncorrelated when averaged with respect to the phase angles $(\theta^{(k)})$ and data symbols $\left(b_I^{(k)}[n], b_Q^{(k)}\right)$ of the various users. Therefore, since they are also jointly Gaussian, $Z_I^{(i)}$ and $Z_Q^{(i)}$ are mutually independent.

By contrast, when the transmitted signal passes through *transponder amplifier nonlinearities* (which introduce AM/AM and AM/PM nonlinear distortion) (see Chapter 9) the cross-correlation of the jointly Gaussian $Z_I^{(i)}$ and $Z_Q^{(i)}$ is not zero, and thus they are not mutually independent. Then a different formula should be used.

Let the outputs of the two branches of the ith correlator be

$$Z_I = D_I + I_I + N_I \quad \text{and} \quad Z_Q = D_Q + I_Q + N_Q$$

where we have suppressed the superscript (i). We assume that the satellite transponder amplifier works in the linear region, and thus no nonlinear distortion is introduced on the sum of the DS/CDMA signals. The three terms above thus represent the desired signal, other-user interference (plus crosstalk, if any), and AWGN. $E\{I_I\} = E\{I_Q\} = 0$. Thus, Z_I and Z_Q are jointly Gaussian uncorrelated and thus independent random variables. Their normalized mean is 1 (signal) and their variance is (for baseband despreading/respreading) $\sigma^2 = \bar{I}_0^e$, where the normalized power of end-to-end interference $\bar{I}_0^e$ is given by Table 6-5 for the various systems of interest to this chapter.

The Symbol Error Rate (SER) of the M-ary PSK CDMA system is then given by

$$P_{e|m} = 1 - \int_{\phi_m - \frac{\pi}{M}}^{\phi_m + \frac{\pi}{M}} \left[\int_0^{\infty} f(\rho, \phi) \rho d\rho \right] d\phi$$

where, through a polar transformation, the joint distribution of the Gaussian pair (Z_I, Z_Q) becomes

$$f(\rho, \phi) = \frac{1}{2\pi\sigma^2} \exp\left[-\frac{(\rho\cos\phi - \cos\phi_m)^2 + (\rho\sin\phi - \sin\phi_m)^2}{2\sigma^2} \right]$$

In the above expression the power of the desired signals is equal to 1, the noise variance is σ^2 and ϕ_m denotes $\phi_m^{(i)}$, the phase of the M-ary PSK symbol transmitted during the 0th time interval by the ith user; $\phi_m \in \{(2m-1)\pi/M,\ m = 1, 2, \ldots, M\}$. This is because the range of angles $\mathcal{R}_m = \left[\phi_m - \frac{\pi}{M}, \phi_m + \frac{\pi}{M}\right]$ corresponds to the decision region where ϕ_m is selected (decided upon), given that ϕ_m is transmitted. The above pdf $f(\rho, \phi)$ takes a different form when there is bias in the terms I_I and

I_Q above (their means are not 0), and when their correlation $E\{I_I I_Q\} \neq 0$; these are the conditions encountered when transponder nonlinearities introduce AM/AM and AM/PM distortion.

It is known in the literature [5] that the above integral can be approximated in closed form by

$$P_{e|m} \approx 2Q\left(\frac{\sin(\frac{\pi}{M})}{\sigma}\right)$$

independent of ϕ_m. Thus, the final expression is

$$P_e \approx 2Q\left(\frac{\sin(\frac{\pi}{M})}{\sqrt{\bar{I}_0^e}}\right) = 2Q\left(\sqrt{\mathrm{SNR}_t}\right)$$

where

$$\mathrm{SNR}_t = \frac{\sin^2(\pi/M)}{\bar{I}_0^e}$$

is the effective total signal-to-noise ratio that includes AWGN noise, crosstalk and other-user interference; $\bar{I}_0^e$ is as in Table 6.5. This quantity corresponds to the familiar cellular CDMA systems term $E_b/(N_0 + I_0)$.

The BER $P_{e,b}$ is evaluated from the SER P_e (of the uncoded system) via the bounds

$$\frac{P_e}{\log_2 M} \leq P_{e,b} \leq P_e$$

6.5.2 BER Evaluation for Concatenated Turbo/RS SE-CDMA

The final stages of our analysis was a combination of analysis and simulation. We used the expressions for the normalized power of total end-to-end interference ($\bar{I}_0^e$) in Table 6.5 and for the approximation of the BER in Section 6.1 above to evaluate the uncoded BER for the FO, MO and SO/SE-CDMA systems of interest. Then we adjusted the E_b to account for the use of outer RS codes; this was accomplished by multiplying E_b by the coding rate (15/16) of the RS code to obtain E_c the (coded) symbol energy. This was then substituted in the expression about $\bar{I}_0^e$ and in the effective $\mathrm{SNR}_t = \sin^2(\pi/M)/\bar{I}_0^e$. This value was used in the simulation results about the Turbo inner coding scheme. It is like simulating the performance of the Turbo codes for an AWGN channel with the above effective signal-to-noise ratio.

Once the simulation results for BER_i of Turbo codes were obtained by linear extrapolation from the BER curves of [3] and [4], we used the expression $m \cdot \mathrm{BER}_i = p_s$ to compute an upper bound on the probability of a symbol error for the RS outer code, and from that the codeword error probability of the binary RS code (of given codeword length and number code rate with errors-only minimum distance decoding) from the familiar RS codeword error probability formulas. The latter RS codeword error probability was also used as an upper bound for the final value of the BER of the system of interest.

6.6 Performance Results

The performance of the FO/SE-CDMA, MO/SE-CDMA and SO/SE-CDMA systems of Table 6.2 (Section 6.2) was evaluated in terms of the end-to-end BER. The basic transmission rate was 64 Kbps, and the rest of the parameters are as shown in Table 6.2. For FO/SE-CDMA $L_b = R_{c2}/R_{c1} = 4$ (length of Walsh beam codes) $L_u = R_{c1}/R_s s = 60$ and (length of QR user codes). For MO/SE-CDMA $L_b = R_{c2}/R_{c1} = 2$ and $L_u = 60$ as well. For SO/SE-CDMA $L_b = 1$ (no overspreading) and $L = L_u = 80$. The full load (capacity) of these systems is $K = 60$ users per beam for the FO and MO/SE-CDMA and $K = 80$ for SO/SE-CDMA, in the sense that this many orthogonal (QR) codes are avilable for reuse within each beam; however, the SO/SE-CDMA cannot operate with 80 users at acceptable BERs, as we will see.

The channel was AWGN (clear-sky conditions); refer to the discussion in Section 6.2.1 for extension of the analysis to Rician SATCOM channels. The performance measure in all comparisons of this section is the end-to-end BER at the CPE user receiver. It is assumed that only baseband despreading/respreading (without demodulation/remodulation or FEC decoding/re-encoding) takes place on-board the GEO satellite; the Turbo (inner) and RS (outer) decoding occurs at the user receiver. Two polarizations are assumed to be used in the FO and MO/SE-CDMA systems.

The code rates of the Turbo code and the RS codes are as shown in Table 6-1. The Turbo codes of [3] (code rate 1/2) and of [4] (for code rate 2/3 and 1/3) were used. The RS(256,240) codes use minimum distance decoding.

We assumed that $E_b^u/N_0 = E_b^d/N_0$. Also the impact of propagation losses, transmit and receive antenna gains, output backoff (OBO) for transponder nonlinearities, and rain fade losses that affect the typical SATCOM link budget (in db) were assumed to balance each other. We assumed this so that we can focus on the effects of other-user intra- and inter-beam interference (and of AWGN). Our evaluated E_b/N_0 is end-to-end (since our normalized interference power in Table 6-5 is end-to-end, assuming that the rest of the link budget is 0 dB. If it is not 0 dB, the positive (or negative) margin can be used to increase (or decrease) the required E_b/N_0 that we provide in our plots.

The normalized average interference from a single user in a first-tier beam (or a second-tier beam) that is used in Table 6.5 are evaluated for uplink and downlink from the double integrals of sections 6.3.3, respectively, for the antenna pattern parameters provided in Appendix 6A. These values turn out to be

$$\bar{I}_1^u = 0.116 \quad \text{and} \quad \bar{I}_2^u = 1.226 \times 10^{-3}$$

$$\bar{I}_1^d = 0.133 \quad \text{and} \quad \bar{I}_2^d = 1.512 \times 10^{-3}$$

In Figure 6.6 we show the BER versus the required signal-to-AWGN ratio E_b/N_0 (in dB) for $K = 1$, 15, 30, 45, and 60 users per beam. FO/SE-CDMA with 8-PSK, rate 2/3 Turbo inner code and RS(256,240) outer code is used. Notice that for E_B/N_0 in the range of 4.8–5 dB BERs lower than 10^{-6} can be achieved for all K. For smaller values of K it is the AGWN noise that dominates the interference in this system (since only interference from two (on average) second-tier adjacent beams is present) and in this (low K) range it is the use of 8-PSK that results in high values of the BER.

In Figure 6.7 the BER versus E_b/N_0 is shown for $K = 1$, 5, 15, 30, 45 and 60 users per beam. MO/SE-CDMA with QPSK, rate 1/2 Turbo inner code and RS(256,240) outer code is used. Notice that now the BER gracefully degrades as the number of users K per beam increases. Here for small K, AWGN does not dominate (there is now interference from 0.5 (on average) first-tier adjacent beams and from 4.5

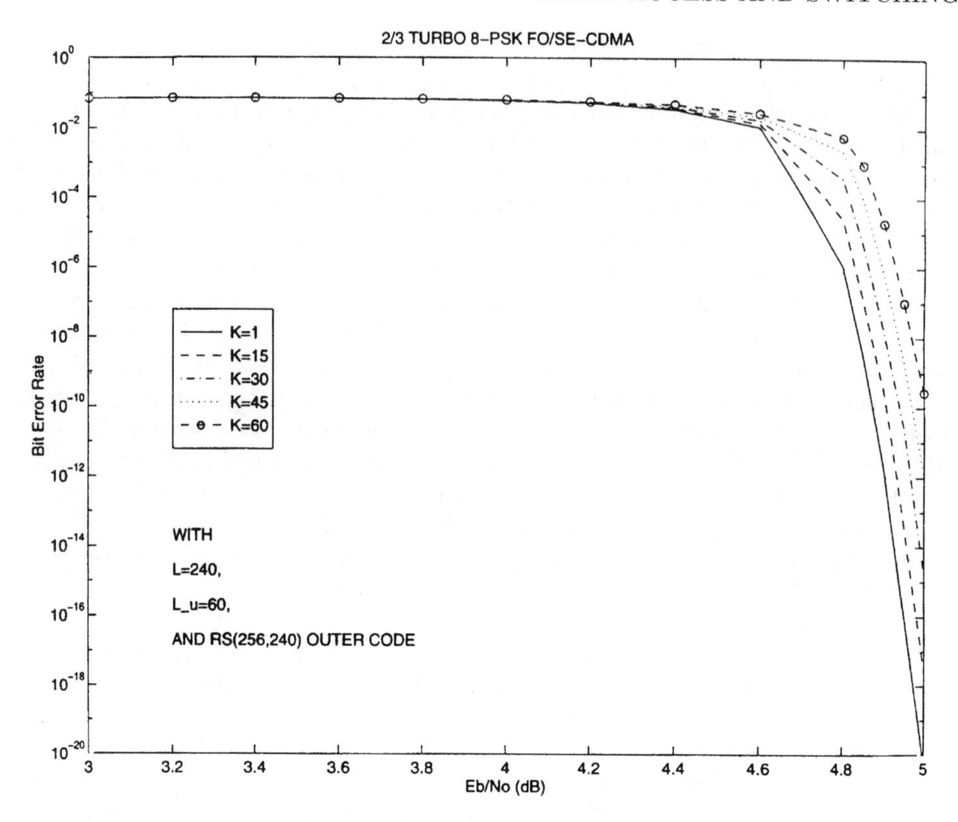

Figure 6.6 Bit Error Rate versus E_b/N_0 for the FO/SE-CDMA with 8-PSK, 2/3 inner turbo code and RS(256,240) outer code. Plotted for different number of users per beam, K.

second-tier adjacent beams) and the modulation is QPSK which has a better BER performance than 8-PSK (for the same code rate). Notice that for E_B/N_0 in the range of 3.5–3.75 dB BERs lower than 10^{-6} can be achieved for $K = 45$ and 60 users per beam.

In Figure 6.8 we show the BER versus E_b/N_0 for $K = 1$, 5, 15, 30, 45, and 60 users per beam for a SO/SE-CDMA system with QPSK, rate 1/2 Turbo inner code and RS(256,240) outer code. Notice that considerable higher E_B/N_0s are required to achieve BERs of 10^{-6} or lower: 7.5 dB for $K = 45$ and 13 dB for $K = 60$. (Both these numbers are below the full load of $K = 80$.) This system is more interference limited than FO or MO/SE-CDMA every user is now subject to interference from more first-tier adjacent beams (3.5 on average), and from five second-tier adjacent beams.

6.7 Conclusions

In this chapter we have presented several aspects of the traffic channels in the SS/CDMA system pertaining to the modulation, spreading and coding. We conducted a detailed interference analysis of SE/CDMA alternative schemes where the partial beam isolation provided via the overspreading mechanism and polarization in the FO

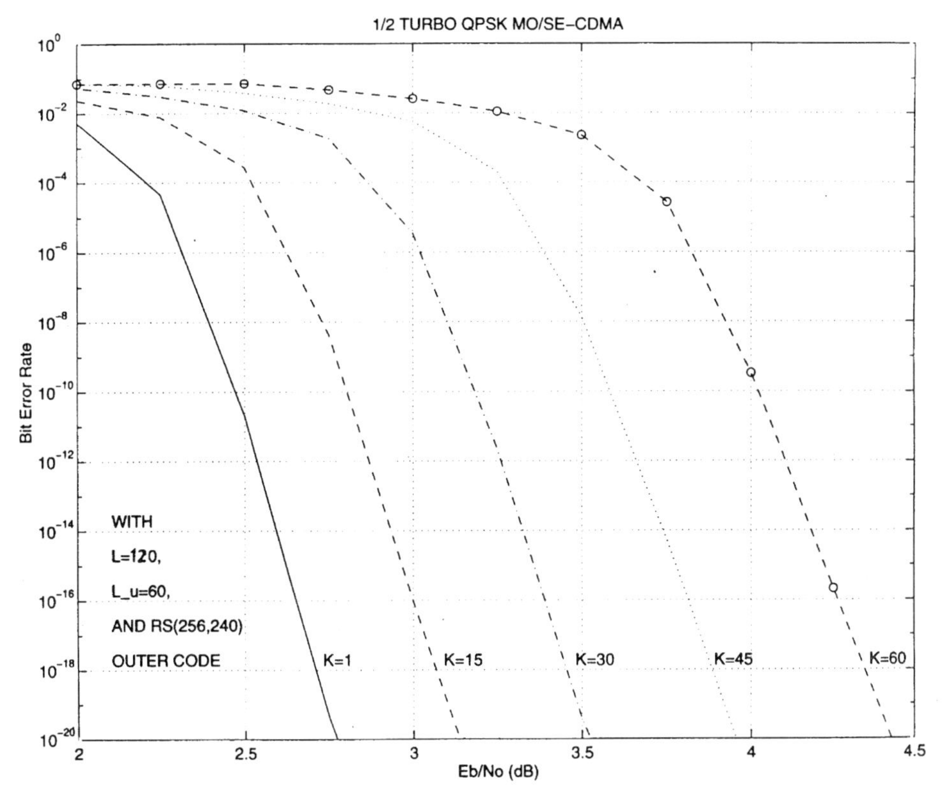

Figure 6.7 Bit Error Rate versus E_b/N_0 for the MO/SE-CDMA with QPSK, 1/2 inner Turbo code and RS(256,240) outer code. Plotted for different number of users per beam, K.

and MO/SE-CDMA systems was compared to the more traditional SO/SE-CDMA implementation. We established that FO and MO/SE-CDMA schemes can provide low BERs (below 10^{-6}) suitable for multimedia applications with low requirements E_b/N_0 (in the range of 4–6 dB); even lower BERs (lower than 10^{-8} or 10^{-10}) can be achieved with a mild increase in the required E_b/N_0.

Key in our performance evaluation are the assumptions of (i) an AWGN SATCOM channel (clear-sky conditions), (ii) completely synchronous operation, (iii) coherent demodulation, and (iv) operation in the linear region (no nonlinear distortion due to amplifier AM/AM and AM/PM distortion). Under these assumptions, the performance of the system that we analyzed represents an upper bound on what is achievable under realistic conditions (actually a lower bound on the BER or the required E_B/N_0).

We have been working on relaxing all four of the above assumptions, and our work in these areas appears the following chapters. In particular, in Chapter 8 we analyze the use of SAD (Symbol-Aided-Demodulation) for Rician channels; this addresses both issues (i) and (iii) above, since the Rician fading channel model is suitable for GEO SATCOM, and SAD is shown to be a viable alternative to fully coherent demodulation. In Chapter 9 we deal with the impact of AM/AM and AM/PM distortion (introduced by the transponder amplifier) on the sum of DS/CDMA signals. In Chapter 2 we

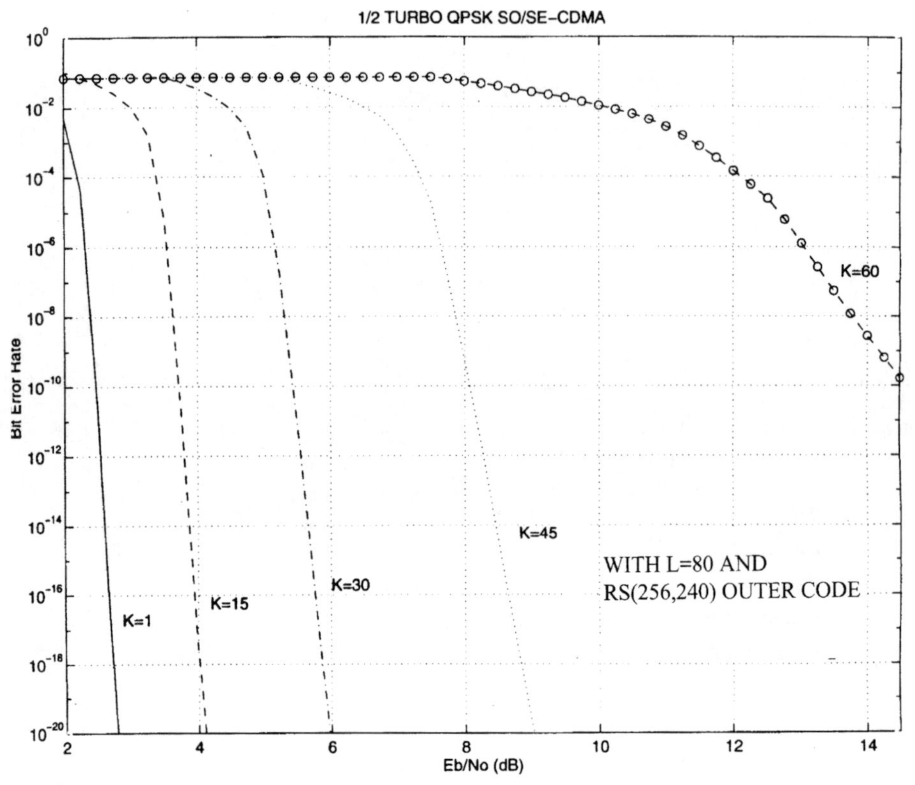

Figure 6.8 Bit Error Rate versus E_b/N_0 for the SO/SE-CDMA with QPSK, 1/2 inner Turbo code and RS(256,240) outer code. Plotted for different number of users per beam, K.

evaluated the impact of time-jitter, and in Chapter 7 we present a scheme for on-board code synchronization and tracking that addresses issue (ii) of the synchronous operation above.

References

[1] E. Geraniotis and D. Gerakoulis 'Bit Error Rate Evaluation of a Spectrally Efficient CDMA Scheme for Geostationary Satellite Communications' Submitted to the *International Journal of Satellite Communications*, 1999.
[2] R. De Gaudenzi, P. Giannetti and M. Luise 'Advances in CDMA Transmission for Mobile and Personal Communications' *Proceedings of the IEEE*, January 1996, pp. 13–37.
[3] Berrou, Glavieux and Thitimajshima 'Near Shannon Limit Error-Correcting Coding and Decoding: Turbo Codes (I)' ICC, 1993, pp. 1064–1070.
[4] LeGoff, Glavieux and Berrou 'Turbo Codes and High Spectral Efficiency Modulation' ICC, 1994, pp. 645–649.
[5] J. Proakis *Digital Communications*, McGraw Hill, 3rd Ed., New York, NY, 1995.

Appendix 6A: Antenna Patterns and Beam Geometry

A.1 Antenna Power Gain Functions
Typically these functions are of the form

$$G(\phi) = G_0 \left[\frac{2J_1(x)}{x}\right]^2 \quad \text{where} \quad x = \frac{\pi d}{\lambda}\sin(\phi)$$

d is the antenna diameter ($d = 2a$ where a is the dish radius), $\lambda = c/f_c$ is the wavelength of transmission related to the center frequency f_c of the system, and the speed of light c, and ϕ is the pointing angle from the user to the satellite or from the satellite to the user. G_0 denotes the center of the beam or boresight gain. The function $J_1(x)$ is the modified Bessel function of order 1; for general n this function is given by the expression $J_n(x) = \frac{1}{\pi}\int_0^{\pi}\cos[x\sin(\theta) - n\theta]d\theta$. For the specific satellite circular reflector antenna, the antenna power gain is

$$G(\phi) = G_0 \left[\frac{C \cdot \frac{2J_1(x)}{x} + \frac{1-C}{n+1} \cdot \frac{2^{n+1}(n+1)!J_{n+1}(x)}{x^{n+1}}}{C + \frac{1-C}{n+1}}\right]^2$$

where $C = 10^{-10/20} = 0.316228$ is the pedestal (edge illumination level), n is the parabolic taper order (taken to be $n = 2$), $\lambda = 0.010169m$ (for $f_c = 29.5GHz$), the antennae 5 dB (power) point is $\theta_{5\text{dB}} = 0.815^o/2$, and $G_0 = 10^{47.8/10}$ (center of the beam or boresight gain of 47.8 dB). The dish radius is $a = 0.529055$ m. The dish efficiency is thus $\eta = G(0)/(2\pi a/\lambda)^2 = 0.563946$.

A.2 User Locations and Satellite Pointing Angles
The relationships between the satellite pointing angles to the users and the distances from the center of the footprints of the beams are given by

$$\tan\phi_0^{(j)} = \frac{r_0^{(j)}}{D_0}\,, \quad \tan\phi_j^{(j)} = \frac{r_j^{(j)}}{D_0}\,, \quad \tan\phi_j^{(0)} = \frac{r_j^{(0)}}{D_0}$$

The interfering user, assumed to belong to beam j, is located in one of the beams adjacent to (surrounding) beam 0 (the reference beam) according to Figure 6.5(a). If beam j is a first-tier adjacent beam, then as can be seen in Figure 6.5(b) (where $\phi_0 = \phi_0^{(j)}$, $\phi_j = \phi_j^{(j)}$, $r = r_0^{(j)}$ and $r' = r_j^{(j)}$), the following equation provides the distance of the interfering user from the center of beam j:

$$r_j^{(j)} = \sqrt{[r_0^{(j)}]^2 + (2R)^2 - 2(2R)r_0^{(j)}\cos\theta_0^{(j)}}$$

for $R \le r_0^{(j)} \le 3R$ and $-\frac{\pi}{6} \le \theta_0^{(j)} \le \frac{\pi}{6}$

Similarly, if the interfering user resides in a second-tier adjacent beam j to beam 0, its distance to the center of beam j is

$$r_j^{(j)} = \sqrt{[r_0^{(j)}]^2 + (4R)^2 - 2(4R)r_0^{(j)}\cos\theta_0^{(j)}}$$

For $\frac{\pi}{12} \le \theta_0^{(j)} \le \frac{\pi}{12}$

We will also need the distance of the user of interest from the center of beam j, given by

$$r_j^{(0)} = \sqrt{[r_0^{(0)}]^2 + (2R)^2 - 2(2R)r_0^{(0)} \cos\theta_0^{(0)}}$$

for a first-tier adjacent beam j, and by

$$r_j^{(0)} = \sqrt{[r_0^{(0)}]^2 + (4R)^2 - 2(4R)r_0^{(0)} \cos\theta_0^{(0)}}$$

for a second-tier adjacent beam j, where the distance of the user of interest from the center of beam 0 and the associated planar angle take the values

$$0 \leq r_0^{(0)} \leq R \text{ and } -\pi \leq \theta_0^{(0)} \leq \pi$$

In the above models both the user of interest (in beam 0) and the interfering user (in beam j) can reside in any arbitrary location within these beams; it is this generality that forces us to introduce all of the above geometric parameters.

A.3 Random Location Model for Beam User Population

Every user can reside with equal probability in any location within the circle of radius R. This implies a specific distribution for the radius and the angle of each user's location. For example, for the user of interest with polar coordinates $\left(r_0^{(0)}, \theta_0^{(0)}\right)$ that satisfy $0 \leq r_0^{(0)} \leq R$ and $-\pi \leq \theta_0^{(0)} \leq \pi$ as seen above, the following is valid:

$$P\left\{\text{user of interest is in the area } \left[r_0^{(0)}, r_0^{(0)} + dr_0^{(0)}\right] \times \left[\theta_0^{(0)}, \theta_0^{(0)} + d\theta_0^{(0)}\right] \text{ of beam } 0\right\}$$
$$= \frac{r_0^{(0)}}{\pi R^2} dr_0^{(0)} d\theta_0^{(0)} = \frac{2r_0^{(0)}}{R^2} dr_0^{(0)} \cdot \frac{1}{2\pi} d\theta_0^{(0)}$$

Therefore, the probability density function (pdf) of being in the above coordinates is

$$f_{r,\theta}\left(r_0^{(0)}, \theta_0^{(0)}\right) = f_r\left(r_0^{(0)}\right) f_\theta\left(\theta_0^{(0)}\right) f_r\left(r_0^{(0)}\right) = \frac{2r_0^{(0)}}{R^2}$$

for $0 \leq r_0^{(0)} \leq R f_\theta\left(\theta_0^{(0)}\right) = \frac{1}{2\pi}$ and $-\pi \leq \theta_0^{(0)} \leq \pi$

The above density definitions pertain to a single user uniformly distributed in the circle of radius R. Let $\bar{\lambda}$ denote the user density (i.e. number of users per unit area). Then in the above density calculation $\bar{\lambda} = K/(\pi R^2)$ for K users in the area, and we can write

$$\tilde{f}_{r,\theta}\left(r_0^{(0)}, \theta_0^{(0)}\right) dr_0^{(0)} d\theta_0^{(0)} = \bar{\lambda} \cdot r_0^{(0)} dr_0^{(0)} \cdot d\theta_0^{(0)} = K \cdot \frac{2r_0^{(0)}}{R^2} dr_0^{(0)} \cdot \frac{1}{2\pi} d\theta_0^{(0)}$$

7

Network Access and Synchronization

7.1 Overview

As we have described in Chapters 3 and 6, the SS/CDMA uses orthogonal CDMA in both uplink and downlink. The orthogonal CDMA can reject the interference between the user traffic channels and thus maximize the system capacity. However, the use of orthogonal CDMA in the uplink requires a network-wide synchronization of all satellite receptions (global synchronization). The accuracy of the synchronization at steady-state and the speed at which synchronization is acquired depends on the propagation environment, i.e. the channel condition, the mobility of the end user, the propagation delay, etc. There are several examples in which synchronous CDMA (orthogonal or quasi-orhtogonal) attempted for use in the uplink or inbound channel. One such example is presented in reference [1], in which synchronous CDMA is proposed for mobile satellite applications. In another example presented in reference [2], orthogonal CDMA is recommended for the inbound and outbound links in terrestrial wireless applications. In this reference it is also shown (by simulation) that the required synchronization jitter from a reference time should not exceed 10% of the chip length for orthogonal CDMA operation. This result has also been verified analytically in Chapter 2. Such a requirement can be achieved easily if the CDMA system has a relatively narrow band and low user mobility. The system presented in reference [3] has a chip rate of 0.65 Mc/s (or a chip length of $T_c = 1.538$ μs, corresponding to a propagation distance of 460 meters), and the cell radious is 230 meters. The synchronization subsystem to meet the above requirement may then be simple. On the other hand, a wideband orthogonal CDMA requires a substantial effort in acquiring and maintaining synchronization, especially in a mobile environment. The above referenced systems, however, assume that a synchronization subsystem is in place without presenting one. In this chapter we present such a synchronization subsystem. This work was originally presented in references [4] and [5], and is a new approach for providing synchronization in an uplink orthogonal CDMA system. The proposed system design and procedures can achieve both access and synchronization in a geostationary satellite orthogonal CDMA for fixed service communications.

As we described in Chapter 3, the multibeam satellite common interface provides signaling control and user traffic channels within each satellite beam. The control channels are used for the acquisition of the user traffic channel. The downlink control channels are, the Pilot, the Sync and the Paging broadcast channels, and are identified by PN and orthogonal codes. In the uplink there is an asynchronous access channel which has an assigned beam PN-code. The access channel operation is based on a Spread-Spectrum Random Access (SSRA) protocol described in the next section. The traffic channels are defined by the user orthogonal code and the beam orthogonal and/or PN codes.

The basic steps of the network synchronization are the initial acquisition of the satellite downlink control channels, the access channel code acquisition, the system-wide synchronization of all traffic channels and the process of retaining and tracking the network sync once synchronization has achieved. Since uplink transmissions are asynchronous, the main part of this process is the synchronization of all uplink traffic channels. This is required in order to align all uplink orthogonal codes to a reference time upon arrival at the satellite despreaders, and thus provide orthogonality between the traffic channels. This alignment, however, may not be ideal, but it is required that the time offset of each signal from the reference time does not exceed 10% of the chip length. The factors which prohibit perfect synchronization include the long satellite propagation delay, the propagation delay variation due to satellite slow drift motion, as well as channel conditions such as rain fade, etc. The synchronization system also has to consider that the complexity of the on-board signal processing is limited by the available mass and power of the satellite.

The proposed synchronization system, although it is designed and evaluated for this particular application, may also be adapted and used in other applications (terrestrial or satellite) which have orthogonal CDMA for uplink access. Limited user mobilty may also be allowed, depending on the CDMA spreading rate and the properties of the codes used. A particular set of quasi-orthogonal codes may be less sensitive to allignment jitter. For example, preferentially phased Gold codes may allow timing jitter of up to 50% of the chip length; see Chapter 2.

The system evaluation is focused on the performance of the access channel code acquisition and the performance of the tracking control loop. The access channel carries control messages from the end user to the satellite, while at the same time provides the timing delay (PN phase offset) of the user code for the purpose of synchronizing the uplink orthogonal codes of the traffic channel. The proposed code acquistion is a serial/parallel scheme adjusted to meet the packet delay requirements. Related work on the subject is found in references [6] and [7]. The analysis of the tracking loop examines the loop stability and its steady state error. Other work related to the tracking control can be found in references [8] and [9]. The proposed feedback tracking-control loop, however, is quite different, since its tracking part resides in the satellite receiver while the control part resides in the user's transmitter. Thus, the tracking loop model also includes the satellite propagation delay.

This chapter is organized as follows. The synchronization procedures and the system design are presented in the next section. In Section 7.3 we provide the access channel performance evaluation, and in Section 7.4 the performance Tracking Contol Loop.

7.2 System Description

The system architecture of the satellite switched CDMA system is illustrated in Figure 7.1. In describing the synchronization procedures of the system, we first identify

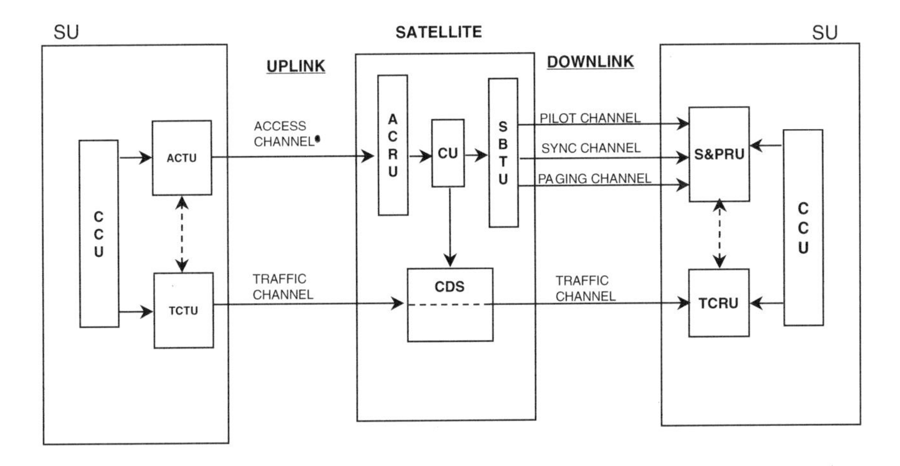

ACRU:	Access Channel Receiver Unit	SBTU:	Satellite Broadcast Transmitter Unit
ACTU:	Access Channel Transmitter Unit	S&PRU:	SYNC & Paging Receiver Unit
CCU:	Call Control Unit	SU:	Subscriber Unit
CDS:	Code Division Switch	TCRU:	Traffic Channel Receiver Unit
CU:	Control Unit	TCTU:	Traffic Channel Transmitter Unit

Figure 7.1 System architecture.

the PN and orthogonal codes of each channel that we use in the process. These codes include the PN code $g_p(t)$, defining the downlink pilot signal which has a rate of R_c and is transmitted continuously in the frequency band of the downlink control channels, ($g_p(t)$ is transmitted with given phase offset Δ_i corresponding to satellite beam i). The uplink access channel in beam i is identified by the PN code $a_i(t)$, which has a chip rate of R_c and operates in an assigned frequency band. Traffic channels are defined by the user orthogonal codes W_k $(k = 1, 2, ...)$, the beam PN-codes $g_i(t)$ $(i = 1, 2, ..)$ and the beam orthogonal codes W_i $(i = 1, 2, 3, 4)$. The traffic channel spreading operation shown in Figure 3.12 of Chapter 3. The beam codes W_i have a chip rate of $R_c (R_c = 4R_{c1})$. Beam PN-code g_i and user orthogonal code W_k have a rate of R_{c1}. All uplink traffic channel codes are required to arrive synchronously at the satellite despreaders in order to maintain the orthogonality among users within the beam, as well as among beams. That is, the starting time of all orthogonal codes should be aligned upon arriving at the satellite despreader.

7.2.1 Synchronization Procedures

The process that provides network-wide synchronization consist of a number of steps, described below:

1. *Initial Synchronization*: upon power-on the Subscriber Unit (SU) acquires synchronization to the Pilot PN sequence using the serial search acquisition circuit in the Sync and Paging Receiver Unit (S&PRU). See Figure 7.1. The S&PRU in the SU will then acquire the corresponding Sync channel in the beam. Based on the system information supplied by the Sync channel, the SU

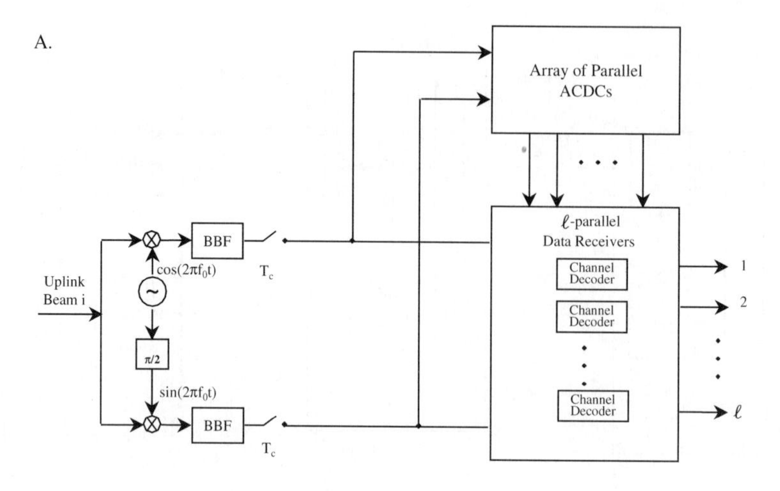

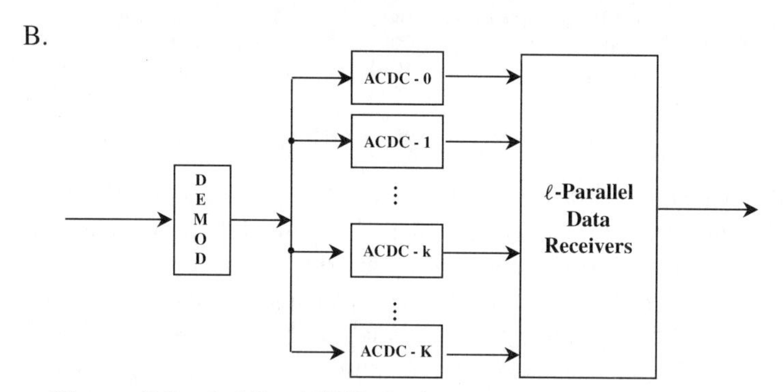

Figure 7.2 A. The ACRU. B. An array of ACDC in parallel.

will receive the orthogonal code W_i of the Paging channel (in the downlink) and the PN code a_i of the corresponding Access channel (in the uplink). The SU then acquires and monitors the Paging channel.

2. *Access Channel Acquisition*: the SU will then make an access attempt in the Access channel. The first message transmitted by the SU_{ki} (SU k in beam i) and received successfully at the Access Channel Receiver Unit (ACRU) will be used to establish the time delay (phase offset) $\Delta\tau_{ki}$ from the reference arrival time (τ_o), i.e. $\Delta\tau_{ki} = \tau_{ki} - \tau_o$. This message may arrive at the satellite despreader at any possible phase offset of the sequence $a_i(t)$. An array of K parallel Access Channel Detection Circuits (ACDC) is then used to cover all phase offsets of the code a_i (as described in Section 7.3) in order to acquire and despread the code. $\Delta\tau_{ki}$ may provide a resolution of T_c (one chip length), $T_c/2$, $T_c/4$ (T_c/ℓ is called the Chip-Cell, $\ell = 1, 2$ *or* 4); that is, $\Delta\tau_{ki} = x_{ki}T_c/4$. The value of $\Delta\tau_{ki}$ will then be sent back to the SU via a Paging channel.
3. *Traffic Channel Acquisition*: the SU will use the value of $\Delta\tau_{ki}$ to establish coarse synchronization to the satellite Reference Arrival Time. This is done by advancing or delaying x_{ki} chip cells the starting point of the code from its

original position at the successful message transmission. Then, the SU aligns each orthogonal and PN code of the uplink traffic channel to the code a_i and begins transmission. (The traffic channel orthogonal and PN codes W_k, W_i, g_i in the uplink, and W_ℓ, W_j, g_j in the downlink, are supplied to the SU by the on-board control unit.)

4. *Fine Sync Control*: after the SU begins transmitting on the traffic channel, a feedback tracking loop will provide fine alignment of the uplink codes with the reference arrival time at the satellite despreaders. This feedback loop extends between the satellite to the SU, and is described in detail in Section 7.2.2. Its transient and steady state response is derived analytically in Section 7.4.
5. *Sync Retention Control*: after a steady state has been reached, another Sync control circuit will be used to retain the fine Sync attained in the previous step. This circuit consists of the downlink (traffic channel) tracking circuit and uplink SYNC control described in Section 7.2.2.

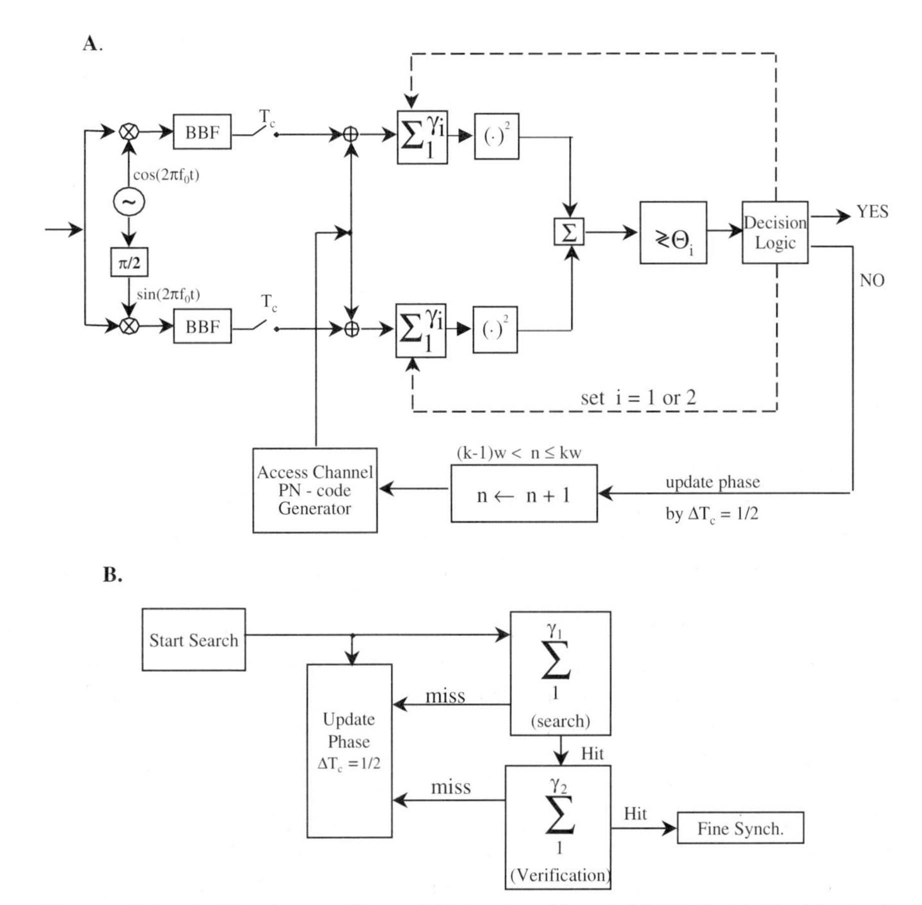

Figure 7.3 A. The Access Channel Detection Circuit (ACDC), B. Double dwell decision logic.

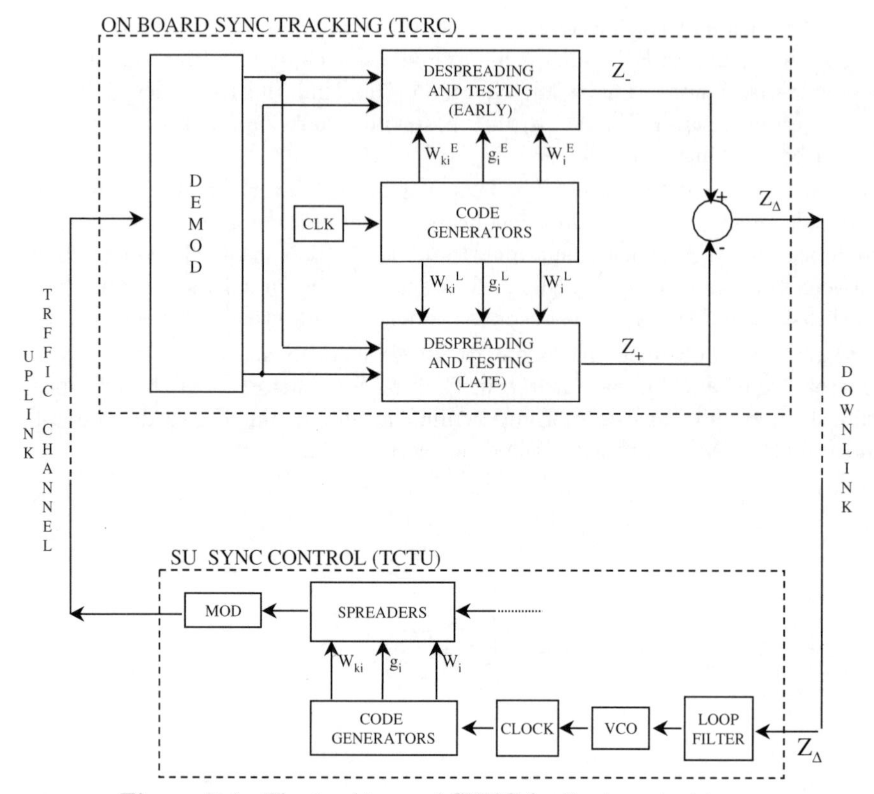

Figure 7.4 The tracking and SYNC feedback control loop.

7.2.2 System Design

In step 1 of the above procedure, the synchronization requirements to the Sync and Paging Receiver Unit (S&PRU) are provided by the Pilot PN code, which is acquired using a serial search acquisition circuit (in the S&PRU). (The Pilot PN-code is a common cover (beam) code for all other downlink control channels which are defined with known orthogonal codes.) In step 2 of the procedure, the access channel provides coarse synchronization for the orthogonal uplink traffic channel. This is an additional function of the access channel which comes at no extra cost. Its main function is to provide access for call set-up signaling messages. The access channel operates as an asynchronous random access channel. Its transmissions obey the Spread Spectrum Random Access (SSRA) protocol. According to SSRA, there is one PN code $a_i(t)$ for all users in beam i. Each user may begin transmitting a message at any time instant (unslotted channel). Each message consists of a preamble (containing no data) and the message information data field. The transmitted preamble signal will arrive at the receiver at any phase offset of the PN-code. Signals arriving at the receiver more than one chip apart will be distinguished and received. Messages that have (uncorrected) errors due to interference or noise will be retransmitted randomly after the time out interval, while messages that are successfully received will be acknowledged. The Access Channel Receiver Unit (ACRU), shown in Figure 7.2-A, consists of a noncoherent detector, an array of parallel Access Channel Detection Circuits (ACDC) and a pool of parallel data decoders. The array of parallel ACDCs, shown in Figure 7.2-B, provides a combination of parallel with serial acquisition circuits. Each ACDC, shown in Figure 7.3-A, searches for synchronization of the message by correlating over a window of w chips during the message preamble. The serial search method utilizes a typical double dwell algorithm, shown in Figure 7.3-B. Given L chips, the length of PN code a_i, and K as the number of ACDCs, the window size will then be $w = L/K$ $(1 \leq w \leq L)$. For example, if $L = 1204$ chips and $K = 16$, then $w = 64$ chips. The correlation process takes place during the message preamble. The actual number of parallel ACDCs K is determined by the required length of the preamble interval. In the serial search (double dwell) method, the length of the dwell time γ_1 and γ_2, as well as the thresholds (Θ_1 and Θ_2), are determined so that the requirements for the false alarm and detection probabilities are met. Also, the access channel is assumed to operate at low traffic load in order to offer a high probability of successful message transmission with the first attempt (see the performance analysis in Section 7.3).

The proposed mechanism for fine sync tracking control, used in step 4, is shown in Figure 7.4. It consists of the on-board SYNC-Tracking circuit, the downlink feedback path, the SU SYNC control circuit and uplink traffic channel timing jitter control. The on-board tracking consists of an Early-Late gate that provides the timing jitter Z_Δ. The timing jitter value Z_Δ will be inserted in a message and sent to the Call Control Unit (CCU) in the SU via the paging channel. The SU SYNC control circuit will then take Z_Δ as input to make the timing adjustment on the uplink traffic channel. The Early or Late despreading circuits may rely on the highest chip rate beam code W_i, i.e. $W_i(t \pm \Delta T_c)$ (the other codes g_i and W_k have a chip length of $T_{c2} = 4T_c$). Hence, the design of the proposed tracking loop differs from the typical design, since the timing adjustment takes place at the transmitter (SU), not the receiver. This is nessassary in order to align the transmitted orthogonal code to the reference time (at the satellite) at which all other transmissions have been aligned with. This tracking loop, however, introduces delays both in the feedback (downlink) as well as in the forward (uplink) path. This delay is equal to the satellite round trip propagation delay, which is about 250 ms. The delay also varies slowly because the satellite has a drift motion of about 2.5 meters/sec. The performance evaluation of this tracking loop have been provided in Section 7.4.

The last step of the process is required in order to maintain the fine synchronization achieved in the previous step without making use of the the on-board sync tracking circuit. The on-board tracking circuit will become available (after a steady state is reached) for reuse in another call, and thus reduce the on-board hardware. The sync retention control circuitry consists of the downlink traffic channel tracking circuit and the SYNC control circuit, shown in Figures 7.5 and 7.6. As shown in Figure 7.6, the feedback signal Z_Δ of the tracking circuit of the downlink traffic channel will also feed the input of the SU Sync control circuit. Hence, any change in the satellite propagation delay with respect to the established timing, by $\Delta\tau_p$ (resulting from satellite drift motion) will be indicated at the downlink traffic channel tracking circuit. The $\Delta\tau_p$ timing jitter will then be used (by the SU Sync control circuit) to compensate for the uplink transmission by advancing or delaying by $\Delta\tau_p$ using the SYNC control circuit.

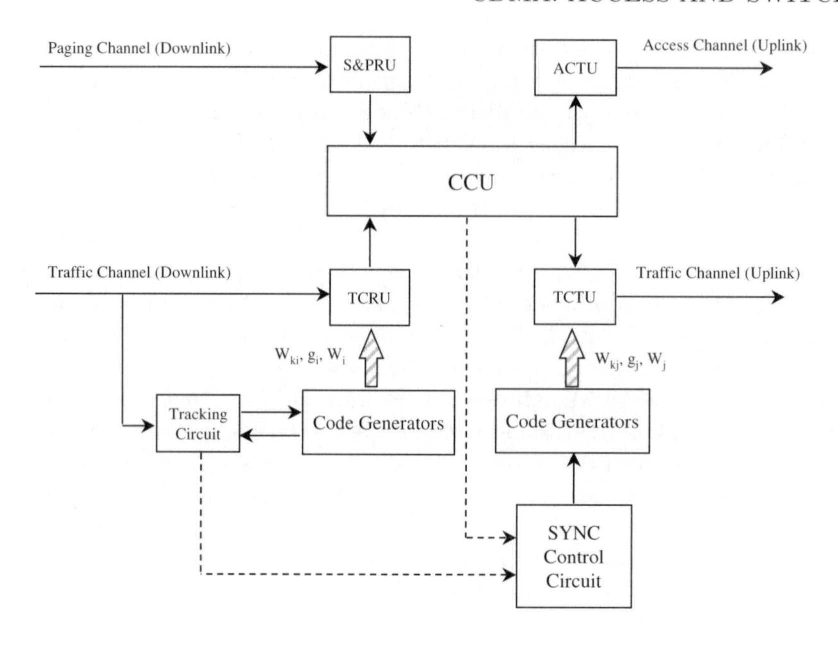

ACTU:	Access Channel Transmitter Unit
S&PRU:	SYNC and Paging Receiver Unit
TCRU:	Traffic Channel Receiver Unit
TCTU:	Traffic Channel Transmitter Unit
CCU:	Call Control Unit

Figure 7.5 The SU tracking and SYNC control circuits.

7.3 Access Channel Performance

Considering the long round trip satellite propagation delay, the main performance requirement of the access channel is to provide a high probability of success with the first transmission attempt. The probability of a successful message transmission depends (a) on the probability of PN code acquisition during the message preamble, (b) on the probability of message collision, and (c) on the probability of no bit errors in the message after channel decoding (called the retention probability).

(a) The performance analysis presented in Section 7.3.1 determines the design parameters for a serial/parallel aquisition circuit which maximizes the probability of successfully acquiring (P_{acq}) within the preamble interval. These parameters determine the minimum preamble interval and the optimum window size ω for a given code length of L chips and known interference noise conditions. The probability P_{acq}, called the *aquisition confidence*, is given by

$$P_{acq} = Pr[T_{acq} \leq T_h] = \int_0^{T_h} f_{T_{acq}}(\tau)d\tau = F_{T_{acq}}(T_h)$$

where T_{acq} is the aquisition time for the window size of ω chips and T_h is the minimum allowed length of the message preamble which satisfies the aquisition confidence. The probability distribution $f_{T_{acq}}$ or the cumulative distribution $F_{T_{acq}}$ functions of the acquisition time have been derived in Section 7.3.1.

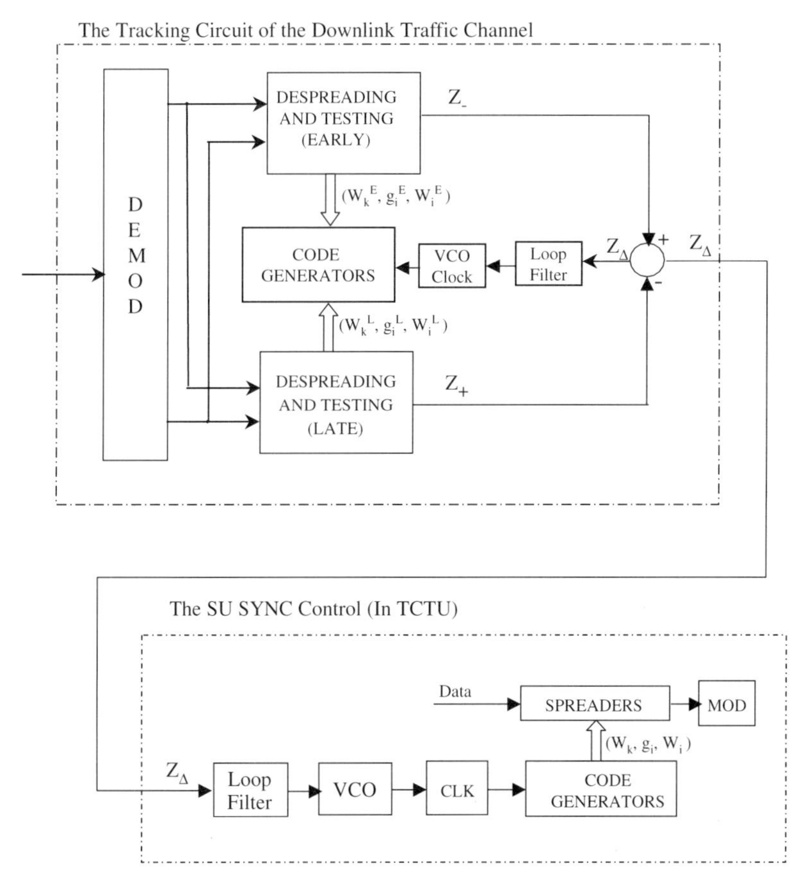

Figure 7.6 The SYNC retention tracking control circuit.

(b) Collision of two or more packets will occur if they are overlapping and have the same PN code phase offset when they arrive at the despreader. This is based on the assumption that the channel is unslotted (continuous time) and a single PN code has been used for all users in the channel. Also, we assume that all packets arrive at the despreader with approximately equal power. The probability that i packets collide given that k_o packets are overlapping at any time instant is given by

$$P_{coll}(i|k_o) = \binom{k_o}{i} (1/L)^i (1 - 1/L)^{k_o - i} \quad for \;\; 2 \leq i \leq k_o$$

where $1/L$ is the probability that i packets have exactly the same phase offset of the PN code. The number of all possible phase offsets is assumed to be L, equal to the length of the code (in terms of the number of chips). (If the phase offset is less than a chip we assume that collision takes placc.)

The probabilty that k_o packets overlap is given by

$$P(k_o) = \frac{(2t_p G)^{k_o}}{k_o!} e^{-2t_p G}$$

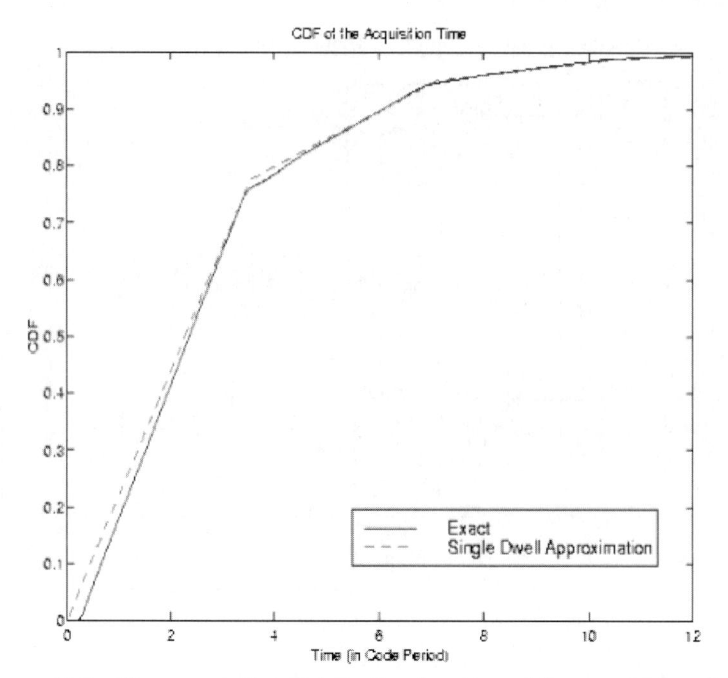

Figure 7.7 The cumulative distribution function of the acquisition time.

In the above expression, t_p is the packet length and G is the total offered traffic load which includes both the newly arrived and retransmitted packets (two or more packets will overlap if they arrive in the interval $2t_p$).

The probability of collision will then be

$$P_{coll} = \sum_{k_o \geq 2} \sum_{i=2}^{k_o} P_{coll}(i|k_o)P(k_o)$$

(c) The probability of packet retention, P_{ret}, is the probability of having no errors in the packet's information field after FEC. That is, if the packet length is n bits, then $P_{ret} = (1 - P_e)^n$, where P_e is the bit error probability.

The probability of successful packet transmission is then given by

$$P_{succ} = P_{acq}(1 - P_{coll})P_{ret}$$

In the above equation, however, we have assumed that there is always a receiver available to decode the data in the packet. If there are ℓ parallel data receivers available (as shown in Figure 7.2-B), then the probability of not finding an available one is $P_{un}(\ell) = \sum_{k_o \geq \ell} P(k_o)$, where $P(k_o)$ is the probability of having k_o packets overlapping at the reception at a given load G (given above). Then, $P'_{succ} = P_{succ}[1 - P_{un}(\ell)]$.

7.3.1 Packet Acquisition Performance Analysis

The PN code aquisition is based on the serial/parallel model shown in Figures 7.2 and 7.3. The PN sequence of length L is divided into K subsequences of ω chips each. There are K double dwell serial search circuits operating in parallel over each of these subsequences or windows. The PN phase offset of each arrived packet in a given window will be detected by the corresponding parallel searching circuit. In general, the acquisition performance can be improved if the outputs of these parallel circuits are processed jointly, therefore the performance of any of the circuits assuming that they work independently can serve as an upper bound.

In order to determine the false alarm and detection probabilities, we first assume that each chip is further divided into l cells during the search. This will give a final pull-in uncertainty of $\frac{T_c}{2l}$ for the code tracking. For each searching window, a total of $\nu = l\omega$ cells will be tested, among which $\nu - 2l$ can lead to false alarms. The false alarm probabilities for noncoherent reception under unfaded AWGN channels can be written as (see [10])

$$P_{Fi} = \int_{\Theta_i}^{\infty} p_{0i}(z)dz = e^{-\Theta_i/V_i}, \quad i = 1, 2$$

where i is the dwell index and Θ_i is the corresponding threshold. $V_i = \gamma_i I_o$ is the equivalent noise spectral density, with γ_i being the dwell (accumulation) time and

$$I_0 = N_0 + 2m_\psi \sum_{j=1}^{K_a} E_c(j).$$

where $E_c(j)$ is the energy per chip for user j and K_a is the number of simultaneous packet receptions; m_ψ is an interference constant depending on the chip waveform. For the $2l$ cells within one chip of the correct timing, we need to determine the detection probabilities. The worst case corresponds to sampling times that differ from the correct (peak) time by

$$\tau_j = \frac{j - (l + \frac{1}{2})}{l} T_c, \quad j = 1, 2, \ldots, 2l$$

For each of these sampling times, the detection probabilities are given by

$$P_{Dij} = \int_{\Theta_i}^{\infty} p_{1ij}(z)dz = \int_{\Theta_i/V_i}^{\infty} e^{-(x+\mu_{ij})} J_0(2\sqrt{\mu_{ij} x})dx, \quad i = 1, 2; \ j = 1, 2, \ldots, 2l \tag{7.1}$$

with

$$\mu_{ij} = \gamma_i \frac{E_c(k)}{I_0} R^2(\tau_j)$$

where $R(\tau_j)$ is the correlation between the chip waveform and the receiver chip filter with time offset τ_j.

The reduced state diagram (in reference [10]) now has the branch transfer functions

$$H_0(z) = z^{\gamma_1}(1 - P_{F1}) + z^{\gamma_1+\gamma_2} P_{F1}(1 - P_{F2}) + z^{\gamma_1+\gamma_2+\gamma_p} P_{F1} P_{F2}$$

$$H_D(z) - \sum_{j=1}^{2l} P_{D1j} P_{D2j} z^{\gamma_1+\gamma_2} \prod_{i=1}^{j-1} [(1 - P_{D1i}) z^{\gamma_1}]$$

$$H_M(z) = z^{2l\gamma_1} \prod_{j=1}^{2l} (1 - P_{D1j}) + \sum_{j=1}^{2l} P_{D1j}(1 - P_{D2j}) z^{\gamma_1+\gamma_2} \prod_{i=1}^{j-1} [(1 - P_{D1i}) z^{\gamma_1}]$$

where $H_0(z)$ corresponds to the transfer function of the branches emerging from the $\nu-2l$ nodes without the presence of the signal and $H_D(z)$ and $H_M(z)$ are the detection and mistransfer functions respectively emerging from the super-state representing the cells within one chip of the correct timing; γ_p is the penalty for a false alarm at the second dwell. The total transfer function, assuming that all cells in the searching window are equally likely to become the starting cell, is given as follows:

$$U(z) = \frac{H_0(z)H_D(z)\left[1 - H_0^{\nu-2l}(z)\right]}{(\nu - 2l)\left[1 - H_0(z)\right]\left[1 - H_M(z)H_0^{\nu-2l}(z)\right]}$$

The mean and variance of the acquisition time can therefore be computed by

$$E[T_{acq}] = \left.\frac{dU(z)}{dz}\right|_{z=1}$$

and

$$Var[T_{acq}] = \left.\left\{\frac{d^2U(z)}{dz^2} + \frac{dU(z)}{dz}\left[1 - \frac{dU(z)}{dz}\right]\right\}\right|_{z=1}$$

These two quantities, however, are not sufficient to evaluate the acquisition confidence for the given preamble length. Fortunately, the operating situation we consider here meets the conditions given in [12]; the approximation therein can therefore be applied to compute the CDF of the acquisition time. The single dwell equivalence of the case considered can be characterized by the following parameters:

$$\nu_e = \nu - 2l + 1$$

$$\gamma_e = \left(\gamma_1 + P_{F1}\frac{1 - P_{F2}}{1 - P_{F1}P_{F2}}\gamma_2\right)\left(\frac{\nu_e - 1}{\nu_e}\right) + \left(\left.\frac{dH_D(z)}{dz}\right|_{z=1} + \left.\frac{dH_M(z)}{dz}\right|_{z=1}\right)\frac{1}{\nu_e}$$

$$\gamma_{pe} = \gamma_p + \frac{1 - P_{F1}}{1 - P_{F1}P_{F2}}\gamma_2$$

$$P_{De} = \left.H_D(z)\right|_{z=1}$$

$$P_{Fe} = P_{F1}P_{F2}$$

where in the equation of $H_M(z)$ the second term is added to account for the additional dwell time introduced by the super-node containing the correct timing. When the search window size $\nu \gg 2l$, the equation of $H_M(z)$ becomes equivalent to (22a) in [12]. The CDF of the acquisition time can then be approximated by [12]

$$F_{T_{acq}}(t) = 1 - (1 - P_{De})^J\left[1 + JP_{De} - \frac{P_{De}t}{\nu_e(\gamma_e + P_{Fe}\gamma_{pe})}\right]$$

where

$$J = \left\lfloor\frac{t}{\nu_e(\gamma_e + P_{Fe}\gamma_{pe})}\right\rfloor$$

Now we assume that the false alarm probabilities P_{Fi}, the penalty γ_p, the window size ν, the chip waveform and E_c/I_0 are given by the system requirements. If we further define an acquisition confidence α so that within the preamble length T_h the probability of acquisition is P_{acq}, according to $\nu_e = \nu - 2l + 1$, we have

$$(1 - P_{De})^J \left[1 + JP_{De} - \frac{P_{De}T_h}{\nu_e(\gamma_e + P_{Fe}\gamma_{pe})}\right] = 1 - P_{acq}$$

This relation can be used to obtain the optimal dwell times γ_1 and γ_2 which will minimize the required preamble length T_h. Due to the integration involved in the evaluation of P_{coll}, this optimization problem cannot be solved analytically. A discrete two-dimensional search, however, can be performed to find the best dwell times. Given the simple form of $E[T_{acq}]$, this search process does not need excessive computation.

During the search, local minima resulted from the discontinuity in the equation of $H_M(Z)$ were observed. Moreover, a smaller T_h does not guarantee a smaller mean acquisition time due to the change of CDF. For example, a longer dwell time can make T_h smaller by reducing the variance, but it also shifts the mean acquisition time towards a larger value. As a result, we have to compromise between these two quantities. The results presented here have the minimal mean acquisition times among the local minima observed.

In order to determine the values of T_h corresponding to the required acquisition confidence, we first derived the CDF of the acquisition time $F_{T_{acq}}$ (see Section 7.3.1). Figure 7.7 shows an approximation of the CDF when $E_c/I_0 = -10\ dB$, $\gamma_1 = 54$ and $\gamma_2 = 137$ chips. Also, the transmission chip waveform employed is a raised cosine with a roll-off factor 0.1, and the receiver uses a matched filter with the same waveform.

7.3.2 Packet Acquisition Performance Results

In Tables 7-1 and 7-2, we present the serial/parallel acquisition circuit performance results for code lengths $L = 1024$, 512 and for $K = 32$ and 16 parallel circuits. The false alarm probabilities used are $P_{F1} = 0.01$, $P_{F2} = 0.1$ and the corresponding penalty γ_p is equal to the PN code period L. The mean acquisition time $E[T_{acq}]$ varies from 0.3 to 1.4 ms, depending on the values of the (E_c/I_0) (chip energy to interference ratio), L and K. The dwell accumulation values γ_1 and γ_2 are optimized for each case. The required preamble length T_h is provided in each case in terms of the number of code lengths $(\times L\ )$ and in msec, assuming the chip rate is $R_c = 9.8304$ Mc/s. The T_h values given in the tables represent the minimum preamble length required to achieve an acquisition confidence of 95%. As shown, the minimum required preamble length T_h varies from 0.73 to 3.65 ms for $L = 1024$ and from 0.16 to 1.83 ms for $L = 512$, depending on the channel conditions (E_c/I_0) and the number of parallel ACDCs (K). The packet delays introduced by these preamble lengths are then feasible and acceptable, even with delay sensitive traffic.

Assuming that the acquisition confidence $P_{acq} = 0.95$ and the bit error rate is 10^{-5} (after FEC), we have also evaluated and plotted the probability of successful packet transmission P_{succ} versus the offered load (packets/sec) for packet lengths of $n = 256$ and $n = 512$ encoded bits (or 128 and 256 information bits assuming FEC rate 1/2) with $\ell = 1, 2, 3$ parallel data receivers (channel decoders). The period of the encoder is 512. The plot is shown in Figure 7.8. As shown, the P_{succ} is near 0.95 for a wide range of packet loads (up to 10 packets/sec), when the packet length is 256 symbols and with two or more channel decoders.

Table 7.1 Acquisition performance for $L = 1024$.

K	E_c/I_0 (dB)	γ_1 (Chips)	γ_2 (Chips)	$E[T_{acq}]$ (ms)	T_h ($\times L$)	T_h (ms)
32	−10	54	137	0.3069	7.1002	0.7396
32	−12	85	243	0.4845	11.1816	1.1648
32	−14	135	371	0.7643	17.6459	1.8382
16	−10	54	141	0.5959	14.1483	1.4738
16	−12	85	243	0.9383	22.2378	2.3165
16	−14	135	371	1.4811	35.1115	3.6576

Table 7.2 Acquisition performance for $L = 512$.

K	E_c/I_0 (dB)	γ_1 (Chips)	γ_2 (Chips)	$E[T_{acq}]$ (ms)	T_h ($\times L$)	T_h (ms)
32	−10	54	137	0.3695	7.0937	0.1611
32	−12	87	183	0.2537	11.2253	0.5847
32	−14	138	290	0.4015	17.7618	0.9251
16	−10	54	137	0.3044	14.0784	0.7333
16	−12	86	204	0.4811	22.2688	1.1599
16	−14	136	332	0.7609	35.1989	1.8334

7.4 Performance of the Tracking Control Loop

In this section we examine the tracking loop stability and steady state error. The tracking loop performance is based on the analysis given in references [10] and [11], which is also outlined in Section 7.4.1. The tracking loop circuitry is shown in Figure 7.9. The mean and variance of the discriminator output has been derived for a discrete time model of the loop. Figure 7.10 shows the loop discriminator gain versus the timing error. The model of the above tracking circuit is shown in Figure 7.11-A. In this model the time unit is the duration of one channel symbol, which is the period of accumulation. Assuming operation in the linear region of the curve, the loop has been approximated by a linear model, shown in Figure 7.11-B.

There are four issues of importance in the practical system design: stability, convergence speed, steady state performance, and feedback bandwidth. In the following we discuss each of these, and their influence on each other.

The steady state timing error (derived in Section 7.4.1) is given by

$$\tau_s = \lim_{z \to 1}(1 - z^{-1})\frac{\tau(z)}{T_c} = \frac{c}{A\alpha F(z)}$$

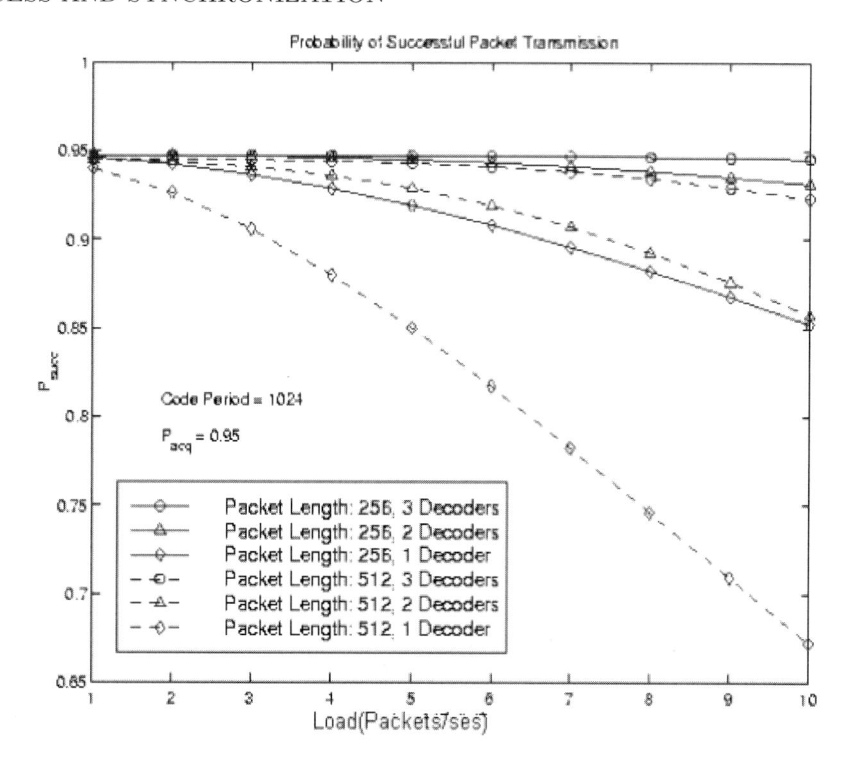

Figure 7.8 The probability of successful packet transmission.

where $F(z)$ is the loop filter and c is the normalized Doppler shift. The variance of the timing error $\sigma_s^2 = Var(\tau/T_c)$ is given by (see Section 7.4.1)

$$\sigma_s^2 = \frac{(V_0 + n_0)}{2\pi} \int_{-\pi}^{\pi} \left| \frac{\alpha F(e^{j\omega}) e^{j\omega(-d-1)}}{1 - e^{-j\omega} + A\alpha F(e^{j\omega}) e^{j\omega(-d-1)}} \right|^2 d\omega$$

The value of $A\alpha$ is usually very small, and as we can see from the above expression, the steady state variance increases as α increases. This implies that given a loop filter, there is an optimal value of α. We note, however, that the interference variance V_0 is itself a function of the steady state error.

So far, we have considered only one accumulator with length L. The length of the accumulation, however, cannot be larger than the proccessing gain because of the data symbol length. This implies very large feedback bandwidth consumption (larger than the traffic channel), which is unreal. In attempting to solve this problem, we built a second accumulator on the satellite to accumulate and average Z_Δ. Assuming the length of the second accumulation is N, every component of the tracking loop now works N times slower. As we observed in the previous equations, we notice that only three parameters are affected by this down-sampling: the Doppler constant c, the delay step d, and the variance of the timing error σ_s^2. The Doppler constant is now replaced by $c' = cN$; the delay step is replaced by $d' = d/N$. The variance of V_o is also divided by N because of the i.i.d. property of the interference from symbol to symbol. σ_s^2/N

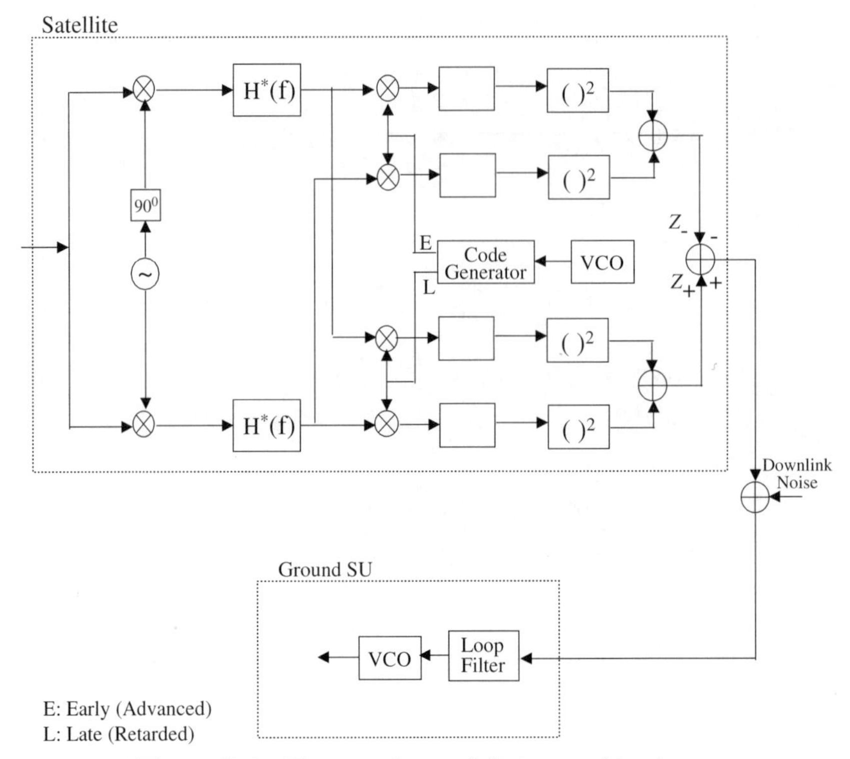

Figure 7.9 The noncoherent full-time tracking loop.

is thus replaced by

$$\sigma_s'^2 = \frac{\sigma_s^2}{N}$$

From from the final value of τ_s we can see that if we want to maintain the same steady state timing error after the down-sampling, α needs to be N times larger. Using the equation of ${\sigma_s}^2$ and the new value of α, we obtain a higher steady state variance from $\sigma_s'^2 = {\sigma_s}^2/N$. Therefore we conclude that finer sampling (smaller N) achieves better steady state performance which, however, requires a wider feedback bandwidth.

The final goal of this performance section is to establish the feasibility of the tracking feedback delay loop with given SS/CDMA system parameters and requirements. Hence, we consider the following system parameters: the chip rate $R_c = 9.8304$ Mc/sec ($T_c = 1/R_c = 1.0173 \times 10^{-7}$ sec). The orthogonal (quadratic residue) code of length ($L = 60$) (one step in the discrete model is equal to 6.1035×10^{-6} sec). The system is assumed to be fully loaded, which means that the number of users $K = 60$. The longest round trip delay is 0.26 sec, which makes the delay step d as large as 42,598. The Doppler shift caused by the satellite drift is 40 ns/s. When normalized with T_c, the Doppler constant c is 2.4×10^{-6}. A raised cosine waveform with a roll-off factor of 0.1 is utilized as the chip waveform. As a result, the waveform factor is $m_\psi = 9.8 \times 10^{-3}$ (see Section 7.4.1). The received signal-to-noise ratio is $E_s/N_0 = 6$ dB. The maximum chip offset from the satellite oscillator is required to be within $\pm\frac{T_c}{10}$. For the early-late correlators, we assume $\Delta T_c = \frac{T_c}{4}$; this puts a constraint of $\frac{T_c}{4}$ on

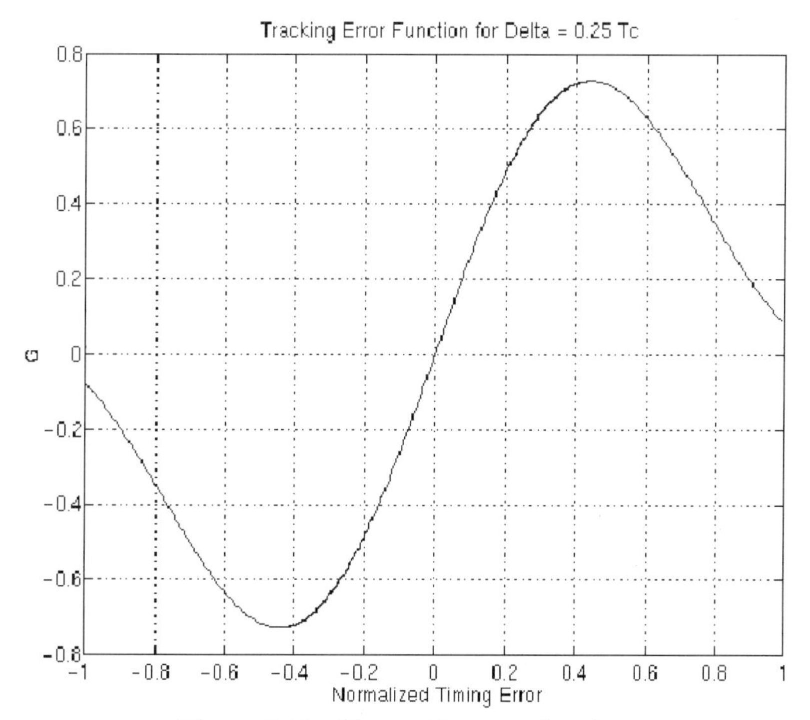

Figure 7.10 The tracking error function.

the acquisition pull-in condition. The equivalent gain γ of the linear model obtained from $\gamma = dG(\tau/T_c)/d(\tau/T_c)|_{\tau=0}$ is 2.6762.

In the following evaluation, we assume that the feedback information is well encoded so that the downlink noise can be ignored. Taking into consideration the accuracy of the feedback information, and the fact that the feedback bandwidth is limited, we reach the decision of setting the down-sampling factor as $N = 1000$. Scaling of d, c and σ_s^2 is done as described above. The assumption that the timing error does not change much in N symbols is justified, since the new normalized Doppler shift is only 2.4×10^{-3} chips. Due to the large delay d, solving the characteristic functions such as $(1 - z^{-1})^2 + A\alpha B(z)z^{-d-1}$ (derived in Section 1.4.1) is not possible. Fortunately, we can use the Jury Stability Test (see [15]) with computer aided search. The result shows that it is not possible to have a loop filter of the form $F(z) = \frac{B(z)}{1-z^{-1}}$ (see Section 1.4.1) without a steady state error. This means that we'll always have a steady state error $\tau_s = \frac{c}{A\alpha F(z)}$ (see Section 1.4.1). Therefore, a scaling adjustment is required when the steady state is reached. This is possible since τ_s is approximately known.

A simple form of loop filter is taken for evaluation:

$$F(z) = \frac{1}{1 + \beta z^{-1}}$$

Computer search shows that β can only be in the range of $(-1, 1)$ so the system is stable. The maximum allowable values of $A\alpha$ for different values of βs are shown in Figure 7.12. We define a cost function as $\tau_s + 2\sigma_s$, which means we have more than 95% of confidence that the timing error will be smaller than $\tau_s + 2\sigma_s$. A typical relationship

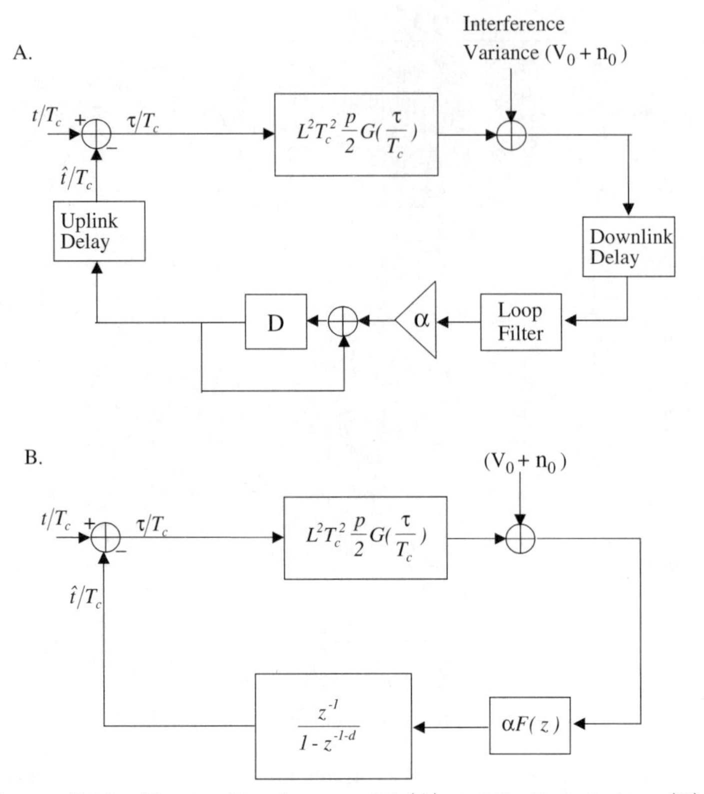

Figure 7.11 The tracking loop model (A) and its linearization (B).

between τ_s and σ_s is shown in Figure 7.13. With the timing error requirement being $T_c/10$, it is shown that the curve of $\tau_s + 2\sigma_s$ has two crossings with $T_c/10$. Considering the time needed for convergence after acquisition pull-in, we always want to maximize the loop gain. Therefore, it is better to choose the crossing point with larger α. In order to get the global optimization of α and β which minimize the convergence time with the timing error requirement matched, we can use the linear model to simulate for each β and the corresponding α (the second crossing point). Hence, the optimal values of α and β for which the convergence time is the shortest can then be determined.

7.4.1 Analysis of the Code Tracking

The mean and variance of the discriminator output of the noncoherent code tracking circuit shown in Figure 7.11, given below, are derived in reference [10]:

$$E[Z_\Delta] = L^2 T_c^2 \frac{p}{2}(R_+^2 - R_-^2)$$

$$\begin{aligned} Var[Z_\Delta] &\leq Var(Z_-) + Var(Z_+) \\ &= 8(\sigma_N^2 + \sigma_I^2)^2 + 8L^2T_c^2 p(R_+^2 + R_-^2)(\sigma_N^2 + \sigma_I^2) \end{aligned}$$

where p is the transmission power, L is the number of chips per accumulation and T_c is the chip duration. R_+ is the partial correlation between the (chip) matched filter of

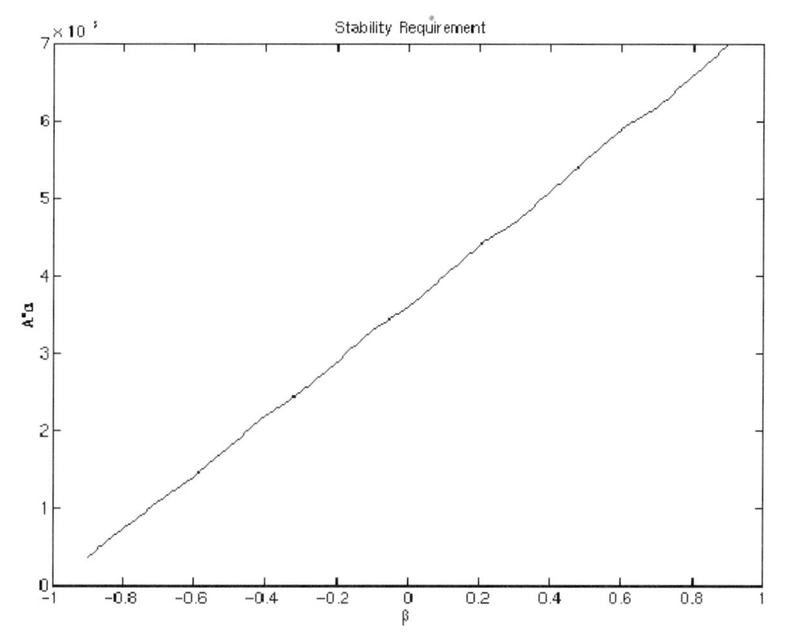

Figure 7.12 The tracking control stability requirement.

the late gate and the signal of interest (see reference [14]). σ_N^2, σ_I^2 are the interferences due to thermal noise and other users, respectively:

$$\sigma_N^2 = LT_c\frac{N_0}{4}, \quad \sigma_I^2 = LK\frac{p}{2}T_c^2 m_\psi, \quad \text{and} \quad m_\psi = \frac{1}{2}\frac{1}{2\frac{T_c}{10}}\sum_{\substack{i=-\infty \\ i\neq 0}}^{\infty}\int_{iT_c-\frac{T_c}{10}}^{iT_c+\frac{T_c}{10}} R^2(t)dt$$

where $R(t)$ is the convolution of the chip waveform and the matched filter (normalized, i.e. $R(0) = 1$).

The model of the above tracking circuit is shown in Figure 7.11-B. In this model the time unit is the duration of one channel symbol, which is the period per accumulation. We define the gain of the early-late discriminator as $E[Z_\Delta(\tau)]$, which is a function of the normalized relative timing error τ/T_c. The relative timing error τ is the timing difference between the incoming signal and the local code generator,

$$E[Z_\Delta(\tau)] = L^2T_c^2\frac{p}{2}G(\tau/T_c)$$

where

$$G(\tau/T_c) = R^2\left(\frac{\tau-\Delta}{T_c}\right) - R^2\left(\frac{\tau+\Delta}{T_c}\right)$$

We also define the interference variance as

$$V_0 \leq Var[Z_\Delta]$$

The Voltage Controlled Oscillator (VCO) in the terminal is modeled as an accumulator, since the absolute timing of the local code generator is modified according to all previous (scaled) timing errors. When the pull-in condition from the acquisition

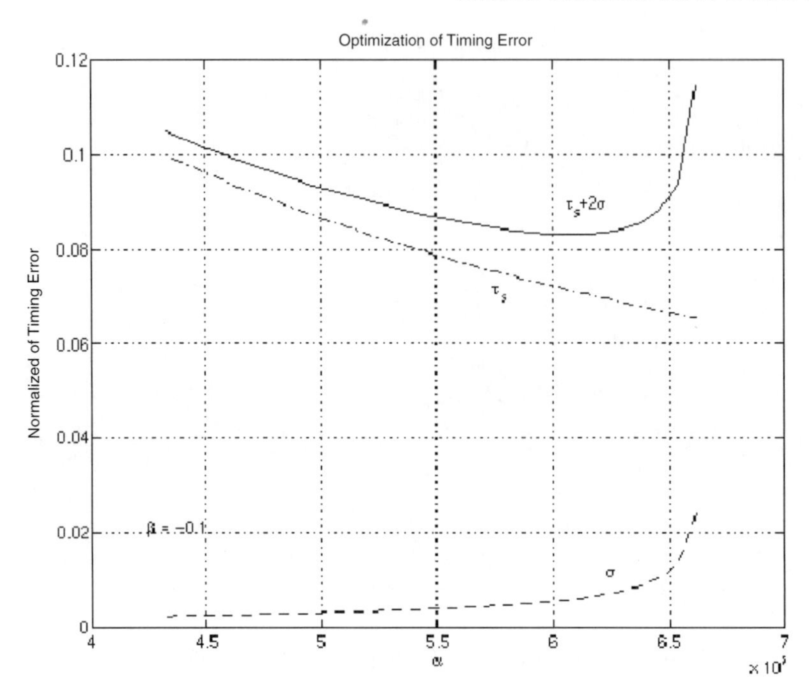

Figure 7.13 Minimization of the timing error.

stage is good, the relative timing error is very small. In this case, the tracking loop is operating in the linear region (see reference [13]). We can then model the loop by a linear model (shown in Figure 7.11-B), in which γ is defined by

$$\gamma = \left. \frac{dG(\tau/T_c)}{d(\tau/T_c)} \right|_{\tau=0}$$

and $F(z)$ is the transfer function of the loop filter. Let the normalized pull-in timing error be τ_0/T_c; it can be represented by a step function whose generating function is $(\tau_0/T_c)/(1-z^{-1})$. In addition to this fixed timing error, we have the error caused by the satellite drift. During the tracking stage, this drift can be assumed to have fixed direction and fixed speed. Its z-transform can be written as $cz^{-1}(\frac{1}{1-z^{-1}})^2$, where c is the normalized Doppler shift. The input to the linear model is then the sum of these two timing error terms. Finally, the generating function of the relative timing error is

$$\frac{\tau(z)}{T_c} = \left[\frac{\tau_0}{T_c} \left(\frac{1}{1-z^{-1}} \right) + cz^{-1} \left(\frac{1}{1-z^{-1}} \right)^2 \right] \frac{1}{1+H(z)}$$

$$H(z) = \frac{L^2 T_c^2 \frac{p}{2} \gamma \alpha F(z) z^{-1}}{1-z^{-d-1}}$$

with d equal to the round trip propagation delay. If the loop filter is chosen such that its bandwidth is sufficiently narrow, V_0 can be assumed to be white Gaussian distributed. In steady state when the timing error is τ_s, V_0 is given by

$$V_0 = 8\left(\sigma_N^2 + \sigma_I^2\right) \left[\sigma_N^2 + \sigma_I^2 + L^2 T_c^2 \frac{p}{4} \left(G(\tau/T_c) \right) \right]$$

The variance of the timing error is obtained from the linear model as

$$Var(\tau/T_c) = \frac{V_0 + n_0}{(L^2 T_c^2 \frac{p}{2}\gamma)^2} \frac{1}{2\pi} \int_{-\pi}^{\pi} \left| \frac{H(e^{j\omega})}{1 + H(e^{j\omega})} \right|^2 d\omega$$

where n_0 is the downlink noise and $A = L^2 T_c^2 \frac{p}{2}\gamma$. Using the equations of $\tau(z)/T_c$ and $H(z)$ above, we obtain the generating function of the relative timing error:

$$\frac{\tau(z)}{T_c} = \frac{1}{(1 - z^{-1})} \frac{\frac{\tau_0}{T_c} - \left(\frac{\tau_0}{T_c} - c\right) z^{-1}}{1 - z^{-1}\,(1 - A\alpha F(z) z^{-d})}$$

In order to have steady state timing error $\tau_s = 0$, we require that the loop filter in the form

$$F(z) = \frac{B(z)}{1 - z^{-1}}$$

where $B(z)$ can be any ratio of polynomials and d is feedback time delay. The characteristic function of $\frac{\tau(z)}{T_c}$ is then

$$(1 - z^{-1})^2 + A\alpha B(z) z^{-d-1}$$

The loop filter $F(z)$, in general, will drive the system unstable when the feedback delay is large. For this reason, we must include nonzero steady state timing error in our design consideration. If $F(z)$ does not have 1 as a pole, the characteristic function of $\frac{\tau(z)}{T_c}$ will be

$$(1 - z^{-1})(1 - z^{-1} + A\alpha F(z) z^{-d-1})$$

By the Final Value Theorem (see reference [15]), the steady state timing error will be given by

$$\tau_s = \lim_{z \to 1} (1 - z^{-1}) \frac{\tau(z)}{T_c} = \frac{c}{A\alpha F(z)}$$

which decreases as α increases.

Using the expression of $Var(\tau/T_c)$ above, the variance of the timing error $\sigma_s^2 = Var(\tau/T_c)$ will be given by

$$\sigma_s^2 = \frac{(V_0 + n_0)}{2\pi} \int_{-\pi}^{\pi} \left| \frac{\alpha F(e^{j\omega}) e^{j\omega(-d-1)}}{1 - e^{-j\omega} + A\alpha F(e^{j\omega}) e^{j\omega(-d-1)}} \right|^2 d\omega.$$

7.5 Conclusion

In this chapter we have presented and evaluated the network synchronization for an orthogonal CDMA satellite system. The objective of providing sychronization of all uplink orthogonal code traffic channels, as shown, can be achieved with a procedure which involves the uplink random access channel for coarse code acquisition and the use of an innovative feedback tracking control loop for fine synchronization. The access channel code acquisition scheme is based on a parallel/serial design which is optimized in terms of minimizing the acquisition time and maximizing the acquisition confidence for a given signal-to-noise ratio. Performance analysis indicates that packets may be transmitted successfuly over the access channel with a probability near 0.95 when the traffic load is up to 10 packets/sec and for a given set of system parameters. Performance analysis of the tracking loop has also been performed in order to determine the stability and optimum loop design. Due to long round trip satellite propagation delay, the loop response has a steady state error which can be corrected by a scaling adjustment. Thus, the requirement of synchronizing each orthogonal code to the reference time within 10% of the chip length can be achieved.

References

[1] R. De Gaudenzi, F. Giannetti and M. Luise 'Advances in Satellite CDMA Transmission for Mobile and Personal Communications' *Proceedings of the IEEE*, Vol. 84, No. 1, January 1996, pp. 18–39.

[2] D. T. Magill, F. D. Natali and G. P. Edwards 'Spread-Spectrum Technology for Commercial Applications' *Proceedings of the IEEE*, Vol. 82, No. 4, April 1994, pp. 572–584.

[3] J. K. Omura and P. T. Yang 'Spread-Spectrum S-CDMA for Personal Communication Services' MILCOM'92, 1992, pp. 11.3.1–5.

[4] D. Gerakoulis, H.-J. Su, and E. Geraniotis 'Network Access and Synchronization Procedures of a CDMA Satellite Communication System' To appear in the *International Journal of Satellite Communications*, 2000–2001.

[5] D. Gerakoulis 'Method of Synchronizing Satellite Switched Communication System' U.S. Patent No. 5,838,669; November 17 1998.

[6] W. Zhuang 'Noncoherent Hybrid Parallel PN Code Acquisition for CDMA Mobile Communications' *IEEE Trans. on Vehicular Tech.* Vol. 45, No. 4, November 1996, pp. 643–656.

[7] R. R. Rick and L. B. Milstein 'Parallel Acquisition of Spread-Spectrum Signals with Antenna Diversity' *IEEE Trans. on Commun.* Vol. 45, No. 8, August 1997, pp. 903–905.

[8] W. R. Braun 'PN Acquisition and Tracking Performance in DS/CDMA Systems with Symbol-Length Spreading Sequences' *IEEE Trans. on Commun.* Vol. 45, No. 12, December 1997, pp. 1595–1601.

[9] S.-L. Su and N.-Y. Yen 'Performance of Digital Code Tracking Loops for Direct-Sequence Spread-Spectrum Signals in Mobile Radio Channels' *IEEE Trans. on Commun.* Vol. 45, No. 5, May 1997, pp. 596–604.

[10] A. J. Viterbi *CDMA : Principles of Spread Spectrum Communications* Addison-Wesley, Massachusetts, 1995.

[11] H.-J. Su, P. Li, E. Geraniotis and D. Gerakoulis 'Code Tracking Loop Performance for an Orthogonal CDMA Uplink SATCOM System' *Conference Proceedings, IEEE ISCC'98*, Athens, Greece, 1998.

[12] V. M. Jovanovic 'On the Distribution Function of the Sread-Spectrum Code Acquisition Time' *IEEE Journal on Selected Areas in Commun.* Vol. 10, No. 4, May 1992, pp. 760–769.

[13] R. L. Peterson, R. E. Ziemer and D. E. Borth *Introduction to Spread Spectrum Communications.* Prentice Hall, New Jersey, 1995.

[14] M. P. Pursley 'Spread Spectrum Multiple Access Communications' *Multi-User Communications Systems*, G. Longo Ed., Springer-Verlag, 1981, pp. 139–199.

[15] G. F. Frankin, J. D. Powell and M. L. Workman *Digital Control of Dynamic Systems, 2nd Ed.* Addison-Wesley, Massachusetts, 1990.

8

Carrier Recovery for 'Sub-Coherent' CDMA

8.1 Overview

In this chapter we examine possible methods of carrier recovery for the SE-CDMA presented in Chapter 6. In particular, we propose, evaluate and compare two techniques; namely Symbol-Aided Demodulation (SAD) and the Pilot-Aided Demodulation (PAD). The performance analysis of each scheme (SAD and PAD) includes both Rician and Rayleigh multipath fading channels, and thus are also useful (in addition to the satellite) in terrestrial mobile applications. Both schemes are promising alternatives to differentially coherent demodulation for scenarios characterized with uncertainties in the carrier phase that make coherent demodulation unfeasible. The frequency selective fading (multipath), the Doppler phenomenon due to user mobility and/or to satellite drift motion, and the temperature variation and ventilation conditions at the sites of the various local oscillators that generate the transmitted signals cause the carrier phase uncertainty.

Coherent demodulation requires the extraction of a reliable (perfect) phase reference from the received signal. A traditional alternative is the differentially coherent demodulation that uses the phase of the previous bit (symbol) as a reference, but requires almost 3dB (for M-ary PSK modulation in AWGN channels, it is less than that for BDPSK) of additional signal-to-noise (E_b/N_0) in order to achieve the same bit error rate as coherent demodulation. This problem is more severe in DS/CDMA systems, which are limited by other-user interference: the additional cost in dBs of differentially coherent over coherent demodulation increases linearly with the number of users in the system, so as to render the fully-loaded multi-user system impractical [1]. Recently, SAD [2] and PAD [3] have been considered a form of '*sub-coherent*' demodulation. In the proposed SAD and PAD schemes, estimates of the channel multipath phases and amplitudes are extracted by smoothing and interpolation of the transmitted known bits in the SAD scheme or the pilot in the PAD scheme. The SAD (or PAD) performance then consists of evaluating the additional Signal-to-Noise Ratio (SNR) needed by either scheme to achieve the same Bit Error Rate (BER) as the coherent demodulation.

In this chapter we first present the system model and the design issues of the SAD scheme in Section 8.2, and its BER analysis (for the uncoded system) in Section 8.3. Then in Section 8.4 we present the system model, the design and the BER analysis (for the uncoded system) of the PAD scheme. The BER analyses of the coded systems are presented in Section 8.5. The coded system is based on a proposed new iterative decoding algorithm. The performance of the coded system of both schemes has been evaluated via simulations. The performance results are presented in Section 8.6.

8.2 Symbol-Aided Demodulation

8.2.1 System Model

In symbol-aided demodulation, known symbols are multiplexed with data bearing symbols. The known symbols are multiplexed with the data symbols at a constant ratio, so that one known symbol is followed by $J-1$ unknown data symbols. This ratio implies a loss in the throughput of $1/J$. At the receiver the known symbols are used to estimate the channel for other sampling points.

The system is as shown in Figure 8.1-A. The transmitted signal for the first user is given by

$$s_1(t) = A \sum_{k=-\infty}^{\infty} b_1(k)a_1(t)p(t-kT)$$

where $b_1(k)$ is the binary data sequence, $a_1(t)$ is the spreading code, which is a periodic sequence of unit amplitude positive and negative rectangular pulses (chips) of duration T_c, $T = NT_c$ is the symbol duration, and N is the processing gain.

The jth code pulse has amplitude $a_i^j = a_i(t)$ for $jT_c \leq t \leq (j+1)T_c$, and $p(t)$ is a unit energy pulse in the interval $0 \leq t \leq T$. The received signal is

$$r(t)\sum_{l=1}^{L} c_{1l}(t-\tau_{1l})s_{1l}(t-\tau_{1l}) + \sum_{m=2}^{K_u}\sum_{l=1}^{L} c_{ml}(t-\tau_{ml})s_{ml}(t-\tau_{ml}) + n(t)$$

where L is the number of paths, $n(t)$ is the AWGN with power spectral density N_0 in the real and imaginary parts, and K_u is the number of users. The channel complex gain $c_{ml}(t)$ represents the Rayleigh or Rician fading for the lth path of the mth user, with an autocorrelation function [4]

$$R_c(\tau) = \sigma_g^2 \left(\frac{K}{1+K} + \frac{1}{1+K} J_0(2\pi f_D \tau) \right)$$

where K is the ratio between the line of sight power and the scattered power, and the paths are assumed independent and with identical distributions.

The output of the normalizing matched filter, representing the finger of the rake receiver, for the first path of the first user, with impulse response $a_1(-t)p^*(-t)/(\sqrt{N_0})$, and assuming $\tau_{11} = 0$, will be given by

$$r_{11}(k) = u_{11}(k)b_1(k) + \sum_{l=2}^{L} u_{1l}(k)I_{1l}(k)e^{j\phi_{1l}(k)}$$
$$+ \sum_{m=2}^{K_u}\sum_{l=1}^{L} u_{ml}(k)I_{ml}(k)e^{j\phi_{ml}(k)} + n_{11}(k)$$

where the Gaussian noise samples $n_{11}(k)$ are white with unit variance, and complex symbol gain $u_{ml}(k)$ has mean

$$E[u_{ml}(k)] = \sqrt{\gamma_{ml}^s}\sqrt{\frac{K}{K+1}}$$

and variance

$$\sigma_{u_{ml}}^2 = \gamma_{ml}^s \frac{1}{K+1}$$

where the average SNR for path l of user m is given by

$$\gamma_{ml}^s = \frac{E_{ml}^s}{N_0}$$

and $I_{ml}(k)$ is the interference from path l of user m to path 1 of user 1. For the SAD scheme

$$E_{ml}^s = E_{ml}^b \frac{J-1}{J}$$

where E_{ml}^b is the energy per bit, and

$$\gamma_{ml}^b = \frac{E_{ml}^b}{N_0}$$

8.2.2 Design of Modulator and Demodulator

There are several issues that must be taken into consideration for the proper design of the symbol-aided modulation/demodulation system.

The Rate of Aid-Symbols

A proper choice of the value of J is of paramount importance for SAD system design. Increasing J will result in increasing the throughput, but at the same time it will increase the processing delay and the carrier-phase estimation error in both the known symbols and the data symbols.

Guidelines for the choice of J are given below. The value of J is determined from the bandwidth, rate of fading or of Doppler, or in general, from the rate of change of the phenomenon that introduces the uncertainty (and change) in the carrier phase. If we assume that the fading rate (or other rate of change) is R_f, then the sampling period T_{fs} and sampling rate $R_{fs} = 1/T_{fs}$ of the channel observations must satisfy the Nyquist condition

$$R_{fs} = \frac{1}{T_{fs}} \geq 2R_f$$

For notational convenience, define

$$J_{\max} = \frac{R_s}{2R_f}$$

where R_s is the symbol rate. $J_{\max}$ corresponds to the sampling of the fading phenomenon at exactly the Nyquist rate, presented in a more convenient form

A.

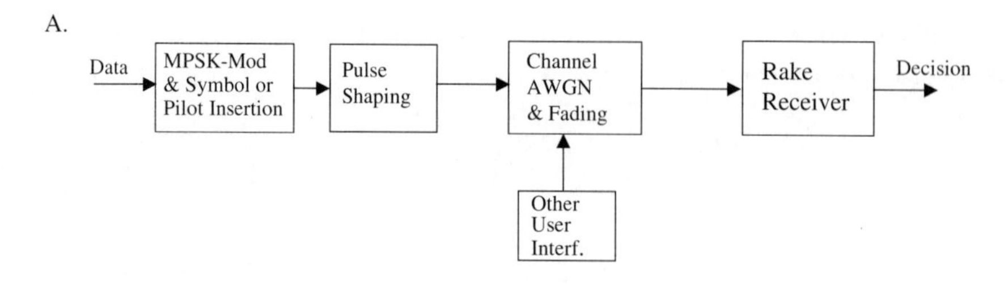

B.

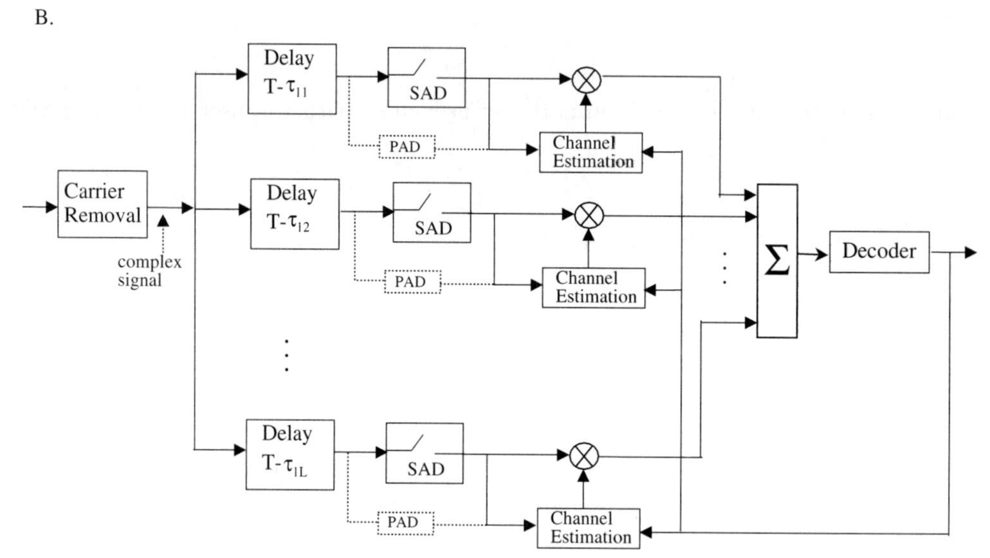

Figure 8.1 A. The SAD/PAD CDMA system B. A rake receiver for SAD (or PAD in dotted lines).

corresponding to the rate which known symbols are inserted for a given data and fading rates in order to fully capture the variation in the fading (or other phenomenon). We expect that

$$R_f << R_s$$

that is, the rate of change of the carrier phase is much slower than the symbol rate of the system. For example, we may have $R_s = 64$ kbps (kHz) while $R_f = 64$ or 128 Hz (or a value in the range of 30 Hz to 200 Hz). Denoting the symbol duration as $T_s = 1/R_s$, we have that $T_f >> T_s$, and define J as the ratio T_{fs}/T_s, i.e.

$$J = [T_{fs}/T_s] = [R_s/R_{fs}] \leq [R_s/(2R_f)] = J_{\max}$$

Therefore, $J_{\max}$ corresponds to sampling at exactly the Nyquist rate, and $J \leq J_{\max}$ corresponds to oversampling; for example $J = J_{\max}/4$ corresponds to sampling at four times the Nyquist rate, while $J = J_{\max}/8$ corresponds to sampling at eight times the Nyquist rate.

The idea is to use a sufficiently small J so that oversampling at rate

$$R_{fs} = (J_{\max}/J) \cdot (2R_f)$$

captures the change in the phenomenon and reduces the noise in the estimates of the phase (by a factor of $J_{\max}/J$ through smoothing, as we will see next) but still maintains the throughput loss (equal to $1/J$) within acceptable values.

A simple analysis of the SAD technique was presented in reference [5]. This is an approximate analysis assuming perfect filtering, but it provides an intuitive understanding of the problem and it helps identify the optimum J. It is done by simply taking into consideration the power loss due to reference insertion, expressed as

$$L_r \approx \frac{J+1}{J} = 1 + \frac{1}{J}$$

and the amount of increase in the noise due to the noisy reference (assuming perfect filtering and interpolation) which is given by

$$L_n \approx 1 + \frac{J}{J_{\max}}$$

Thus the total loss compared to coherent system is given by

$$L_t(dB) = L_r(dB) + L_n(dB)$$

The optimum choice of J given $J_{\max}$ can be obtained by calculating the minimum achievable loss; we can easily get

$$J_{opt} \approx \sqrt{J_{\max}}$$

$$L_t(J_{opt}) \approx (1 + \frac{1}{\sqrt{J_{\max}}})^2$$

Then the conclusion is that the performance of any SAD system can be no better than L_t dBs below (worse than) the performance of a coherent system (which assumes the perfect knowledge of the fading phase). For example, for $J_{\max} = 50$, $J_{\text{opt}} \approx 7$, and $L_t \approx 1.15$dB.

For the SAD scheme, the demodulation will delay the data symbols by JM symbols, where M is half the order of the smoothing filter. Clearly, decreasing J will produce a shorter delay, but as we mentioned, it will decrease the throughput, and the estimation error will be increased.

The Smoothing Filter

The bandwidth of the smoothing filter is another important issue. This filter is a digital filter that estimates $u_{ml}(k)$ of the unknown symbol samples. Decreasing J (which is equivalent to oversampling) will enable the filter to better estimate $u_{ml}(k)$ by removing more noise, and allow easier tracking of the relatively slower fading.

Two approaches are addressed here. The first is to derive the optimal Wiener filter for every unknown data point within the frame of length J, which means that filtering and interpolation are done simultaneously (in a sliding window manner). The second approach is to use a single filter for filtering all the known symbols, which is also a Wiener filter, and then to linearly interpolate the resulting output in order to obtain all unknown data symbols. The difference in performance between the two approaches is evaluated below.

8.3 BER Analysis for SAD

The first step for calculating the performance is to calculate the interference (see $r_{11}(k)$ in Section 8.2.1). The best way to proceed is to calculate $I_{ml}(k)$ for a given code selection, and hence calculate the mean square power of the interference averaged over τ_{ml}. A very good approximation is to follow references [6] and [1], and to assume a random signature sequence of length N. This approximation is very accurate if the system uses long (period) codes like the IS-95 system [3], and has sufficiently large N and K_u. Following references [6] or [1], we can calculate

$$E\{I_{ml}(k)\} = 0$$
$$E\{I_{ml}^2(k)\} = \frac{2}{3N}$$

In this section the analysis of the SAD system will be presented for both the optimum Wiener filter and linear interpolation case.

8.3.1 Optimum Wiener Filtering

The best performance that can be expected from the SAD technique (for a given filter length) can be obtained from Wiener filters. We will obtain its performance in this section for multipath Rician fading channels. Cavers [2] was the first to perform this analysis for Rayleigh fading channels. The phase reference of the lth path of the mth user for the unknown symbols is obtained from

$$v_{ml}(k) = \mathbf{h}^{\dagger}(k)\mathbf{r}_{ml} = \sum_{i=-M}^{M} h^*(i,k) r_{ml}(iJ)$$

where the dagger denotes conjugate transpose, and $\mathbf{r}_{ml}$ is the vector formed from $r_{ml}(iJ)$, the samples of the output of a matched filter of a finger of the rake receiver, $-M \le i \le M$ is the index of the known (SAD) symbols, and $1 \le k \le J-1$. Note that there will be $J-1$ different filters used.

The Wiener filter equation will be given by

$$\tilde{R}\mathbf{h}(k) = \mathbf{w}(k)$$

where $\tilde{R}$ is the autocorrelation matrix of size $2M+1$ defined by

$$\tilde{R} = \frac{1}{2}E[\mathbf{r}_{ml} r_{ml}]$$

and the $J-1$ vectors are

$$\mathbf{w}(k) = \frac{1}{2}E[u_{ml}^*(k)\mathbf{r}_{ml}]$$

The channel is Rician as described by $R_c(\tau)$ in Section 8.2.1. Perfect power control is assumed, such that γ_{ml}^b is constant for all users and is denoted by γ_{bL}. It is assumed that all the paths are identical, and so

$$\gamma_{bL} = \frac{\gamma_b}{L}$$

where $\gamma_b = L\gamma_{ml}^b$ is the total average SNR for every bit from all the paths.

Now we can obtain $\tilde{R}$ and $\mathbf{w}(k)$ from

$$\begin{aligned} R_{ij} &= \frac{\gamma_{bL}}{K+1}\frac{J-1}{J}\left[K + J_0\left(\pi\frac{(i-j)J}{J_{\max}}\right)\right] \\ &+ \left[\gamma_{bL}\frac{J-1}{J}\frac{2(K_u * L-1)}{3N} + 1\right]\delta_{i,j} \\ w_i(k) &= \frac{\gamma_{bL}}{K+1}\frac{J-1}{J}\left[K + J_0\left(\pi\frac{(iJ-k)}{J_{\max}}\right)\right] \end{aligned}$$

where $\delta_{i,j}$ is the Kronecker delta. The Rake receiver is shown in Figure 8.1-B, which is the maximal ratio combiner with noisy reference. From reference [7] we can calculate the probability of error using

$$\begin{aligned} Pe &= Q_1(a,b) - I_0(ab)e^{[-\frac{1}{2}(a^2+b^2)]} + \frac{I_0(ab)e^{[-\frac{1}{2}(a^2+b^2)]}}{[2/(1-\mu)]^{2L-1}}\sum_{k=0}^{L-1}\binom{2L-1}{k}\left(\frac{1+\mu}{1-\mu}\right)^k \\ &+ \frac{e^{[-\frac{1}{2}(a^2+b^2)]}}{[2/(1-\mu)]^{2L-1}} \times \sum_{n=1}^{L-1} I_n(ab) \sum_{k=1}^{L-1-n}\binom{2L-1}{k} \\ &\cdot\left[\left(\frac{b}{a}\right)^n\left(\frac{1+\mu}{1-\mu}\right)^k - \left(\frac{a}{b}\right)^n\left(\frac{1+\mu}{1-\mu}\right)^{2L-1-k}\right] \end{aligned}$$

where

$$a^2 = \frac{L}{2}\left|\frac{E\{r\}}{\sigma_r} - \frac{E\{v\}}{\sigma_v}\right|^2 \quad \text{and} \quad b^2 = \frac{L}{2}\left|\frac{E\{r\}}{\sigma_r} + \frac{E\{v\}}{\sigma_v}\right|^2$$

$$Q_1(a,b) = \int_b^{\infty} xe^{[-\frac{1}{2}(a^2+x^2)]}I_0(ax)dx \quad \text{and} \quad \mu = \frac{\sigma_{rv}^2}{\sigma_v\sigma_r}$$

where for the SAD scheme,

$$\begin{aligned} \sigma_{rv}^2 &= \underline{w}^{\dagger}(k)\underline{h}(k) - \frac{K}{K+1}\frac{J-1}{J}\gamma_{bL}S_h(k) \\ \sigma_v^2 &= \underline{w}^{\dagger}(k)\underline{h}(k) - (E\{v\})^2 \\ \sigma_r^2 &= \gamma_{bL}\frac{J-1}{J}\frac{1}{K+1} + 1 + \gamma_{bL}\frac{J-1}{J}\frac{2[K_u * L-1]}{3N} \\ E\{v\} &= \sqrt{\frac{K}{K+1}\frac{J-1}{J}\gamma_{bL}S_h^2(k)} \\ E\{r\} &= \sqrt{\frac{K}{K+1}\frac{J-1}{J}\gamma_{bL}} \end{aligned}$$

where $S_h(k) = \sum_{i=-M}^{M} h(i,k)$, while for coherent demodulation we have $v_{ml}(k) = u_{ml}(k)$

$$\sigma_{rv}^2 = \frac{1}{K+1}\gamma_{bL}$$
$$\sigma_v^2 = \frac{1}{K+1}\gamma_{bL}$$
$$\sigma_r^2 = \gamma_{bL}\frac{1}{K+1} + 1 + \gamma_{bL}\frac{2[K_u * L - 1]}{3N}$$
$$E\{v\} = \sqrt{\frac{K}{K+1}\gamma_{bL}}$$
$$E\{r\} = \sqrt{\frac{K}{K+1}\gamma_{bL}}$$

and for differential modulation we have $v_{ml}(k) = r_{ml}(k-1)$

$$\sigma_{rv}^2 = \frac{1}{K+1}\gamma_{bL} J_0\left(\frac{\pi}{J_{\max}}\right)$$
$$\sigma_v^2 = \gamma_{bL}\frac{1}{K+1} + 1 + \gamma_{bL}\frac{2[K_u * L - 1]}{3N}$$
$$\sigma_r^2 = \gamma_{bL}\frac{1}{K+1} + 1 + \gamma_{bL}\frac{2[K_u * L - 1]}{3N}$$
$$E\{v\} = \sqrt{\frac{K}{K+1}\gamma_{bL}}$$
$$E\{r\} = \sqrt{\frac{K}{K+1}\gamma_{bL}}$$

Filtering Followed by Interpolation

The second approach is to design a single filter to filter the known samples and then linearly interpolate the output to estimate the channel at the unknown samples. A Wiener filter can still be used to maximize the effective SNR at $k = iJ$. Following the same Wiener optimization approach as before, we can obtain $\mathbf{h}$ for $k = 0$ and use it.

Following the filter, the interpolator linearly interpolates the estimates of the carrier (or fading) inphase and quadrature components of the data symbols between each two successive known symbols (0 and J). We consider the case of $1 \le k \le J-1$ without any loss of generality. We have

$$v(k) = \frac{k}{J}v(J) + \frac{J-k}{J}v(0)$$

The expression for the probability of error P_e (given above) will be used again to calculate the performance, where now

$$\sigma_{rv}^2 = \sum_{i=-M}^{M} h(i)\gamma_{bL}\frac{J-1}{J}\frac{1}{K+1}\left[K + \frac{J-k}{J}\right.$$
$$\left. \cdot J_0\left(\pi\frac{iJ-k}{J_{\max}}\right) + \frac{k}{J}J_0\left(\pi\frac{(i+1)J-k}{J_{\max}}\right)\right]$$
$$- \frac{K}{K+1}\frac{J-1}{J}\gamma_{bL}S_h(k)$$

$$\sigma_v^2 = \sum_{i=-M}^{M}\sum_{j=-M}^{M} h(i)h(j)\gamma_{bL}\frac{J-1}{J}\frac{1}{K+1}\left(\left(\frac{J-k}{J}\right)^2\right.$$
$$\left. + \left(\frac{k}{J}\right)^2\right)\left[K + J_0\left(\pi\frac{(i-j)J}{J_{\max}}\right)\right]$$
$$+ h(i)h(j)\gamma_{bL}\frac{J-1}{J}\frac{1}{K+1}\frac{J-k}{J}\frac{k}{J}\left[2K\right.$$
$$\left. + J_0\left(\pi\frac{(i-j-1)J}{J_{\max}}\right) + J_0\left(\pi\frac{(i-j+1)J}{J_{\max}}\right)\right]$$
$$+ h(i)h(j)\left(\left(\frac{J-k}{J}\right)^2 + \left(\frac{k}{J}\right)^2\right)\left[1\right.$$
$$\left. + \gamma_{bL}\frac{J-1}{J}\frac{2(K_u * L-1)}{3N}\right]\delta_{i,j} + h(i)h(j)\frac{J-k}{J}\frac{k}{J}$$
$$\cdot\left[1 + \gamma_{bL}\frac{J-1}{J}\frac{2(K_u * L-1)}{3N}\right](\delta_{i,j-1} + \delta_{i,j+1})$$
$$- (E\{v\})^2$$

The other parameters $(E\{v\}, \sigma_r^2, E\{r\}, S_h(k))$ are like those given in the previous section for the optimum Wiener filter case.

8.4 Pilot-Aided Demodulation

8.4.1 System Model

The transmitted signal for user 1 is given by

$$s_1(t) = A\sum_{k=-\infty}^{\infty}((A_p a_1^p(t) + b_1(k)a_1(t))p(t-kT)$$

where A_p is the pilot amplitude, $a_1^p(t)$ is the pilot spreading code, and all the other parameters are as described for the SAD scheme. $a_1^p(t)$ and $a_1(t)$ could easily be made orthogonal through the use of code concatenation. The orthogonality will be maintained for every path because both codes will pass through the same channel.

The received signal will be given by

$$r(t) = \sum_{l=1}^{L} c_{1l}(t - \tau_{1l}) s_{1l}(t - \tau_{1l}) + \sum_{m=2}^{K_u} \sum_{l=1}^{L} c_{ml}(t - \tau_{ml}) s_{ml}(t - \tau_{ml}) + n(t)$$

The output of the normalizing matched filter, representing the finger of the rake receiver, for the first path of the first user, with impulse response $a_1(-t)p^*(-t)/(\sqrt{N_0})$ assuming equal energy pulses and BPSK modulation, will be given by

$$r_{11}(k) = u_{11}(k)(b_1(k) + I_{11_p}(k)) + \sum_{l=2}^{L} u_{1l}(k) I_{1l}(k) e^{j\phi_{1l}(k)}$$
$$+ \sum_{m=2}^{K_u} \sum_{l=1}^{L} u_{ml}l(k) I_{ml}(k) e^{j\phi_{ml}} + n_{11}(k)$$

where the Gaussian noise samples $n_{11}(k)$ are white with unit variance. $I_{11_p}(k)$ is the interference from the pilot signal to the data signal from the same path. As mentioned above, each user's data and pilot codes are assumed orthogonal, and so $I_{11_p}(k) = 0$.

For the pilot-aided scheme, for fair comparison, the energy per bit E_b will be the sum of the pilot energy E_p and data energy E_d, which means $E_b = E_p + E_d$. In the following, we will denote the power in the pilot as a fraction of the power of the data signal, and so we can write $E_p = PE_b$.

The complex symbol gain $u_{ml}(k)$ has mean

$$E[u_{ml}(k)] = \sqrt{\gamma_{ml}^d} \sqrt{\frac{K}{K+1}}$$

and variance

$$\sigma_{u_{ml}}^2 = \gamma_{ml}^d \frac{1}{K+1}$$

where

$$\gamma_{ml}^d = \frac{E_{ml}^d}{N_0}$$

and $I_{ml}(k)$ is the interference from path l of user m to path 1 of user 1, including the interference from both the data and pilot signals. The same expression can be obtained for the pilot fingers of the rake, but with E_p replacing E_d. The output of the first finger of the first user pilot Rake will be denoted by $r_{11p}(k)$.

8.4.2 Design of Modulator and Demodulator

There are several issues that must be taken into consideration in the proper design of a pilot-aided modulation/demodulation system.

Filter Length

The filter length is of great importance for the performance of the PAD scheme. If the fading is very slow relative to the data rate, an averaging filter could be used; this filter will give equal weight to each sample. If, on the other hand, the fading is not very slow, or it is required to have a long filter, a Wiener filter could be used, and it should be designed as explained before for the SAD scheme, but $\tilde{R}$ and $\mathbf{w}$ will be given by

$$\begin{aligned} R_{ij} &= \frac{\gamma_{pL}}{K+1}\left[K + J_0\left(\pi\frac{(i-j)}{J_{\max}}\right)\right] \\ &\quad + \left[\gamma_{dL}\frac{2(K_u * L - 1)}{3N} + \gamma_{pL}\frac{2(K_u * L - 1)}{3N} + 1\right]\delta_{i,j} \\ w_i &= \frac{\sqrt{\gamma_{pL}\gamma_{dL}}}{K+1}\left[K + J_0\left(\pi\frac{i}{J_{\max}}\right)\right] \end{aligned}$$

Again, it is assumed that all the paths are identical; we have

$$\gamma_{pL} = \frac{E_p}{N_0} = \frac{\gamma_p}{L}$$

$$\gamma_{dL} = \frac{E_d}{N_0} = \frac{\gamma_d}{L}$$

where γ_p and γ_d are the average SNRs corresponding to every bit from all the paths for the pilot and data, respectively.

The Ratio of Powers

The ratio of the power of the pilot to the power of the signal is the other parameter that should be studied carefully. The choice of this parameter is very similar to the choice of the parameter J in SAD scheme. Increasing this power will give a better estimate, but the overall performance may be worse. There will be an optimum level for this power that can be obtained with a similar argument to that shown in Section 8.2.2.

Again, this is an approximate analysis assuming perfect filtering, but it provides an intuitive understanding of the problem, and figuring out the optimum P. It is done by simply taking into consideration the power loss due to pilot insertion, expressed as

$$L_r \approx 1 + P$$

and the amount of increase in the noise due to the noisy reference (assuming perfect filtering and interpolation), which is given by

$$L_n \approx 1 + \frac{1}{PJ_{\max}}$$

Thus, the total loss compared to coherent system is given by

$$L_t(dB) = L_r(dB) + L_n(dB)$$

The optimum choice of P given $J_{\max}$ can be obtained by calculating the minimum achievable loss; we can easily get

$$P_{opt} \approx \sqrt{\frac{1}{J_{\max}}}$$

$$L_t(J_{opt}) \approx \left(1 + \frac{1}{\sqrt{J_{\max}}}\right)^2$$

Note the similarity of this result to the one given for the SAD scheme. Ideally, the two schemes will have the same performance. The question is, in practical situations where there is only finite length filtering and other user interference, which one of the two schemes will be better. The other question is how well will the two schemes fare in iterative decoding environment.

8.4.3 BER Analysis for PAD

Equation (8.1) could be used to evaluate the performance of PA scheme with

$$\begin{aligned}
\sigma_{rv}^2 &= \underline{w}^\dagger \underline{h} - \frac{K}{K+1}\sqrt{\gamma_{dL}\gamma_{pL}} S_h \\
\sigma_v^2 &= \underline{w}^\dagger \underline{h} - (E\{v\})^2 \\
\sigma_r^2 &= \gamma_{bL}\frac{1}{K+1} + 1 + \gamma_{dL}\frac{2[K_u * L - 1]}{3N} + \gamma_{pL}\frac{2[K_u * L - 1]}{3N} \\
E\{v\} &= \sqrt{\frac{K}{K+1}\gamma_{pL} S_h^2} \\
E\{r\} &= \sqrt{\frac{K}{K+1}\gamma_{dL}}
\end{aligned}$$

where $S_h = \sum_{i=-M}^{M} h(i)$.

8.5 The Coded SAD and PAD Systems

In this section we will consider the effect of iterative decoding schemes. We consider the case where the known symbols in the SAD scheme and the pilot symbols in the PAD schemes are uncoded. A block diagram of the rake-receiver/decoder is shown in Figure 8.1-B. It was shown that this method has a great advantage over that where each process is done separately. In this section we describe a scheme for iterative decoding and channel estimation for both the SAD and PAD schemes.

Interest in iterative decoding was ignited by the introduction of Turbo codes [8]. The superb performance of these codes stimulated the use of the same concept for other modules in the receiver. Hagenauer [9] called this structure the turbo processing principle, and argued that it can be used to improve the performance of all receiver modules. The optimum receiver for any communication system should be one big combined maximum likelihood estimator, that takes into account all of the information and processes it. For current complex systems it is not feasible to do that. Traditionally, all receiver modules have worked separately, and information has been lost when passing from one module to the other. In addition, every module does not make use of information supplied by preceding modules. Turbo processing introduces a partial solution to this problem. All modules in the receiver are designed to be Soft-Input/Soft-Output (SISO), to minimize the loss when passing information from one module to the other. The other problem is solved by feeding back the output of the last module as an input to the first one, which will exhaust all of the information used. In this section we will describe a method to use the decoder and the channel estimator to form an iterative decoding pair.

8.5.1 Coded SAD

A block diagram of the receiver is shown in Figure 8.1-B. In Section 8.2.1 the equation for $r_{11}(k)$ defines the output of the matched filter. Let us now define the energy per coded symbol

$$E_s = \sigma_u^2 = \frac{(J-1)E_b r_c}{J}$$

where r_c is the coding rate. A new scheme for iterative decoding channel estimation is presented in references [10] and [11]. At the first iteration, the estimate for the channel is obtained form the known symbols. Following the first iteration, the data and known symbols are used to obtain the channel estimate. The filter used from the start of the second iteration will be defined as before, with $\tilde{R}$ and $\mathbf{w}$ given by

$$\begin{aligned} R_{ij} &= \frac{\gamma_{sL}}{K+1}\frac{J-1}{J}\left[K + J_0\left(\pi\frac{(i-j)}{J_{\max}}\right)\right] \\ &+ \left[\gamma_{sL}\frac{J-1}{J}\frac{2(K_u * L-1)}{3N}+1\right]\delta_{i,j} \\ w_i &= \frac{\gamma_{sL}}{K+1}\frac{J-1}{J}\left[K + J_0\left(\pi\frac{i}{J_{\max}}\right)\right], \qquad i,j = -M,..,-1,1,..,M \end{aligned}$$

Before the first iteration, the channel estimate will be calculated, as in the uncoded system from only the known symbols. Following the first iteration, the reliability information $L(k)$ at the output of the decoder will be used to calculate the probability of the symbol (data or code). The reliability information $L(k)$ will be given by

$$L(k) = \sum_{l=1}^{L} L_{cl}(k) + L_e(k)$$

The sign of $L(k)$ is an estimate of $b(k)$, which is now either a data or code bit, while the magnitude $|L(k)|$ is the reliability of this estimate. $L_{cl}(k)$ is the channel log likelihood value depending on the received symbols at the output of every matched filter for the first user, corresponding to one of the L paths, and is given by

$$\begin{aligned} L_{cl}(k) &= ln\left[\frac{p(r_{1l}(k)|x(k)=+1)}{p(r_{1l}(k)|x(k)=0)}\right] \\ L_{cl}(k) &= 4\frac{E_s}{N_0}Re(r_{1l}(k)\hat{c}_l^*(k)) \end{aligned}$$

where $Re(.)$ denotes the real part, and * denotes the complex conjugate. $\hat{c}(k)$ represents the channel estimate, which is calculated from the known symbols only before the first iteration. The signals $L_c(k)$ enter the decoder before the first iteration, and at the output of the decoder there will be available $L_e(k)$, which is the extrinsic information for both the data and code bits.

Following the first iteration, the probability of the bits is calculated according to

$$p(x(k)=1) = e^{L(k)}/(1+e^{L(k)})$$

and the channel estimate is adapted after every iteration according to

$$\hat{c}_l(m) = \sum_{l=-M, l\neq 0}^{M} h(l) r_{1l}(l+m)[2p(l+m)-1]$$

where $p(k) = 1$ for known symbols. The new $L_{cl}(k)$'s are then calculated and a new iteration is performed.

8.5.2 Coded PAD

A similar scheme as that used for SAD will be used here. Before the first iteration, the pilot symbols are the only ones used to estimate the channel. The channel estimate is calculated using the filters obtained in Section 8.4.2, but now γ_{dL} represents, the energy per coded bit instead of per data bit.

Following the first iteration, we form the signal

$$r_{11t}(k) = (2p(k)-1)\sqrt{1-P} r_{11}(k) + \sqrt{P} r_{11p}(k)$$

This signal will maximize the information known about symbol k of the first path of the first user. At moderate (operating) SNRs, after the first iteration the probabilities p are close to either 0 or 1; this will allow designing of the new channel estimation filter that uses the signal $r_{11t}(k)$. If p's are not very reliable, the solution is to design an adaptive filter that takes into consideration the actual values of p, and this is changed after every iteration. Simulations were performed, and it was shown that the adaptive filtering alternative will not improve the performance, and that the fixed filter performance will converge to the adaptive filter after two or three iterations. Again $\gamma_p = P\gamma_s$, where now γ_s represents the energy per coded bit. Now to design the filter. We have

$$\begin{aligned} R_{ij} &= \frac{\gamma_{sL}}{K+1}\left[K + J_0\left(\pi \frac{(i-j)}{J_{\max}}\right)\right] \\ &\quad + \left[((K_u * L - 1) * \gamma_{sL})\frac{2}{3N} + 1\right]\delta_{i,j} \\ w_i &= \frac{\gamma_{sL}}{K+1}\left[K + J_0\left(\pi \frac{i}{J_{\max}}\right)\right] \end{aligned}$$

The resulting filter h will be used to calculate the channel estimate as

$$\hat{c}(m) = \sum_{l=-M, l\neq 0}^{M} h(l) r_{11t}(l+m)$$

and the iterations are repeated in a similar way to the SAD scheme.

8.6 Performance Results

Figures 8.2-A and -B, show the BER over Rayleigh and Rician fading channels, respectively, when the optimum Wiener filters are used and for different numbers

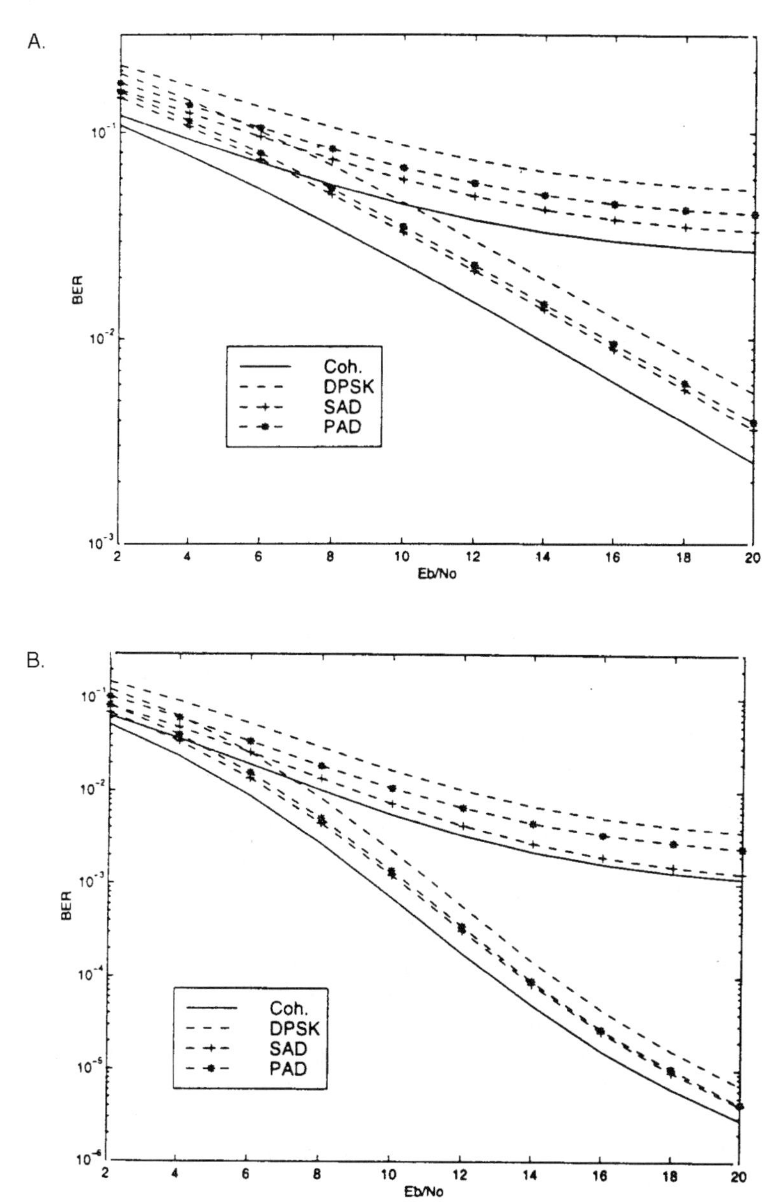

Figure 8.2 BER in (A) Rician ($K = 10$), (B) Rayleigh fading and for one and six users. $J = 7$, $J_{\max} = 50$, $P = 1$, $N = 31$, $L = 1$. Filter length $= 31$ for both schemes.

of users $K_u = 1, 6$, and the processing gain (number of chips per symbol) is $N = 31$. The filters for the SAD and PAD schemes are of length 29. $J = 7$, $P = 1/7$ and $J_{\max} = 50$.

It was noticed and previously mentioned [2] that if the optimum filter is used for every point between two known symbols, there will be approximately no change in the BER values of different symbols along the frame for the SAD scheme. For the PAD scheme there is only one filter used. As can be shown, for $K_u = 1$, the SAD scheme has a small advantage over the PAD scheme, while for $K_u = 6$, the SAD scheme is notably better than the PAD. This can be anticipated as the PAD scheme will have more interference because of the different pilot signals from all the users, which is greater than that of the SAD scheme and increases with the number of users.

Figures 8.3-A and -B show the irreducible BER (i.e. the BER that corresponds to $E_b/N_0 = \infty$, and is thus caused by other-user interference alone) versus the number of users. It is clear that the performance of SAD is still very near that of coherent demodulation, and outperforms the PAD scheme for the same filter length, which is 11 for both schemes. We also noticed that the difference between the BER performance of the SAD for a filter length of 11 and 29 is much smaller than that of the PAD scheme, which means that the SAD scheme achieves its limit with a shorter filter.

In Figures 8.4-A and -B, the performance of suboptimum filtering schemes is presented for two different scenarios. A low data rate scenario with $R_s = 10$ kbps, a fading rate $R_f = 200$ Hz, and Raleigh fading (typical of terrestrial mobile communications) is shown for two different frequencies of SAD symbols: one in twenty ($J = 20$) and one in seven ($J = 7$), respectively. For Figure 8.4-A, the oversampling ratio is $J_{\max}/J = 25/20 = 1.25$ and the throughput loss is $1/J = 0.05$ (5%). Then a high data rate scenario with $R_s = 200$ kbps and a fading rate $R_f = 20$ Hz (typical of GEO satellite communications) is shown in Figure 8.4-B. In this case the SAD insertion frequency is one in one hundred ($J = 100$), the oversampling ratio $J_{\max}/J = 5000/100 = 50$ is high, and the throughput loss $1/J = 0.01$ (1%) is low.

These figures show the performance of coherent, DPSK, optimum SAD (where every point in the frame uses its own unique optimum filter) and of three suboptimal SAD schemes. The first schemes use an optimum Wiener filter for the known symbols, and then interpolates the output linearly this is denoted by 'Opt. filter int.' in the graphs. The second suboptimal scheme is to use an optimized Wiener filter for the midpoint of the frame ($k = 4, 10$ or 50) and use it to filter the known symbols to get a direct estimate of the fading component and use it at every point in the frame; this scheme is easy to implement because the system has to perform the filtering only once for every frame, and is denoted by 'Opt. filter for $k = 4, 10$ or 50' in the graphs. The third option is to use a simple LPF to filter the known symbols, then linearly interpolate the output; we choose it to be a rectangular filter in the time domain, where all $h(i) = 1/(2M + 1)$; this is denoted by 'Rect. filter int.' in the graphs.

Figure 8.4-A shows that DPSK is better than all the SAD options. Only the all-optimal filter is close to DPSK, although still worse than it. We can also notice that the BER of the 'Rect. filter int.' is much worse than all the other schemes.

Figure 8.4-B indicates that all the SAD options have better performance over DPSK.

Figures 8.4-A and -B show that oversampling is very important for the SAD scheme. The optimum J was obtained previously, and it should be followed. It also shows that

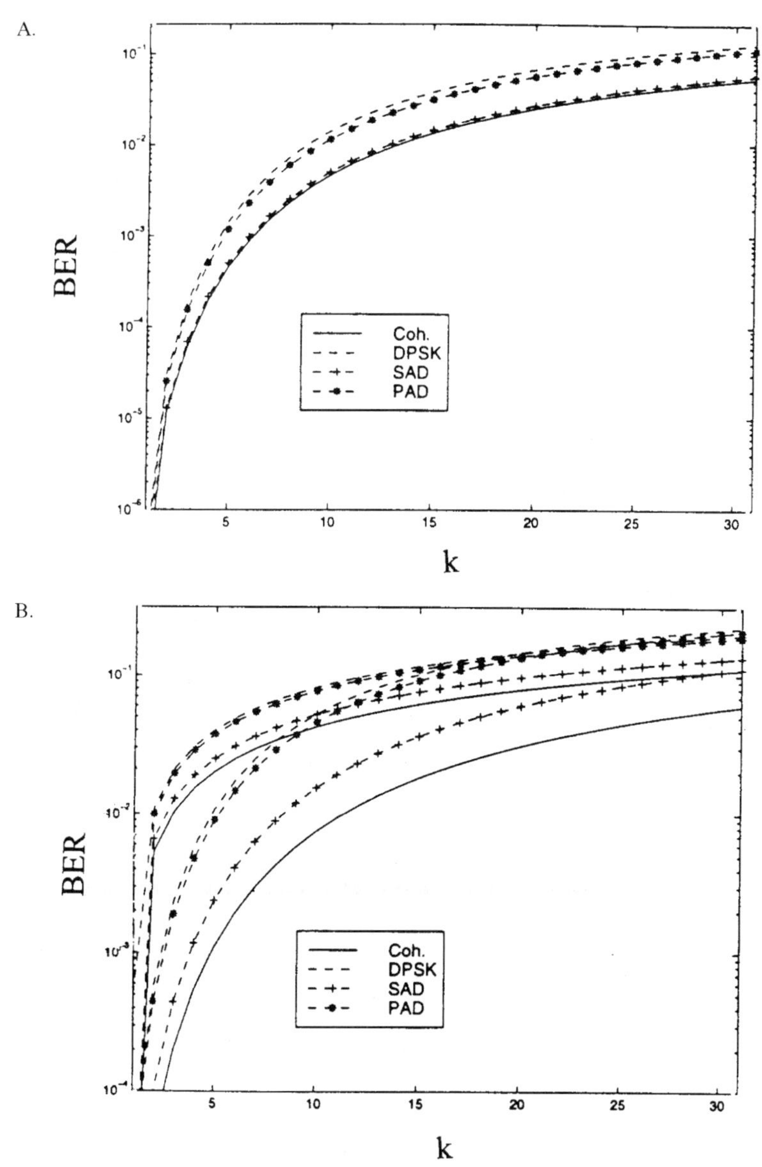

Figure 8.3 BER versus the number of users, for (A) Rician ($K = 10$), (B) Rayleigh fading. $J = 7$, $J_{\max} = 50$, $P = 1/7$, $L = 1, 4$ and SNR very large. Filter order = 11.

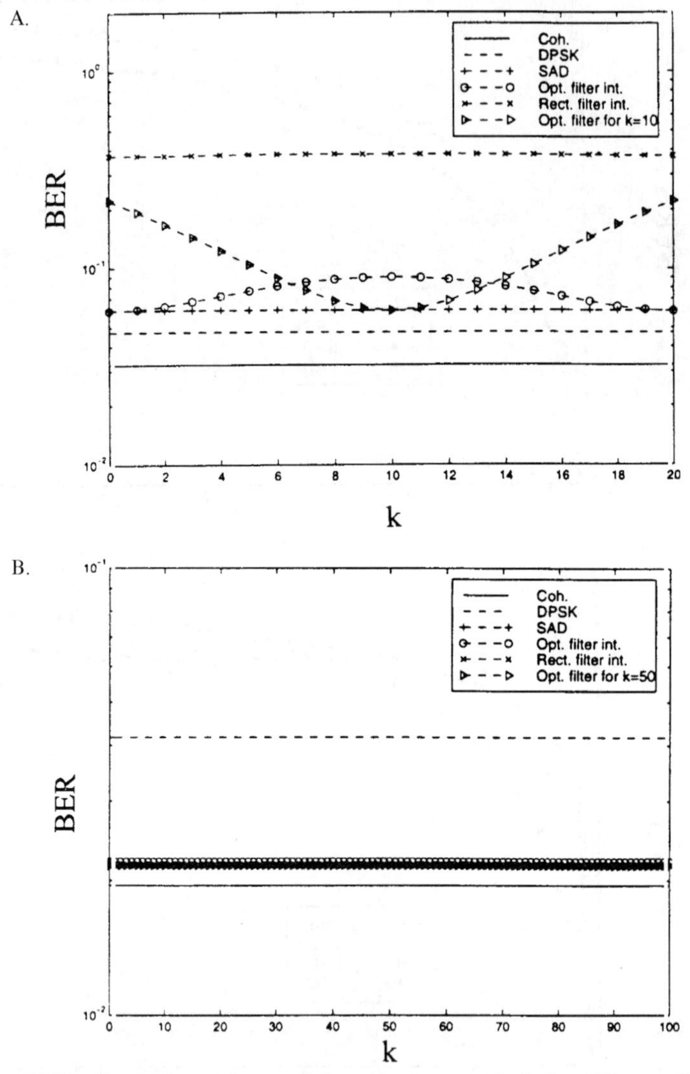

Figure 8.4 BER versus the position of the unknown symbol, for different SAD filtering schemes. (A) Rayleigh fading, $R_s = 10$ kb/s, $R_f = 200$ Hz, $J = 20$, $J_{\max} = 25$, $K_u = 3$, $N = 31$, SNR = 10 dB, (B) Rician ($K = 10$), $R_s = 200$ kb/s, $R_f = 20$ Hz, $J = 100$, $J_{\max} = 5000$, $K_u = 3$, $N = 31$, SNR = 5 dB; Filter order = 11.

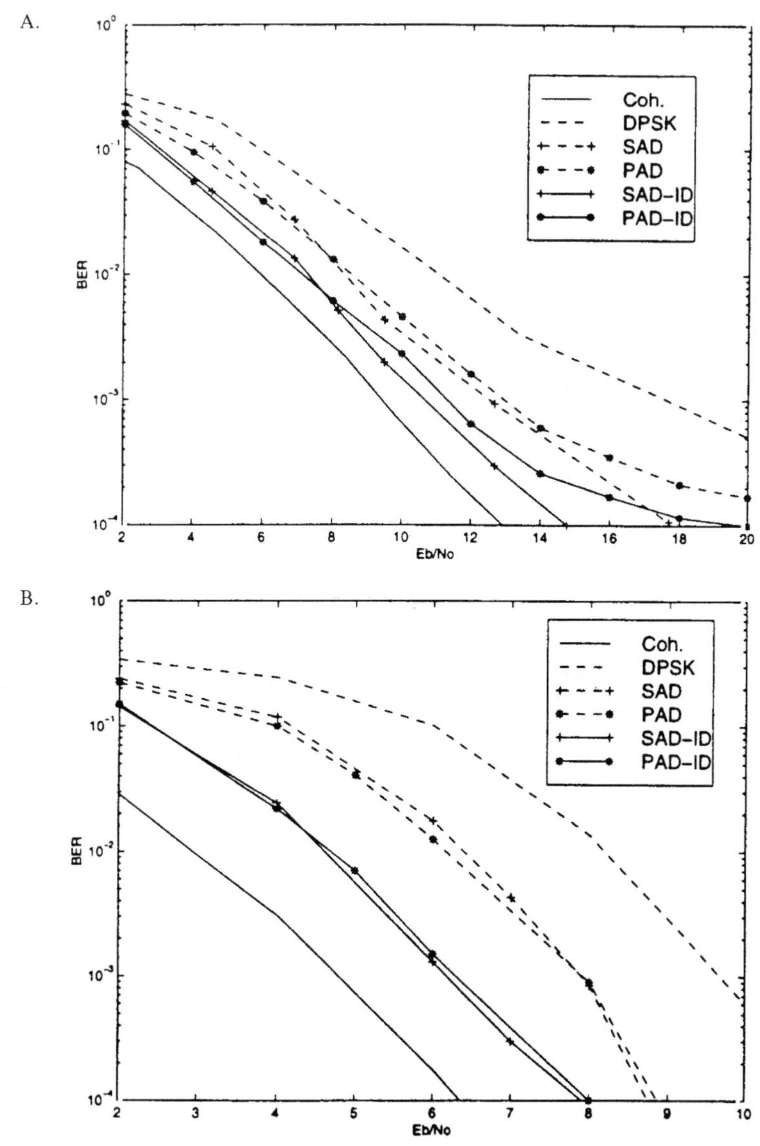

Figure 8.5 BER versus E_b/N_0 for coded signal in Rayleigh fading. (A) $L = 1$, $K_u = 6$, $J = 7$, $J_{\max} = 50$, $P = 1/7$, $N = 31$. (B) $L = 4$, $K_u = 1$, $J_{\max} = 50$, $P = 1/7$, $N = 31$.

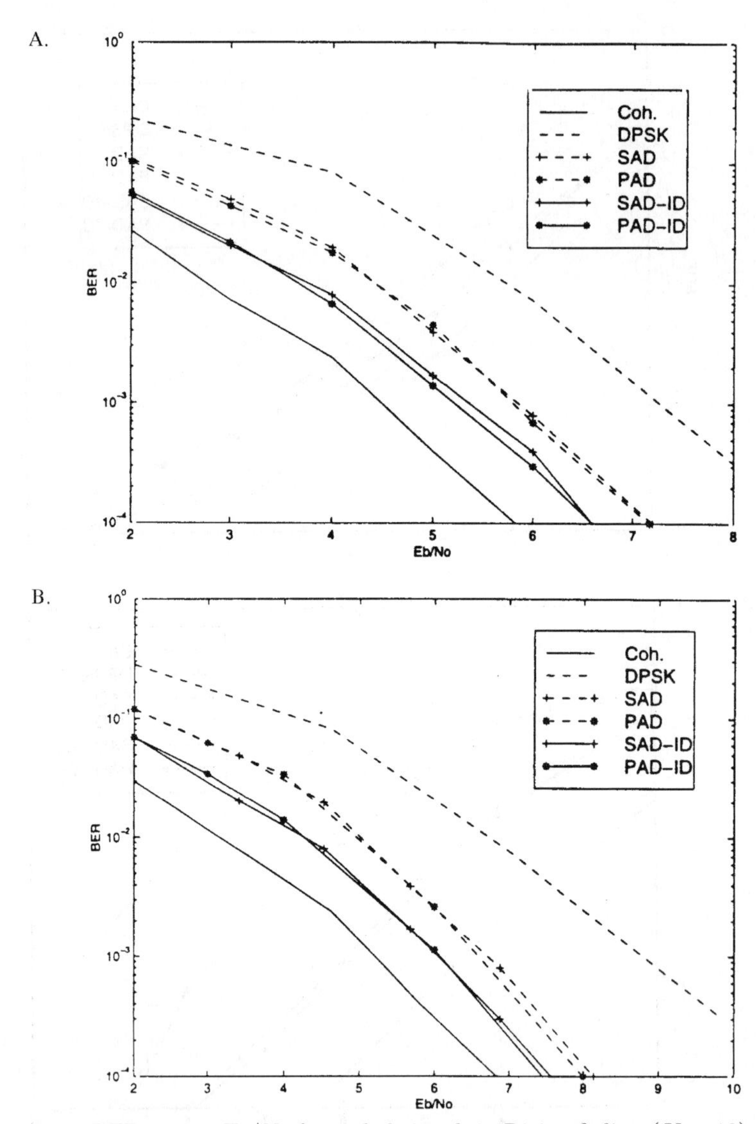

Figure 8.6 BER versus E_b/N_0 for coded signal, in Rician fading ($K = 10$), $L = 1$. (A) $K_u = 1$, $J = 7$, $J_{\max} = 50$, $P = 1/7$, $N = 31$. (B) $K_u = 6$, $J = 7$, $J_{\max} = 50$, $P = 1/7$, $N = 31$.

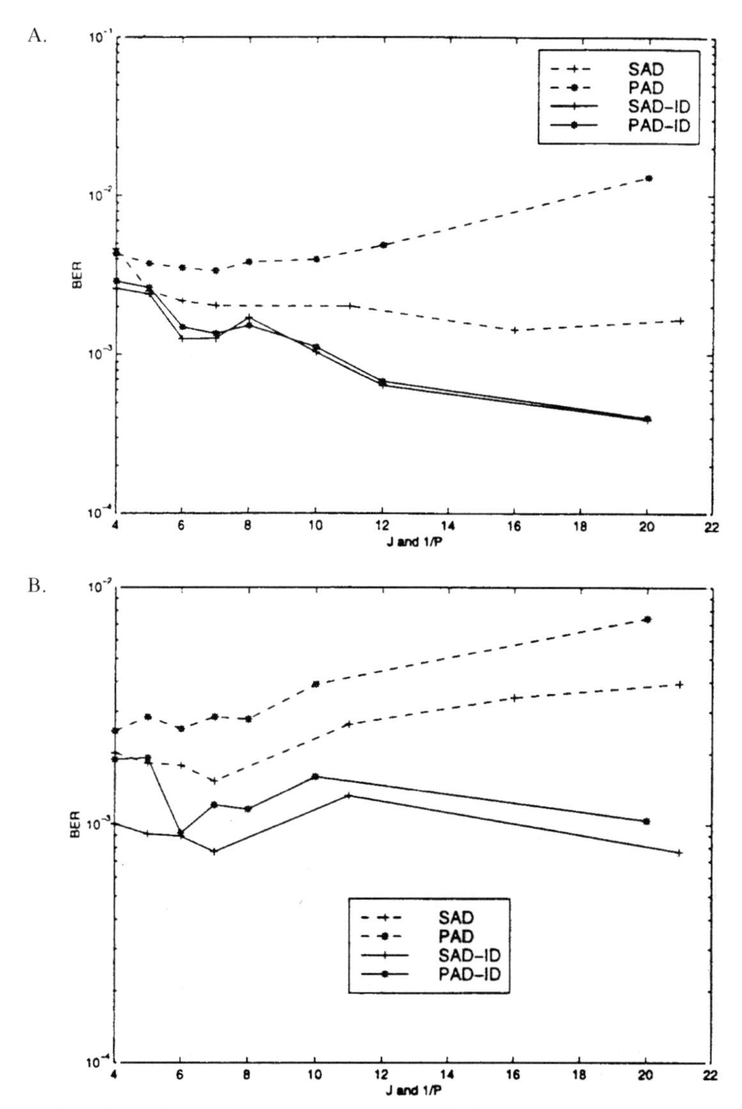

Figure 8.7 The effect of J and P on the BER for (A) Rayleigh fading, $L = 1$, $J_{\max} = 50$, $K_u = 6$, $N = 31$, $E_b/N_0 = 11$ dB; (B) Rician fading, $L = 1$, $J_{\max} = 50$, $K_u = 6$, $N = 31$, $E_b/N_0 = 6$ dB.

the filter design is very important for fast fading Rayleigh channels, while averaging could be used for very slow Rician fading channels.

Figures 8.5-A, -B and 8.6-A, -B indicate the performance of the coded system. In these graphs, the order of the filter used for the SAD scheme is 6, while the PAD scheme uses a filter length of 19. For the SAD-ID and PAD-ID schemes, the filter used following the first iteration is 19.

As can be seen from Figure 8.5-A, the SAD scheme gives better performance than the PAD scheme at high SNR, due to the fact that the interference for the PAD scheme is more than that for the SAD scheme, and so at high SNR the SAD scheme is better. It can also be shown that the iterative decoding algorithm used for both SAD and PAD schemes, and denoted by SAD-ID and PAD-ID, improves the performance of both systems significantly.

For the Rician and 4-path Rayleigh fading it is clear that the SAD and PAD schemes are almost identical. The higher error floor due to other-user interference is also present here but at a lower BER. Due to simulation difficulties, we weren't able to get the exact BER floor for six users, but we increased the number of users until this floor was within our accuracy and the same phenomenon was noticed as for 1-path Rayleigh fading.

Figures 8.7-A and -B shows the effect of changing P and J for SAD and PAD. As shown, for conventional SAD and PAD, the optimum J and P obtained for the uncoded scheme are still optimum.

8.7 Conclusions

We conclude that the choice of filtering of the known symbols in SAD is crucial for Rayleigh channels with a relatively fast fading change rate. If a simple LPF is used, the performance may turn out to be worse than DPSK. This is due to the fact that the channel changes significantly (unlike the Rician fading channel where, due to the LoS path, the changes of the channel are relatively small), which makes it more important to have an optimum filter to track the fading.

Therefore for terrestrial system where Rayleigh fading is the more acceptable model, if the fading is studied well and parameterized carefully, optimum filtering should be used and the insertion frequency of SAD symbols should be moderately high (corresponding to a throughput loss more than 10%), otherwise DPSK should be preferred.

By contrast, for those channel (e.g. GEO satellite links) where Rician fading is a good model, if the direct path is strong (i.e. K is large enough), a simple LPF will be sufficient. For high data rate transmission systems the throughput loss will be negligible.

We also conclude that the SAD scheme is better than the PAD scheme, especially if the number of users is large and the fading is fast. We saw from the numerical results that for the same filter length, the SAD scheme gives better performance, and that the performance was nearly the same with a much lower filter length. It was also noticed that in some applications, when there is strict constraint on the delay, the PAD scheme may be preferable. This may happen if $J_{\max}$ is very high, which will give the optimum SAD scheme a much higher delay than the PAD scheme. If iterative decoding and filtering is used, the first SAD filter could be of a short length and the delay problem may be partially solved. It is also noted that the PAD scheme uses only one filter, while the SAD scheme uses approximately $J/2$ filters, which may be an advantage for PAD scheme, but in return the PAD rake receiver is double as complex as the SAD scheme.

References

[1] E. Geraniotis 'Performance of Noncoherent Direct Sequence Spread Spectrum Multiple Access Communications' *IEEE J. Selected Areas in Commun.*, Vol. SAC-3, September 1985, pp. 687–694.

[2] J. Cavers 'An Analysis of Pilot Symbol Assisted Modulation for Rayleigh Fading Channels' *IEEE Trans. on Vehicular Technology,* Vol. 40, No. 4, November 1991.

[3] A. J. Viterbi *CDMA, Principles of Spread Spectrum Communications.* Addison-Wesley, 1995.

[4] W. Lee *Mobile Communications Engineering*, McGraw-Hill, 1982.

[5] F. Ling 'Method and Apparatus for Coherent Communication in a Spread-Spectrum Communication System' US Patent 5,329,547, 1994.

[6] M. Pursley 'Performance evaluation for phase coded Spread Spectrum Multiple Access Communications, Part I: System Analysis' *IEEE Trans. Commun.*, Vol. COM-25, August 1977, pp. 795–799.

[7] J. G. Proakis *Digital Communications.* McGraw-Hill, 1983.

[8] C. Berrou, A. Glavieux and P. Thitimajshima 'Near Shannon limit error-correcting coding and decoding' *Proc ICC'93*, May 1993.

[9] J. Hagenauer 'The Turbo Principle: Tutorial Introduction and State of the Art' *International Symposium on Turbo Codes and Related Topics*, Brest, France, September 1997, pp. 1–9.

[10] H. ElGamal, M. Khairy and E. Geraniotis 'Iterative Decoding and Channel Estimation of DS/CDMA over Slow Rayleigh Fading Channels' *PIMRC 98*, Boston, MA, 1998.

[11] M. Khairy and E. Geraniotis 'Asymmetric Modulation and Multistage Coding for Multicasting with Multi-Level Reception over Fading Channels' *MILCOM 99*, Atlantic City, NJ, 1999.

[12] M. Khairy and E. Geraniotis 'BER of DS/CDMA Using Symbol-Aided Coherent Demodulation over Rician and Rayleigh Fading Channels' *IEEE ISSSTA*, 1998.

9

Nonlinear Amplification of Synchronous CDMA

9.1 Overview

As we have described in Chapter 3, the SS/CDMA uses orthogonal CDMA for both uplink and downlink transmission. On board the satellite, each CDMA channel is routed to a destination downlink beam by the Code Division Switch (CDS). All channels in the same output port of the CDS are combined and then amplified with a Traveling Wave Tube (TWT) amplifier for downlink transmission. Due to the large number of users in the system, the amplitude of the combined signal has a large variance, which makes its amplification difficult since it may drive the TWT into saturation. This phenomenon also appears in terrestrial wireless systems if the downlink transmission at the base stations is CDMA.

Nonlinear distortions in satellites may also result for other reasons, such as drastically physical changes in the environment, for example, temperature variations and vibration noise. The causes of these distortions are sometimes predictable, such as the significant temperature variations in satellites; or they might be unpredictable, such as the variable aging of local oscillators in harsh environments. These hardware distortions, appearing in the forms of phase noise, spurious phase modulation, frequency offset, filter amplitude and phase ripple, data asymmetry and modulator gain imbalance in the transmitter, as well as nonlinear amplitude and phase distortions in the power amplifier, typically reduce the system performance from a few tenths of dB to as large as ten dB. Among all of them, the nonlinear distortions existing in the power amplifiers contribute to the degradation of the system. While analyses of CDMA systems are based on the assumptions that signal waveforms are ideally linearly retransmitted over the power amplifiers, it has been known that the existence of these nonlinearities impacts upon the real system design.

The effects of nonlinear distortions on CDMA systems can be categorized into two classes – *out-band degradation* and *in-band degradation*. Due to high demand on the frequency bandwidth, stringent regulatory emission requirements have always been enforced to prevent interference with other communication systems. To accommodate more users simultaneously in the designated frequency bandwidth, signals transmitted over wireless channels are always shaped so as to have a compact spectrum within this frequency bandwidth. It also means the out-band emission has to be below the regulated level. However, nonlinear distortions reshape the signals so that they lose their compactness in spectrum which leads to out-band spectral regrowth (see references [1] and [2]). A band-pass filter then has to be utilized before the signal transmission in order to reject this undesired out-band power. The inefficiency of power utilization thus results, and the filtered signal experiences higher intersymbol interference. The BER then increases due to this extra undesired interference. We call this *out-band degradation*, since the degradation is caused by the rejection of out-band power.

The second effect of nonlinear distortions is the *in-band degradation* (see references [3] and [4]). Suppose the power gained from the nonlinear amplifier is totally consumed while signals travel through the channel. This nonlinear amplifier can then be regarded as a nonlinear transformation between the transmitters and receivers. Note that in a fully orthogonal system (i.e. all users are fully synchronized with a set of orthogonal spreading codes), orthogonality is preserved only if the channel is linear. In other words, in a non-fading linear channel, the only interference to the receiver in this system is the thermal noise. No Multiple-Access Interference (MAI) exists. Therefore, with or without the out-band filter, the nonlinear transformation has destroyed part of the orthogonality of the system, which introduces MAI to an originally orthogonal system or more MAI to an originally non-orthogonal system. The result is a higher BER, which further leads to lower system capacity.

Aein and Pickholtz [5] presented a simple phaser model to analyze the Bit Error Rate (BER) performance of an asynchronous CDMA system accessing a RF limiter possessing amplitude-phase conversion (AM/PM) intermodulation effect. Their model considered interference in the form of multiple access noise, a Continuous Wave (CW) tone, or a combination of both. In the above reference, however, the amplitude-to-amplitude conversion (AM/AM) intermodulation effect in the channel was not included. Baer [6] analyzed a two-user PN spread-spectrum system with a hard limiter in the channel. However, the effect of MAI was not considered. Note that both of these papers focused on systems with very few users (one or two), and thus with the major source of interference on the desired user being the self-interference due to nonlinear distortion. Nonlinear amplifiers usually exhibit both AM/PM and AM/AM distortions, in which the hard limiter is not the most accurate model for the AM/AM conversion. Recently, Chen [7] analyzed the effects of nonlinearities on asynchronous systems with MAI, but his results were only limited to the evaluation of the Signal-to-Noise Ratio (SNR).

None of the previous work analyzed the CDMA system performance in terms of BER and system capacity for different data modulation schemes and various spreading codes. Neither did they provide detailed descriptions of the effects of nonlinearities on CDMA systems. The effects of AM/AM and AM/PM distortion on the sum of a number of DS/CDMA with M-ary PSK modulation signals have not been modeled and analyzed in detail before, although simulations have been performed for DS/CDMA signals with BPSK or QPSK modulation. Thus, although it is known that transponder nonlinearities have a significant effect on the performance of DS/CDMA systems, the performance of such systems has never been quantified in a manner that allows us to understand how these AM/AM and AM/PM distortions affect other-user interference, and thus how to mitigate it.

In this chapter, we evaluate the performance of synchronous M-PSK CDMA systems in the presence of nonlinear distortions. Emphasis is placed on the modeling and analytical evaluation of the combined effects of nonlinear distortion and other-user interference on the systems of interest. Besides the complete parametric performance evaluation, our work also prepares the ground for developing novel techniques for Output Back-Off (OBO) mitigation. Mitigation techniques can be used against both the nonlinear distortion and other-user interference generated from nonlinearities.

The application motivating this study is the satellite switched CDMA system presented in Chapter 3. The link access of the SS/CDMA is based on the SE-CDMA, which is a synchronous CDMA as described in Chapters 3 and 6. In this application nonlinear distortion comes from the on-board power amplifier which is a Traveling Wave Tube (TWT). In addition to the satellite applications, the results presented in this chapter are also applicable to base stations in wireless terrestrial networks.

The chapter is organized as follows. After this overview, Section 9.2 describes the model of a synchronous M-PSK CDMA system. The mathematical model for various nonlinear distortions, as well as their effects, are discussed afterwards. System performance evaluated by the Gaussian approximation is presented in Section 9.3. Section 9.4 pertains to the numerical results and simulations.

9.2 System Model

As depicted in Figure 9.1, the input signal $S(t)$ consists of a sum of synchronous CDMA signals from the same satellite beam (or from the same cell-sector). Each beam has K users. The nonlinear amplifier can be seen as a nonlinear pipe, which is especially well known in satellite communications. We further assume the downlink channel is an Additive White Gaussian Noise (AWGN) channel with attenuation equal to the power gained through the nonlinear amplifier. This can be easily achieved since, in the case of unequal attenuation, its effect can be easily incorporated into the parameters of the model of the nonlinear amplifier. This system model can be viewed as a satellite system with high uplink power, which suggests the omission of uplink noise. In this case, the phases of all the local oscillators might be the same, which is just a special case of this model.

9.2.1 *Transmitter*

Suppose each user sends an M-PSK signal with inphase (I) and quadrature (Q) components spread by the respective user code sequence (see Figure 9.2-A). For the kth user, the M-PSK data signal of I and Q components can be represented respectively as

$$b_I^{(k)}(t) = \sum_{i=-\infty}^{\infty} b_I^{(k)}[i]\mathcal{I}_{T_s}(t - iT_s)$$

$$b_Q^{(k)}(t) = \sum_{i=-\infty}^{\infty} b_Q^{(k)}[i]\mathcal{I}_{T_s}(t - iT_s)$$

$$\left(b_I^{(k)}[i], b_Q^{(k)}[i]\right) \in \left\{\left(\sqrt{2p}\cos\frac{(2m-1)\pi}{M}, \sqrt{2p}\sin\frac{(2m-1)\pi}{M}\right),\ m = 1, .., M\right\}$$

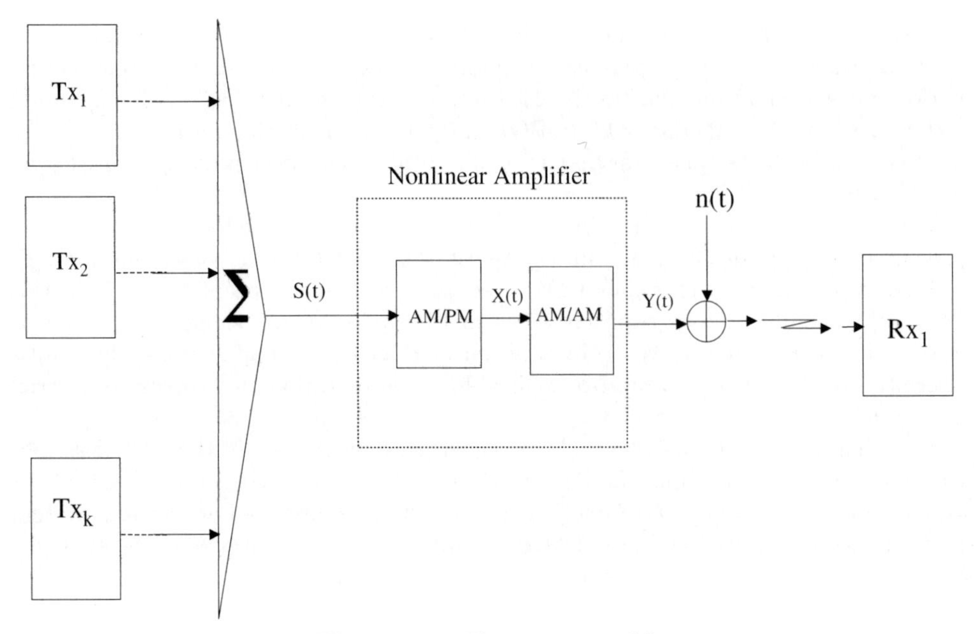

Figure 9.1 The system model.

with equal probability $\frac{1}{M}$. p is the transmitted power, T_s is the symbol period, and $\mathcal{I}_T$ is defined as

$$\mathcal{I}_T(t) = \begin{cases} 1 & \text{if } 0 \leq t \leq T \\ 0 & \text{otherwise.} \end{cases}$$

The spreading code signals of the kth user can be represented as

$$c^{(k)}(t) = \sum_{i=-\infty}^{\infty} c^{(k)}[i]\mathcal{I}_{T_c}(t - iT_c)$$

where T_c is the chip duration and $c^{(k)}[i] \in \{-1, +1\}$. Taking into account the phase of each user's local oscillator, the output signal of user k can be represented as

$$\begin{aligned} S^{(k)}(t) &= b_I^{(k)}(t)c^{(k)}(t)\cos\left(\omega_c t + \theta^{(k)}\right) + b_Q^{(k)}(t)c^{(k)}(t)\sin\left(\omega_c t + \theta^{(k)}\right) \\ &= A^{(k)}(t)\cos(\omega_c t) + B^{(k)}(t)\sin(\omega_c t) \end{aligned}$$

where

$$\begin{aligned} A^{(k)}(t) &= b_I^{(k)}(t)c^{(k)}(t)\cos\theta^{(k)} + b_Q^{(k)}(t)c^{(k)}(t)\sin\theta^{(k)} \\ B^{(k)}(t) &= -b_I^{(k)}(t)c^{(k)}(t)\sin\theta^{(k)} + b_Q^{(k)}(t)c^{(k)}(t)\cos\theta^{(k)} \end{aligned}$$

Since this is a synchronous system, the transmitted signal is the sum of the individual signal with perfect time alignment. Therefore,

$$S(t) = \sum_{k=1}^{K} S^{(k)}(t) = V(t)\cos(\omega_c t - \Phi(t))$$

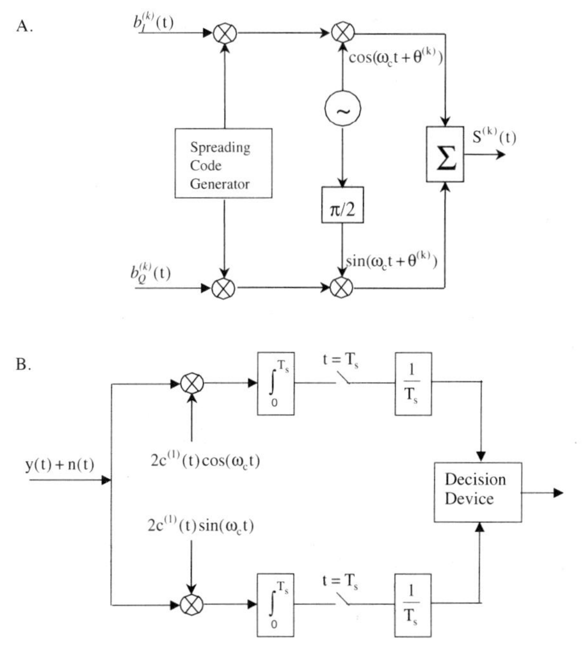

Figure 9.2 The transmitter (A), and receiver (B) models.

where

$$V(t) = \sqrt{\left[\sum_{k=1}^{K} A^{(k)}(t)\right]^2 + \left[\sum_{k=1}^{K} B^{(k)}(t)\right]^2}$$

$$\Phi(t) = \tan^{-1}\left[\frac{\sum_{k=1}^{K} B^{(k)}(t)}{\sum_{k=1}^{K} A^{(k)}(t)}\right]$$

9.2.2 Nonlinear Amplifiers

The most commonly used AM/AM model for nonlinear amplifiers is the Q function (see Figure 9.3-A). Almost all the smooth limiters of interest preserve this shape. The drive power at which the output power saturates is called the *input saturation power*. In Figure 9.3-A, the corresponding baseband input amplitude is ± 2.3. The ratio of input saturation power to desired drive power is called the *input back-off* (IBO). Similarly, the *output saturation power* is the maximum output power of an amplifier. Its corresponding baseband output amplitude in Figure 9.3-A is ± 1. *Output back-off* (OBO) is therefore the ratio of the output saturation power to the actual output power. Increasing IBO or OBO leads to less output power, but reduces the nonlinearities introduced during signal amplification. The trade-off between lower power and more nonlinearities results in the highest effective SNR or, equivalently, the highest effective E_s/N_o and the lowest BER. Note that increasing OBO reduces the efficiency of amplifier power usage, which is particularly undesirable in satellite communications. OBO can thus be regarded as the immunity strength of a communication system to nonlinearity. The higher OBO that is required, the more vulnerable the system is to nonlinear distortion and the less efficient in power usage.

However, since the Q function is not an analytical function, in order to evaluate the system performance, we may only use computer simulation. The hard limiter model, on the other hand, although simple is not an accurate representation of our system.

Furthermore, the hard limiter model will not be able to determine the IBO or OBO of a nonlinear amplifier. In addition to the above two extreme options, i.e. Q function and hard limiter, Chen [7], Forsey [8] and Kunz [9] have proposed different models. Our model is based on both Chen [7] and Kunz [9] for reasons of accuracy and simplicity in the evaluation of the Signal-to-Noise Ratio (SNR) of the CDMA system.

In the particular model we use, AM/AM is represented by a third-order polynomial, which is a memoryless nonlinearity affecting only the amplitude of the input signal. The coefficients of the polynomial are determined by placing the local maximum or minimum at the saturation point of the nonlinear amplifier. In the meantime, AM/PM introduces a phase distortion which is proportional to the square of the envelope of the input signal. In other words, AM/PM: $\Theta[V(t)] = \eta V^2(t)$, $\eta \ll 1$, and AM/AM: $V'(t) = \alpha_1 V(t) + \alpha_3 V^3(t)$, where η, α_1 and α_3 are the corresponding parameters. Then the input signal to the receiver is

$$\begin{aligned} z(t) &= V'(t)\cos\{\omega_c t - \Phi(t) + \Theta[V(t)]\} + n(t) \\ &= [S_I(t) + D_I(t) + n_I(t)]\cos(\omega_c t) + [S_Q(t) - D_Q(t) + n_Q(t)]\sin(\omega_c t) \end{aligned}$$

where

$$n(t) = n_I(t)\cos(\omega_c t) + n_Q(t)\sin(\omega_c t)$$

$$S_I(t) = \left[\alpha_1 + \alpha_3 V^2(t)\right]\sum_{k=1}^{K} A^{(k)}(t)$$

$$D_I(t) = \eta\left[\alpha_1 V^2(t) + \alpha_3 V^4(t)\right]\sum_{k=1}^{K} B^{(k)}(t)$$

$$S_Q(t) = \left[\alpha_1 + \alpha_3 V^2(t)\right]\sum_{k=1}^{K} B^{(k)}(t)$$

$$D_Q(t) = \eta\left[\alpha_1 V^2(t) + \alpha_3 V^4(t)\right]\sum_{k=1}^{K} A^{(k)}(t)$$

9.3 Performance Analysis

The effects of nonlinearities on CDMA signals are evaluated by analyzing their distortions on the first two moments of the received signal. Since there is a large number of users in the system (32 to 64), it is important that the methodology we use is very accurate. The proposed methodology is based on so-called Gaussian Approximation (GA). Note that to achieve accurate results, we apply the GA after the nonlinear effect. In other words, we analyze the first two moments of the signal after the nonlinear distortion instead of analyzing them before the distortion and then nonlinearly transforming them to analyze the performance.

A

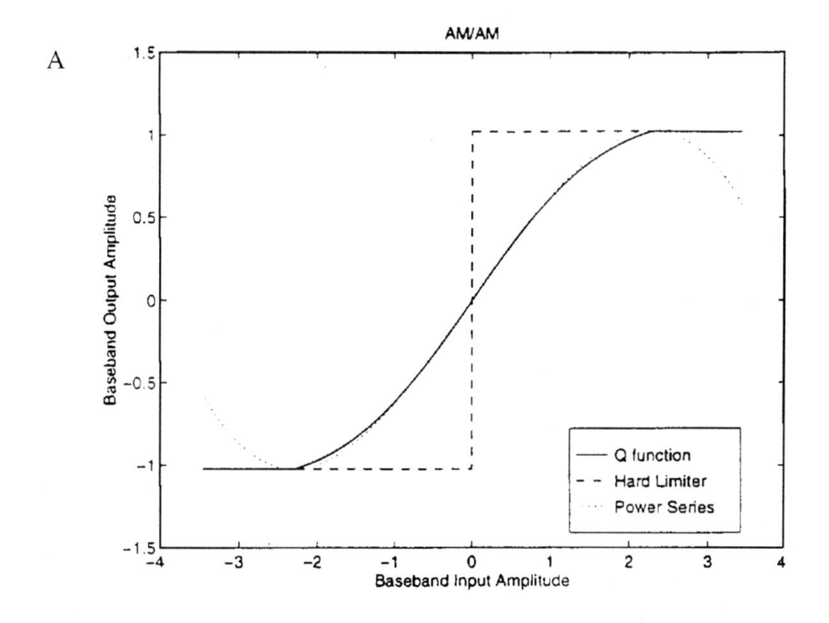

B

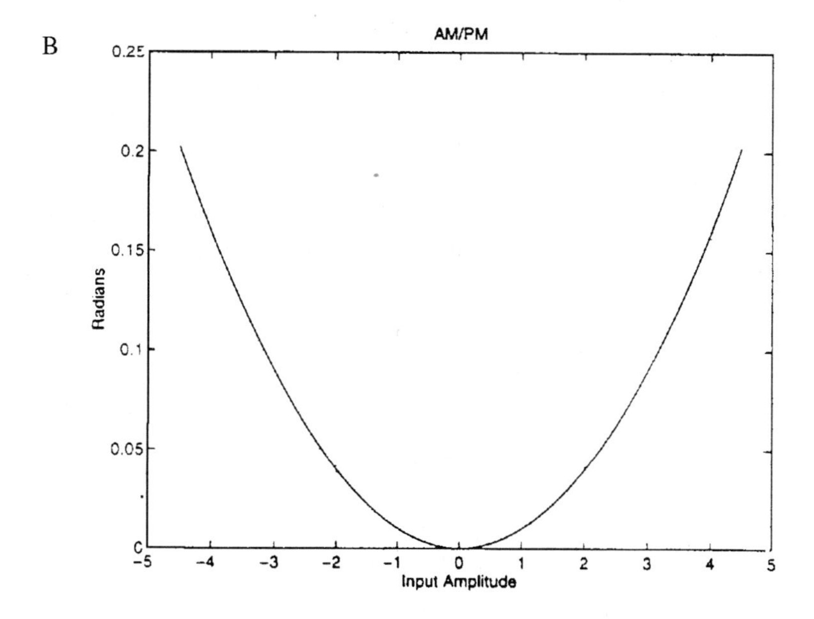

Figure 9.3 (A) The amplitude transfer (AM/AM) curve and (B) the phase transfer (AM/PM) curve.

The distortion on the first moment is sometimes called *constellation warping.* The received constellation points are no longer at their original grids due to the distortion of amplitude and phase. Amplitude distortion changes the distance from the original to the constellation points, while phase distortion changes their angles. Constellation warping leads to the undesired preferences of some of the constellation points, which means that the detection is no longer maximum likelihood. The amplitude shrinkage also reduces the effective E_b/N_o.

The distortion on the second moment is called *cloud forming.* As in SS/CDMA, in most synchronous CDMA systems, Multiple-Access Interference (MAI) is minimized by utilizing orthogonal spreading codes. We assume that users within each beam are distinguished by orthogonal codes, while between beams we use PN-codes. Therefore, since all downlink transmissions are perfectly synchronized, the same-beam MAI shall be zero and the other-beam MAI is a function of the cross-correlation functions of PN-codes. This suggests that the BER performance is a function of the received user power, the thermal noise figure and the other-beam MAI. Note that both the orthogonality and cross-correlation functions are the second moment of code functions. However, in the presence of nonlinear distortions of amplitude and phase, the second-moment distortion generates high-order moments of code functions. These high-order moments destroy the orthogonality between same-beam spreading codes, which leads to nonzero same-beam MAI. The other-beam MAI now also consists of second moments and higher moments of code functions. The nonlinear MAI, composed of both same-beam and other-beam MAIs, becomes a function of the second or even third power of the received user power, the thermal noise figure, the high-order moments of code cross-correlation functions and, worst of all, of the number of active users in the system. System capacity thus reduces due to nonlinearities.

In general, amplitude nonlinearity leads to the generation of harmonics and amplitude cross-modulation. Nonlinear phase characteristics lead to phase cross-modulation. The effects of these two nonlinearities on the sum of CDMA signals will force warping of signal constellations as well as the intra- and inter-beam cross-correlation of signals.

The receiver model is shown in Figure 9.2-B. Without loss of generality, we consider the performance of the first user. Then the output of the in-phase receiver can be represented as

$$\mathbf{Z_I} = \frac{1}{T_s}\int_0^{T_s} y(t)c_I^{(1)}(t)2\cos(\omega_c t)dt$$

The integrator will then reject the high frequency portion of the signal. Hence,

$$\mathbf{Z_I} = \mathbf{S_I} + \mathbf{D_I} + \mathbf{N_I}$$

where

$$\mathbf{S_I} = \frac{1}{T_s}\int_0^{T_s} S_I(t)c_I^{(1)}(t)dt$$

$$\mathbf{D_I} = \frac{1}{T_s}\int_0^{T_s} D_I(t)c_I^{(1)}(t)dt$$

The mean of $\mathbf{N_I}$ is 0 and the variance of $\mathbf{N_I}$ is $\sigma_N^2 = \frac{N_0}{T_s}$. $\frac{N_0}{2}$ is the two-sided power spectral density.

Suppose $b_I^{(1)}[0] = d_I$ and $b_Q^{(1)}[0] = d_Q$. We also assume synchronization is perfect, which means $\theta^{(1)} = 0$. Thus $A^{(1)}[0] = d_I c^{(1)}[n]$, $B^{(1)}[0] = d_Q c^{(1)}[n]$, $n = 0..N-1$, where $A^{(1)}[0]$ represents the value of $A^{(1)}(t)$ during the time interval from $t = 0$ to $t = T_s$. The same rule, i.e. '[0]' representing $t = 0..T_s$, also applies to all other functions. Therefore,

$$
\begin{aligned}
\mathbf{S_I} &= \frac{1}{N}\sum_{n=0}^{N-1} S_I[0]c_I^{(1)}[n] \\
&= \frac{1}{N}\sum_{n=0}^{N-1}\left\{\left(\alpha_1 + \alpha_3 V^2[0]\right)\sum_{k=1}^{K} A^{(k)}[0]\right\} c_I^{(1)}[n] \\
&= \alpha_1 \frac{1}{N}\sum_{n=0}^{N-1}\sum_{k=1}^{K} A^{(k)}[0]c_I^{(1)}[n] + \alpha_3 \frac{1}{N}\sum_{n=0}^{N-1}\sum_{k_1,k_2,k_3=1}^{K} A^{(k_1)}[0]A^{(k_2)}[0]A^{(k_3)}[0]c_I^{(1)}[n] \\
&\quad + \alpha_3 \frac{1}{N}\sum_{n=0}^{N-1}\sum_{k_1,k_2,k_3=1}^{K} A^{(k_1)}[0]B^{(k_2)}[0]B^{(k_3)}[0]c_I^{(1)}[n]
\end{aligned}
$$

Similarly,

$$
\begin{aligned}
\mathbf{D_I} &= \frac{1}{N}\sum_{n=0}^{N-1} D_I[0]c_I^{(1)}[n] \\
&= \frac{1}{N}\sum_{n=0}^{N-1}\left\{\eta\left(\alpha_1 V^2[0] + \alpha_3 V^4[0]\right)\sum_{k=1}^{K} B^{(k)}[0]\right\} c_I^{(1)}[n] \\
&= \eta\alpha_1 \frac{1}{N}\sum_{n=0}^{N-1}\sum_{k_1,k_2,k_3=1}^{K} B^{(k_1)}[0]A^{(k_2)}[0]A^{(k_3)}[0]c_I^{(1)}[n] \\
&\quad + \eta\alpha_1 \frac{1}{N}\sum_{n=0}^{N-1}\sum_{k_1,k_2,k_3=1}^{K} B^{(k_1)}[0]B^{(k_2)}[0]B^{(k_3)}[0]c_I^{(1)}[n] \\
&\quad + \eta\alpha_3 \frac{1}{N}\sum_{n=0}^{N-1}\sum_{\substack{k_1,k_2\\k_3,k_4,k_5=1}}^{K}\Big\{B^{(k_1)}[0]A^{(k_2)}[0]A^{(k_3)}[0]A^{(k_4)}[0]A^{(k_5)}[0] \\
&\quad + 2B^{(k_1)}[0]B^{(k_2)}[0]B^{(k_3)}[0]A^{(k_4)}[0]A^{(k_5)}[0] \\
&\quad + B^{(k_1)}[0]B^{(k_2)}[0]B^{(k_3)}[0]B^{(k_4)}[0]B^{(k_5)}[0]\Big\} c_I^{(1)}[n]
\end{aligned}
$$

The detailed derivations are shown in Appendix 9A. For the case of QPSK, the following results can be obtained for the I component. General results for M-PSK cases are shown in Appendix 9A. The corresponding expression for the Q component is obtained by interchanging I to Q and Q to I in all terms:

$$
\begin{aligned}
Y_I &= \alpha_1 d_I + \alpha_3 d_I^3 + \alpha_3 d_I d_Q^2 + \eta\alpha_1 d_Q^3 + \eta\alpha_1 d_I^2 d_Q + \eta\alpha_3 d_Q^5 + 2\eta\alpha_3 d_I^2 d_Q^3 + \eta\alpha_3 d_I^4 d_Q \\
U_I &= a_2 K^2 + a_1 K + a_0 \\
\sigma_I^2 &= b_3 K^3 + b_2 K^2 + b_1 K + b_0 + \sigma_N^2 \\
\sigma_{IQ}^2 &= 4\alpha_1\alpha_3 d_I d_Q (K-1) R^2 p
\end{aligned}
$$

where

$$
\begin{aligned}
a_2 &= 24\alpha_3\eta d_Q p^2 \\
a_1 &= -60\alpha_3\eta d_Q p^2 + \left(4\alpha_3 d_I + 4\alpha_1\eta d_Q + 12\alpha_3\eta d_Q^3 + 12\alpha_3\eta d_I^2 d_Q\right) p \\
a_0 &= 36\alpha_3\eta d_Q p^2 - \left(12\alpha_3\eta d_I^2 d_Q + 4\alpha_3 d_I + 4\alpha_1\eta d_Q + 12\alpha_3\eta d_Q^3\right) p \\
b_3 &= \left(16\alpha_3^2 R^2 + \frac{19}{3}\alpha_3^2 \hat{R}^2\right) p^3 \\
b_2 &= \left(-60\alpha_3^2 R^2 - 38\alpha_3^2 \hat{R}^2\right) p^3 + \left(12\alpha_3^2 d_Q^2 R^2 + 8\alpha_1\alpha_3 R^2 + 44\alpha_3^2 d_I^2 R^2\right) p^2 \\
b_1 &= \left(72\alpha_3^2 R^2 + \frac{209}{3}\alpha_3^2 \hat{R}^2\right) p^3 \\
&\quad + \left(2\alpha_3^2 d_I^2 - 120\alpha_3^2 d_I^2 R^2 - 20\alpha_1\alpha_3 R^2 - 32\alpha_3^2 d_Q^2 R^2 + 2\alpha_3^2 d_Q^2\right) p^2 \\
&\quad + \left(\alpha_3^2 R^2 d_Q^4 + \alpha_1^2 R^2 + 6\alpha_1\alpha_3 R^2 d_I^2 + 2\alpha_1\alpha_3 R^2 d_Q^2 + 10\alpha_3^2 d_I^2 d_Q^2 R^2 + 9\alpha_3^2 R^2 d_I^4\right) p \\
b_0 &= \left(-28\alpha_3^2 R^2 - 38\alpha_3^2 \hat{R}^2\right) p^3 \\
&\quad + \left(12\alpha_1\alpha_3 R^2 + 76\alpha_3^2 d_I^2 R^2 + 20\alpha_3^2 d_Q^2 R^2 - 2\alpha_3^2 d_Q^2 - 2\alpha_3^2 d_I^2\right) p^2 \\
&\quad + \left(-\alpha_1^2 R^2 - 6\alpha_1\alpha_3 R^2 d_I^2 - 2\alpha_1\alpha_3 R^2 d_Q^2 - 9\alpha_3^2 R^2 d_I^4 - 10\alpha_3^2 d_I^2 d_Q^2 R^2 \right. \\
&\quad \left. -\alpha_3^2 R^2 d_Q^4\right) p
\end{aligned}
$$

σ_I^2 is the variance of the I component interference, and σ_{IQ}^2 is the covariance of the I and Q components. $S_I = \mathbf{S_I}$ and $D_I = \mathbf{D_I}$. $Y_I + U_I = S_I + D_I$. Y_I is the part of $\mathbf{S_I} + \mathbf{D_I}$ from only the first user. All other terms, regarded as a disturbance of the first moment from other users, are represented by U_I. Note that a non-zero-mean cross-correlated Gaussian interference results due to nonlinearities. All the corresponding results for the Q components can be easily obtained by exchanging I and Q and Q and I in the expressions of I the component.

To compute the probability of symbol error, let $\phi_m = \tan^{-1}\frac{Y_I+U_I}{Y_Q+U_Q}$ and m correspond to different d_I and d_Q, where

$$
(d_I, d_Q) \in \left\{ \left(\sqrt{2p}\cos\frac{(2m-1)\pi}{M},\ \sqrt{2p}\sin\frac{(2m-1)\pi}{M} \right),\ m = 1, .., M \right\}
$$

Furthermore, let

$$
\rho_{I,Q} = \frac{E\{(\mathbf{Z_I} - Z_I)(\mathbf{Z_Q} - Z_Q)\}}{\sigma_I \sigma_Q}
$$

For equi-probable M-ary PSK symbols, the probability of error is (see Appendix 9A for details)

$$P_e = \frac{1}{M} \sum_{m=1}^{M} P_{e|m}$$

where

$$P_{e|m} = 1 - \int_{\phi_m - \frac{\pi}{M}}^{\phi_m + \frac{\pi}{M}} \bar{f}(\phi, \phi_m) d\phi$$

$\bar{f}(\phi, \phi_m)$ is defined as

$$\bar{f}(\phi, \phi_m) = \frac{1}{\sqrt{2\pi\sigma_I\sigma_Q}\sqrt{1 - \rho_{I,Q}\sin(2\phi)}} \frac{\cos(\phi - \phi_m) - \rho_{I,Q}\sin(\phi + \phi_m)}{1 - \rho_{I,Q}\sin(2\phi)} \cdot \exp\left\{ - \frac{[1 - \rho_{I,Q}\sin(2\phi_m)] - \frac{[\cos(\phi - \phi_m) - \rho_{I,Q}\sin(\phi + \phi_m)]^2}{1 - \rho_{I,Q}\sin(2\phi)}}{2(1 - \rho_{I,Q}^2)\sigma_I\sigma_Q} \right\}$$

The effects on the spread-spectrum signals after despreading can be summarized as

1. *Cloud forming*: each constellation point becomes a cloud due to linear and nonlinear ISI, as well as intra- and inter-user cross-modulation of the I and Q components, and
2. *Signal warping*: the respective centroid of the clouds (i.e. the received constellation points) are no longer at the original position of the constellation points.

From another point of view, nonlinearities shrink the constellation and transform its power into clouds. So besides the filtered out-of-band power, nonlinearities further reduce the effective signal power.

9.4 Numerical Results

We first consider a system with 64 chips per code. Three different code families (i.e. OG, PG and PN) are investigated along with BPSK, QPSK and 8-PSK data modulation schemes. The AM/AM curve of the nonlinear amplifier is shown as a dotted line in Figure 9.3-A, while the AM/PM curve is shown in Figure 9.3-B. The way to compute the input power, output power and IBO of a nonlinear amplifier is shown in Appendix 9A.5.

Figure 9.4 shows the IBO versus E_s/N_0 and effective E_s/N_0. It shows that the E_s/N_0 (dashed line) increases as IBO decreases, because less IBO means more power is being transmitted. On the other hand, as IBO decreases the effective E_s/N_0 (solid line) deviates from the E_s/N_0 (dashed line) to a lower position beginning from about 8 dB. It even descends after the IBO is less than 3 dB. This behavior reveals that the effects of nonlinearities start emerging at 8 dB. They further deteriorate, and thus dominate the output signals after IBO is less than 3 dB.

Figure 9.5 shows the SER for 32 users, and Figure 9.6 shows that for 64 users. As shown, the effective E_s/N_0 begins lowering, and after 3 dB IBO decreases. This concludes that in this particular system considered, the optimum operating point of the nonlinear amplifier is around 3 dB IBO.

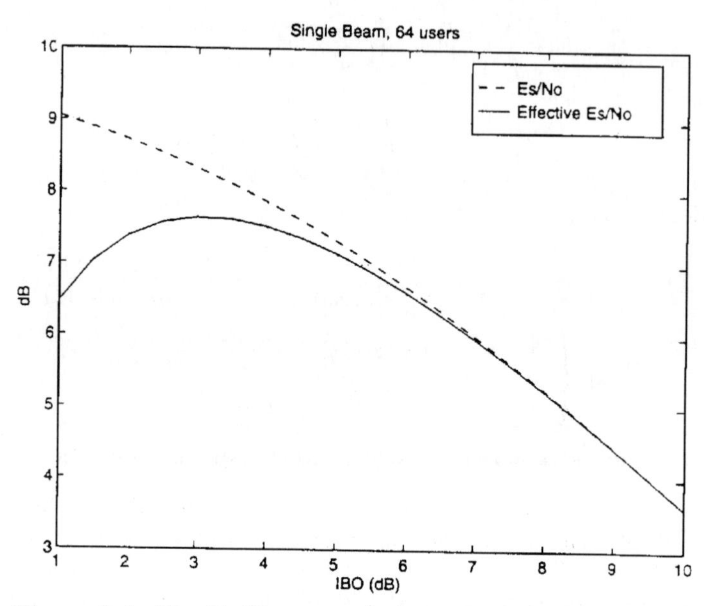

Figure 9.4 The E_s/N_0 versus the Input-Back-Off (IBO) in db.

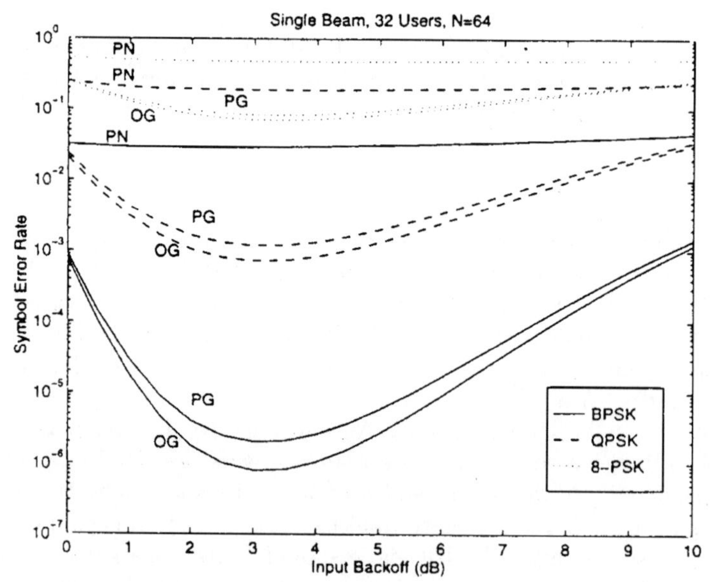

Figure 9.5 The Symbol Error Rate (SER) versus the Input Back-Off (IBO) in db with 32 users and for orthogonal (OG), preferred-phased (PG) and pseudonoise (PN) sequences.

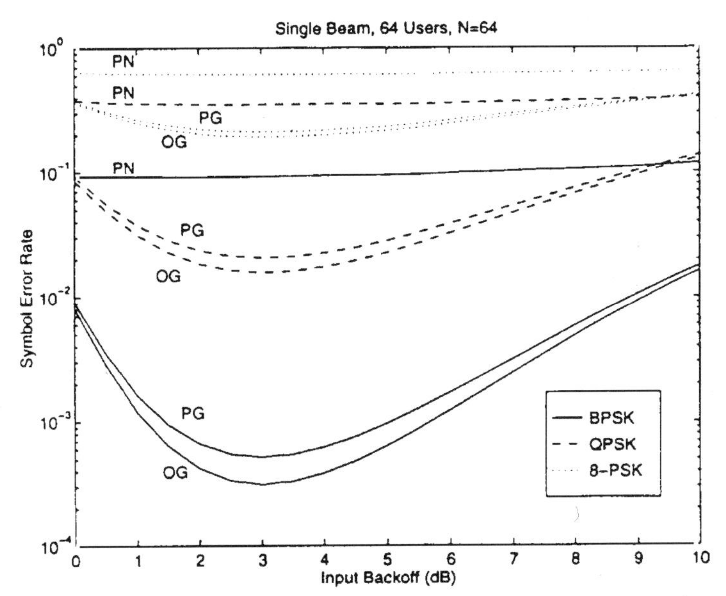

Figure 9.6 The Symbol Error Rate (SER) versus the Input Back-Off (IBO) in db with 64 users and for orthogonal (OG), perferred-phased (PG) and pseudonoise (PN) sequences.

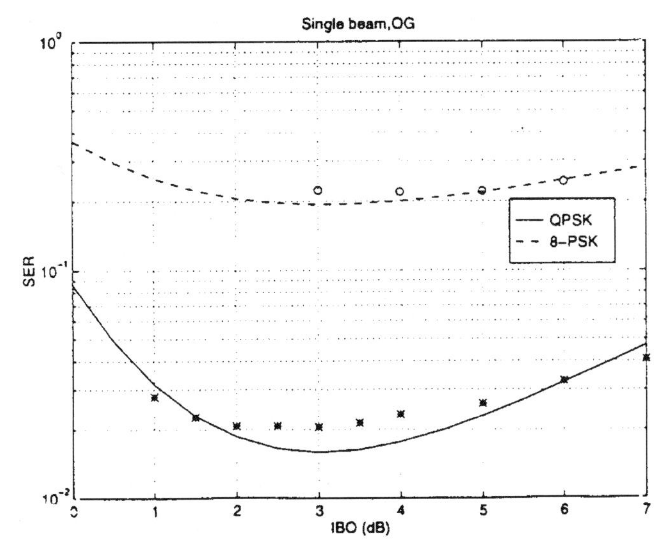

Figure 9.7 Comparisons between analysis and simulation of the SER vs. the IBO for orthogonal (OG) sequences and for QPSK and 8-PSK.

To compare with these analytical results, computer simulations were conducted for systems with QPSK and 8-PSK data modulation schemes and OG spreading code. AM/AM nonlinear distortion is represented as the solid line in Figure 9.3-A. The results are shown in Figure 9.7. Their solid consistency further assures the validity of the analytical results.

According to the analytical and numerical results obtained, several conclusions are made which are summarized as follows:

1. The received constellation points are no longer at their original places.
2. Given an amplifier, for a fixed E_b/N_0 requirement, different number of active users in a system has different Output Back-Off (OBO) values.
3. The value of the OBO as well as the interference and received constellation points are implicit functions of the number of active users.
4. Nonlinearities destroy the orthogonality of the inter-user same-phase components.
5. Nonlinearities leads to a non-zero-mean and cross-correlated jointly Gaussian interference in the I and Q branches of the receiver.

9.5 Conclusions

The performances of synchronous M-PSK CDMA systems with nonlinear distortions were evaluated analytically. The effects of nonlinearities appear more significantly while the back-off is less. The trade-off between power and nonlinearities, which achieves maximal effective E_s/N_0, determines the desired operating point of a nonlinear amplifier. The performance of BPSK data modulation scheme, as expected, is better than QPSK, which is even better than 8-PSK. The OG and PG spreading codes have almost the same performance and both of them outperform the PN spreading code.

References

[1] S.-W. Chen, W. Panton and R. Gilmore 'Effects of Nonlinear Distortion on CDMA Communication Systems' *IEEE MTT-S Digest*, 1996, pp. 775–778.

[2] N. I. Smirnov and S. F. Gorgadze 'An Estimate of the Power Utilization Efficiency of a Nonlinear Tranponder in Data Transmission Systems with Code Division Multiplexing' *Elektrosvyaz*, June 1995, pp. 21–24.

[3] Pen Li and E. Geraniotis 'Effects of Nonlinear Distortion on Synchronous M-PSK DS/CDMA Systems' *Conference on Information Science and Systems*, Baltimore, MD, March 1997, pp. 966–971.

[4] Pen Li and E. Geraniotis 'Performance Analysis of Synchronous M-PSK DS/CDMA Multi-Tier System with a Nonlinear Amplifier' *IEEE Symposium on Computers and Communications*, Alexandria, Egypt, July 1997, pp. 275–279.

[5] J. M. Aein and R. L. Pickholtz 'A Simple Unified Phasor Analysis for PN Multiple Access to Limiting Repeaters' *IEEE Trans. on Commun.*, Vol. COM-30, No. 5, May 1982, pp. 1018–1026.

[6] H. P. Baer 'Interference Effects of Hard Limiting in PN Spread-Spectrum Systems' *IEEE Trans. on Commun.*, Vol. COM-30, No. 5, May 1982, pp. 1010–1017.
[7] L. Chen, S. Ikuta, M. Kominami and H. Kusaka 'An Analysis of Nonlinear Distortion due to TWT in Asynchronous DS-SSMA Communication Systems' *IEEE Spread Spectrum Commun.*, 1994, pp. 279–302.
[8] R. J. Forsey, V. E. Gooding, P. J. McLane and L. L. Campbell 'M-ary PSK Transmission via a Coherent Two-Link Channel Exhibiting AM-AM and AM-PM Nonlinearities' *IEEE Trans. on Commun.*, Vol. COM-26, No. 1, January 1978, pp. 116–123.
[9] W.E. Kunz, J. H. Foster and R. F. Lazzarini 'Travelling-Wave Tube Amplifier Characteristics for Communications' *The Microwave Journal*, March 1967, pp. 41–46.

Appendix 9A: Performance Evaluation

9A.1 Evaluation of the First Moment

Let

$$\mathrm{E}_{\boldsymbol{\theta}}[\cdot] = \frac{1}{(2\pi)^{K-1}} \int_0^{2\pi} \int_0^{2\pi} \cdots \int_0^{2\pi} \cdot d\theta_K \cdots d\theta_3 d\theta_2$$

By knowing the fact that

$$\mathrm{E}_{\boldsymbol{\theta}}\left[A^{(k)}[0]\right] = \begin{cases} A^{(1)}[0] & \text{if } k = 1; \\ 0 & \text{otherwise} \end{cases}$$

$$\mathrm{E}_{\boldsymbol{\theta}}\left[A^{(k_1)}[0]A^{(k_2)}[0]\right] = \begin{cases} A^{(1)}[0]^2 & k_1 = k_2 = 1; \\ \frac{1}{2}\left(b_I^{(k)}[0]^2 + b_Q^{(k)}[0]^2\right) & k_1 = k_2 = k \neq 1; \\ 0 & \text{otherwise} \end{cases}$$

$$\mathrm{E}_{\boldsymbol{\theta}}\left[A^{(k_1)}[0]A^{(k_2)}[0]A^{(k_3)}[0]\right]$$
$$= \begin{cases} A^{(1)}[0]^3 & k_1 = k_2 = k_3 = 1; \\ \frac{1}{2}A^{(1)}[0]\left(b_I^{(k)}[0]^2 + b_Q^{(k)}[0]^2\right) & \text{if one of } k_1, k_2, \text{ or } k_3 = 1 \\ & \text{and the other two are equal to } k; \\ 0 & \text{otherwise} \end{cases}$$

$$\mathrm{E}_{\boldsymbol{\theta}}\left[A^{(k_1)}[0]B^{(k_2)}[0]B^{(k_3)}[0]\right]$$
$$= \begin{cases} A^{(1)}[0]B^{(1)}[0]^2 & k_1 = k_2 = k_3 = 1; \\ \frac{1}{2}A^{(1)}[0]\left(b_I^{(k)}[0]^2 + b_Q^{(k)}[0]^2\right) & k_1 = 1, k_2 = k_3 = k \neq 1; \\ 0 & \text{otherwise} \end{cases}$$

Then

$$\sum_{k_1=1}^{K}\sum_{k_2=1}^{K} \eta A^{(k_1)}[0]A^{(k_2)}[0] = d_I^2 + \frac{1}{2}\sum_{k=2}^{K}\left(b_I^{(k)}[0]^2 + b_Q^{(k)}[0]^2\right)$$

$$\sum_{k_1=1}^{K}\sum_{k_2=1}^{K}\sum_{k_3=1}^{K} \mathrm{E}_{\boldsymbol{\theta}}\left[A^{(k_1)}[0]A^{(k_2)}[0]A^{(k_3)}[0]\right] = A^{(1)}[0]^3 + \frac{3}{2}\sum_{k=2}^{K} A^{(1)}[0]\left(b_I^{(k)}[0]^2 + b_Q^{(k)}[0]^2\right)$$

$$\sum_{k_1=1}^{K}\sum_{k_2=1}^{K}\sum_{k_3=1}^{K} \mathrm{E}_{\boldsymbol{\theta}}\left[A^{(k_1)}[0]B^{(k_2)}[0]B^{(k_3)}[0]\right] = A^{(1)}[0]B^{(1)}[0]^2 + \frac{1}{2}\sum_{k=2}^{K} A^{(1)}[0]\left(b_I^{(k)}[0]^2 + b_Q^{(k)}[0]^2\right)$$

Therefore, from the expression of $S_I(t)$ in Section 9.2.2.

$$\begin{aligned}
\mathrm{E}_{\boldsymbol{\theta}}\left[S_I\right] &= \alpha_1 \frac{1}{N}\sum_{n=0}^{N-1} A^{(1)}[0]c_I^{(1)}[n] \\
&+ \alpha_3 \frac{1}{N}\sum_{n=0}^{N-1}\left\{A^{(1)}[0]^3 + \frac{3}{2}\sum_{k=2}^{K} A^{(1)}[0]\left(b_I^{(k)}[0]^2 + b_Q^{(k)}[0]^2\right)\right\} c_I^{(1)}[n] \\
&+ \alpha_3 \frac{1}{N}\sum_{n=0}^{N-1}\left\{A^{(1)}[0]B^{(1)}[0]^2 + \frac{1}{2}\sum_{k=2}^{K} A^{(1)}[0]\left(b_I^{(k)}[0]^2 + b_Q^{(k)}[0]^2\right)\right\} c_I^{(1)}[n] \\
&= \alpha_1 d_I + \alpha_3 d_I^3 + \alpha_3 d_I d_Q^2 + 2\alpha_3 d_I \sum_{k=2}^{K}\left(b_I^{(k)}[0]^2 + b_Q^{(k)}[0]^2\right)
\end{aligned}$$

The first moment of $\mathbf{D_I}$ can be obtained similarly.

9A.2 Evaluation of the Second Moment

To compute the variance, we rewrite the expression of $S_I(t)$ in Section 9.2.2 as follows:

$$\begin{aligned}
S_I = \alpha_1 &\sum_{k=1}^{K} C(I_1, I_k) b_I^{(k)}[0]\cos\theta^{(k)} \\
&+ \alpha_1 \sum_{k=1}^{K} C(I_1, Q_k)(k) b_Q^{(k)}[0]\sin\theta^{(k)} + \alpha_3 \Psi
\end{aligned}$$

where

$$C(E_i, F_j) = \frac{1}{N}\sum_{n=0}^{N-1} c_E^{(i)}[n] c_F^{(j)}[n], \;\; E, F \in \{I, Q\}$$

$$\Psi = \sum_{k_1=1}^{K}\sum_{k_2=1}^{K}\sum_{k_3=1}^{K}\Big\{a_7(k_1,k_2,k_3)\cos\theta^{(k_1)}\cos\theta^{(k_2)}\cos\theta^{(k_3)}$$
$$+ a_6(k_1,k_2,k_3)\cos\theta^{(k_1)}\cos\theta^{(k_2)}\sin\theta^{(k_3)}$$
$$+ a_5(k_1,k_2,k_3)\cos\theta^{(k_1)}\sin\theta^{(k_2)}\cos\theta^{(k_3)}$$
$$+ a_4(k_1,k_2,k_3)\cos\theta^{(k_1)}\sin\theta^{(k_2)}\sin\theta^{(k_3)}$$
$$+ a_3(k_1,k_2,k_3)\sin\theta^{(k_1)}\cos\theta^{(k_2)}\cos\theta^{(k_3)}$$
$$+ a_2(k_1,k_2,k_3)\sin\theta^{(k_1)}\cos\theta^{(k_2)}\sin\theta^{(k_3)}$$
$$+ a_1(k_1,k_2,k_3)\sin\theta^{(k_1)}\sin\theta^{(k_2)}\cos\theta^{(k_3)}$$
$$+ a_0(k_1,k_2,k_3)\sin\theta^{(k_1)}\sin\theta^{(k_2)}\sin\theta^{(k_3)}\Big\}$$

Also, $a_i\,(k_1,k_2,k_3)$'s are defined as

$$a_7\,(k_1,k_2,k_3) = C(I_{k_1},I_{k_2},I_{k_3},I_1)b_I^{(k_1)}[0]b_I^{(k_2)}[0]b_I^{(k_3)}[0] + C(I_{k_1},Q_{k_2},Q_{k_3},I_1)b_I^{(k_1)}[0]b_Q^{(k_2)}[0]b_Q^{(k_3)}[0]$$
$$a_6\,(k_1,k_2,k_3) = C(I_{k_1},I_{k_2},Q_{k_3},I_1)b_I^{(k_1)}[0]b_I^{(k_2)}[0]b_Q^{(k_3)}[0] - C(I_{k_1},Q_{k_2},I_{k_3},I_1)b_I^{(k_1)}[0]b_Q^{(k_2)}[0]b_I^{(k_3)}[0]$$
$$a_5\,(k_1,k_2,k_3) = C(I_{k_1},Q_{k_2},I_{k_3},I_1)b_I^{(k_1)}[0]b_Q^{(k_2)}[0]b_I^{(k_3)}[0] - C(I_{k_1},I_{k_2},Q_{k_3},I_1)b_I^{(k_1)}[0]b_I^{(k_2)}[0]b_Q^{(k_3)}[0]$$
$$a_4\,(k_1,k_2,k_3) = C(I_{k_1},Q_{k_2},Q_{k_3},I_1)b_I^{(k_1)}[0]b_Q^{(k_2)}[0]b_Q^{(k_3)}[0] + C(I_{k_1},I_{k_2},I_{k_3},I_1)b_I^{(k_1)}[0]b_I^{(k_2)}[0]b_I^{(k_3)}[0]$$
$$a_3\,(k_1,k_2,k_3) = C(Q_{k_1},I_{k_2},I_{k_3},I_1)b_Q^{(k_1)}[0]b_I^{(k_2)}[0]b_I^{(k_3)}[0] + C(Q_{k_1},Q_{k_2},Q_{k_3},I_1)b_Q^{(k_1)}[0]b_Q^{(k_2)}[0]b_Q^{(k_3)}[0]$$
$$a_2\,(k_1,k_2,k_3) = C(Q_{k_1},I_{k_2},Q_{k_3},I_1)b_Q^{(k_1)}[0]b_I^{(k_2)}[0]b_Q^{(k_3)}[0] - C(Q_{k_1},Q_{k_2},I_{k_3},I_1)b_Q^{(k_1)}[0]b_Q^{(k_2)}[0]b_I^{(k_3)}[0]$$
$$a_1\,(k_1,k_2,k_3) = C(Q_{k_1},Q_{k_2},I_{k_3},I_1)b_Q^{(k_1)}[0]b_Q^{(k_2)}[0]b_I^{(k_3)}[0] - C(Q_{k_1},I_{k_2},Q_{k_3},I_1)b_Q^{(k_1)}[0]b_I^{(k_2)}[0]b_Q^{(k_3)}[0]$$
$$a_0\,(k_1,k_2,k_3) = C(Q_{k_1},Q_{k_2},Q_{k_3},I_1)b_Q^{(k_1)}[0]b_Q^{(k_2)}[0]b_Q^{(k_3)}[0] + C(Q_{k_1},I_{k_2},I_{k_3},I_1)b_Q^{(k_1)}[0]b_I^{(k_2)}[0]b_I^{(k_3)}[0]$$

where

$$C(E_{l_1},F_{l_2},G_{l_3},H_{l_3}) = \frac{1}{N}\sum_{n=0}^{N-1} c_E^{(l_1)}[n]c_F^{(l_2)}[n]c_G^{(l_3)}[n]c_H^{(l_4)}[n]$$

where $E, F, G, H \in \{I, Q\}$. Note that $\mathbf{S_I}$ and $\mathbf{D_I}$ are uncorrelated and $\eta \ll 1$. So

$$\begin{aligned}
&\eta(\mathbf{S_I} + \mathbf{D_I})^2 \\
&= \eta \mathbf{S_I}^2 + 2\mathbf{S_I}\mathbf{D_I} + \mathbf{D_I}^2 \\
&\approx \alpha_1^2 (\Upsilon_1 + \Upsilon_2) + 2\alpha_1\alpha_3 (\Upsilon_3 + \Upsilon_4) + \alpha_3^3 \eta \Psi
\end{aligned}$$

where

$$\begin{aligned}
\Upsilon_1 &= \eta \left(\sum_{k=1}^{K} C(I_1, I_k) b_I^{(k)}[0] \cos\theta^{(k)} \right)^2 \\
\Upsilon_2 &= \eta \left(\sum_{k=1}^{K} C(I_1, Q_k) b_Q^{(k)}[0] \sin\theta^{(k)} \right)^2 \\
\Upsilon_3 &= \eta \Psi \sum_{k=1}^{K} C(I_1, I_k) b_I^{(k)}[0] \cos\theta^{(k)} \\
\Upsilon_4 &= \eta \Psi \sum_{k=1}^{K} C(I_1, Q_k) b_Q^{(k)}[0] \sin\theta^{(k)}
\end{aligned}$$

For $1 \le k_0, k_1, k_2, k_3 \le K$

$$\begin{aligned}
\eta\cos\theta^{(k_0)}\cos\theta^{(k_1)}\cos\theta^{(k_2)}\sin\theta^{(k_3)} &= \eta\cos\theta^{(k_0)}\cos\theta^{(k_1)}\sin\theta^{(k_2)}\cos\theta^{(k_3)} = 0 \\
\eta\cos\theta^{(k_0)}\sin\theta^{(k_1)}\cos\theta^{(k_2)}\cos\theta^{(k_3)} &= \eta\cos\theta^{(k_0)}\sin\theta^{(k_1)}\sin\theta^{(k_2)}\sin\theta^{(k_3)} = 0 \\
\eta\sin\theta^{(k_0)}\cos\theta^{(k_1)}\cos\theta^{(k_2)}\cos\theta^{(k_3)} &= \eta\sin\theta^{(k_0)}\cos\theta^{(k_1)}\sin\theta^{(k_2)}\sin\theta^{(k_3)} = 0 \\
\eta\sin\theta^{(k_0)}\sin\theta^{(k_1)}\cos\theta^{(k_2)}\sin\theta^{(k_3)} &= \eta\sin\theta^{(k_0)}\sin\theta^{(k_1)}\sin\theta^{(k_2)}\cos\theta^{(k_3)} = 0
\end{aligned}$$

After some computations, we can get

$$\begin{aligned}
\Upsilon_1 &= \eta \left(\sum_{k=1}^{K} C(I_1, I_k) b_I^{(k)}[0] \cos\theta^{(k)} \right)^2 = d_I^2 + \frac{1}{2} \sum_{k=2}^{K} C(I_1, I_k)^2 b_I^{(k)}[0]^2 \\
\Upsilon_2 &= \eta \left(\sum_{k=1}^{K} C(I_1, Q_k) b_Q^{(k)}[0] \sin\theta^{(k)} \right)^2 = \frac{1}{2} \sum_{k=2}^{K} C(I_1, Q_k)^2 b_Q^{(k)}[0]^2 \\
&\eta \left(\sum_{k=1}^{K} C(I_1, I_k) b_I^{(k)}[0] \cos\theta^{(k)} \right) \\
&\quad \cdot \left(\sum_{k_1=1}^{K} \sum_{k_2=1}^{K} \sum_{k_3=1}^{K} a_7(k_1, k_2, k_3) \cos\theta^{(k_1)} \cos\theta^{(k_2)} \cos\theta^{(k_3)} \right)
\end{aligned}$$

$$= d_I^4 + d_I^2 d_Q^2 + \sum_{k=2}^{K} \left\{ \frac{3}{2} d_I^2 b_I^{(k)}[0]^2 + \frac{3}{2} C(I_1, I_k)^2 d_I^2 b_I^{(k)}[0]^2 + \frac{1}{2} d_I^2 b_Q^{(k)}[0]^2 \right.$$

$$+ C(I_1, I_k, Q_1, Q_k) d_I d_Q b_I^{(k)}[0] b_Q^{(k)}[0] + C(I_1, I_k) C(Q_1, Q_k) d_I d_Q b_I^{(k)}[0] b_Q^{(k)}[0]$$

$$\left. + \frac{1}{2} C(I_1, I_k)^2 d_Q^2 b_I^{(k)}[0]^2 + \frac{3}{8} C(I_1, I_k)^2 b_I^{(k)}[0]^4 + \frac{3}{8} C(I_1, I_k)^2 b_I^{(k)}[0]^2 b_Q^{(k)}[0]^2 \right\}$$

$$+ \sum_{\substack{2 \le k,l \le K \\ k \ne l}} \left\{ \frac{3}{4} C(I_1, I_k)^2 b_I^{(k)}[0]^2 b_I^{(l)}[0]^2 + \frac{1}{4} C(I_1, I_k)^2 b_I^{(k)}[0]^2 b_Q^{(l)}[0]^2 \right.$$

$$\left. + \frac{1}{2} C(I_1, I_k) C(I_1, I_l, Q_k, Q_l) b_I^{(k)}[0] b_I^{(l)}[0] b_Q^{(k)}[0] b_Q^{(l)}[0] \right\}$$

$$\eta \left(\sum_{k=1}^{K} C(I_1, I_k) b_I^{(k)}[0] \cos\theta^{(k)} \right)$$

$$\cdot \left(\sum_{k_1=1}^{K} \sum_{k_2=1}^{K} \sum_{k_3=1}^{K} a_4(k_1, k_2, k_3) \cos\theta^{(k_1)} \sin\theta^{(k_2)} \sin\theta^{(k_3)} \right)$$

$$= \sum_{k=2}^{K} \left\{ \frac{1}{2} d_I^2 b_I^{(k)}[0]^2 + \frac{1}{2} d_I^2 b_Q^{(k)}[0]^2 + \frac{1}{8} C(I_1, I_k)^2 b_I^{(k)}[0]^2 b_Q^{(k)}[0]^2 \right.$$

$$\left. + \frac{1}{8} C(I_1, I_k)^2 b_I^{(k)}[0]^4 \right\}$$

$$+ \sum_{\substack{2 \le k,l \le K \\ k \ne l}} \left\{ \frac{1}{4} C(I_1, I_k)^2 b_I^{(k)}[0]^2 b_I^{(l)}[0]^2 + \frac{1}{4} C(I_1, I_k)^2 b_I^{(k)}[0]^2 b_Q^{(l)}[0]^2 \right\}$$

$$\eta \left(\sum_{k=1}^{K} C(I_1, I_k) b_I^{(k)}[0] \cos\theta^{(k)} \right)$$

$$\cdot \left(\sum_{k_1=1}^{K} \sum_{k_2=1}^{K} \sum_{k_3=1}^{K} a_2(k_1, k_2, k_3) \sin\theta^{(k_1)} \cos\theta^{(k_2)} \sin\theta^{(k_3)} \right)$$

$$= \sum_{k=2}^{K} \left\{ \frac{1}{2} d_I^2 b_Q^{(k)}[0]^2 - \frac{1}{2} C(I_1, I_k, Q_1, Q_k) d_I d_Q b_I^{(k)}[0] b_Q^{(k)}[0] \right\}$$

$$+ \sum_{\substack{2 \le k,l \le K \\ k \ne l}} \left\{ \frac{1}{4} C(I_1, I_k)^2 b_I^{(k)}[0]^2 b_Q^{(l)}[0]^2 \right.$$

$$\left. - \frac{1}{4} C(I_1, I_k) C(I_1, I_l, Q_k, Q_l) b_I^{(k)}[0] b_I^{(l)}[0] b_Q^{(k)}[0] b_Q^{(l)}[0] \right\}$$

$$\eta \left(\sum_{k=1}^{K} C(I_1, I_k) b_I^{(k)}[0] \cos\theta^{(k)} \right)$$

$$\cdot \left(\sum_{k_1=1}^{K} \sum_{k_2=1}^{K} \sum_{k_3=1}^{K} a_1(k_1, k_2, k_3) \sin\theta^{(k_1)} \sin\theta^{(k_2)} \cos\theta^{(k_3)} \right)$$

$$= \sum_{k=2}^{K} \left\{ \frac{1}{2} d_I^2 b_Q^{(k)}[0]^2 - \frac{1}{2} C(I_1, I_k, Q_1, Q_k) d_I d_Q b_I^{(k)}[0] b_Q^{(k)}[0] \right\}$$

$$+ \sum_{\substack{2 \le k,l \le K \\ k \ne l}} \left\{ \frac{1}{4} C(I_1, I_k)^2 b_I^{(k)}[0]^2 b_Q^{(l)}[0]^2 \right.$$

$$\left. - \frac{1}{4} C(I_1, I_k) C(I_1, I_l, Q_k, Q_l) b_I^{(k)}[0] b_I^{(l)}[0] b_Q^{(k)}[0] b_Q^{(l)}[0] \right\}$$

$$\eta \left(\sum_{k=1}^{K} C(I_1, Q_k) b_Q^{(k)}[0] \sin\theta^{(k)} \right)$$

$$\cdot \left(\sum_{k_1=1}^{K} \sum_{k_2=1}^{K} \sum_{k_3=1}^{K} a_6(k_1, k_2, k_3) \cos\theta^{(k_1)} \cos\theta^{(k_2)} \sin\theta^{(k_3)} \right)$$

$$= \sum_{k=2}^{K} \left\{ \frac{1}{2} C(I_1, Q_k)^2 d_I^2 b_Q^{(k)}[0]^2 - \frac{1}{2} C(I_1, Q_k) C(Q_1, I_k) d_I d_Q b_I^{(k)}[0] b_Q^{(k)}[0] \right\}$$

$$+ \sum_{\substack{2 \le k,l \le K \\ k \ne l}} \left\{ \frac{1}{4} C(I_1, Q_k)^2 b_I^{(l)}[0]^2 b_Q^{(k)}[0]^2 \right.$$

$$\left. - \frac{1}{4} C(I_1, Q_k) C(I_1, I_k, I_l, Q_l) b_I^{(k)}[0] b_I^{(l)}[0] b_Q^{(k)}[0] b_Q^{(l)}[0] \right\}$$

$$\eta \left(\sum_{k=1}^{K} C(I_1, Q_k) b_Q^{(k)}[0] \sin\theta^{(k)} \right)$$

$$\cdot \left(\sum_{k_1=1}^{K} \sum_{k_2=1}^{K} \sum_{k_3=1}^{K} a_5(k_1, k_2, k_3) \cos\theta^{(k_1)} \sin\theta^{(k_2)} \cos\theta^{(k_3)} \right)$$

$$= \sum_{k=2}^{K} \left\{ \frac{1}{2} C(I_1, Q_k)^2 d_I^2 b_Q^{(k)}[0]^2 - \frac{1}{2} C(I_1, Q_k) C(Q_1, I_k) d_I d_Q b_I^{(k)}[0] b_Q^{(k)}[0] \right\}$$

$$+ \sum_{\substack{2 \le k,l \le K \\ k \ne l}} \left\{ \frac{1}{4} C(I_1, Q_k)^2 b_I^{(l)}[0]^2 b_Q^{(k)}[0]^2 \right.$$

$$\left. - \frac{1}{4} C(I_1, Q_k) C(I_1, I_k, I_l, Q_l) b_I^{(k)}[0] b_I^{(l)}[0] b_Q^{(k)}[0] b_Q^{(l)}[0] \right\}$$

$$\eta \left(\sum_{k=1}^{K} C(I_1, Q_k) b_Q^{(k)}[0] \sin\theta^{(k)} \right)$$

$$\cdot \left(\sum_{k_1=1}^{K} \sum_{k_2=1}^{K} \sum_{k_3=1}^{K} a_3(k_1, k_2, k_3) \sin\theta^{(k_1)} \cos\theta^{(k_2)} \cos\theta^{(k_3)} \right)$$

$$= \sum_{k=2}^{K} \left\{ \frac{1}{2} C(I_1, Q_k)^2 d_I^2 b_Q^{(k)}[0]^2 + \frac{1}{2} C(I_1, Q_k)^2 d_Q^2 b_Q^{(k)}[0]^2 + \right.$$

$$+\frac{1}{8}C(I_1,Q_k)^2 b_I^{(k)}[0]^2 b_Q^{(k)}[0]^2 \frac{1}{8}C(I_1,Q_k)^2 b_Q^{(k)}[0]^4\Bigg\}$$

$$+\sum_{\substack{2\le k,l\le K\\ k\neq l}}\left\{\frac{1}{4}C(I_1,Q_k)^2 b_I^{(l)}[0]^2 b_Q^{(k)}[0]^2+\frac{1}{4}C(I_1,Q_k)^2 b_Q^{(k)}[0]^2 b_Q^{(l)}[0]^2\right\}$$

$$\eta\left(\sum_{k=1}^{K}C(I_1,Q_k)b_Q^{(k)}[0]\sin\theta^{(k)}\right)$$

$$\cdot\left(\sum_{k_1=1}^{K}\sum_{k_2=1}^{K}\sum_{k_3=1}^{K}a_0(k_1,k_2,k_3)\sin\theta^{(k_1)}\cos\theta^{(k_2)}\cos\theta^{(k_3)}\right)$$

$$=\sum_{k=2}^{K}\left\{\frac{3}{8}C(I_1,Q_k)^2 b_Q^{(k)}[0]^4+\frac{3}{8}C(I_1,Q_k)^2 b_I^{(k)}[0]^2 b_Q^{(k)}[0]^2\right\}$$

$$+\sum_{\substack{2\le k,l\le K\\ k\neq l}}\Bigg\{\frac{3}{4}C(I_1,Q_k)^2 b_Q^{(k)}[0]^2 b_Q^{(l)}[0]^2+\frac{1}{4}C(I_1,Q_k)^2 b_I^{(l)}[0]^2 b_Q^{(k)}[0]^2$$

$$+\frac{1}{2}C(I_1,Q_k)C(I_1,I_k,I_l,Q_l)b_I^{(k)}[0]b_I^{(l)}[0]b_Q^{(k)}[0]b_Q^{(l)}[0]\Bigg\}$$

Therefore,

$$\Upsilon_3=\eta\Psi\sum_{k=1}^{K}C(I_1,I_k)b_I^{(k)}[0]\cos\theta^{(k)}$$

$$=d_I^4+d_I^2d_Q^2+\sum_{k=2}^{K}\Bigg\{2d_I^2b_I^{(k)}[0]^2+2d_I^2b_Q^{(k)}[0]^2+\frac{3}{2}C(I_1,I_k)^2d_I^2b_I^{(k)}[0]^2$$

$$+C(I_1,I_k)C(Q_1,Q_k)d_Id_Qb_I^{(k)}[0]b_Q^{(k)}[0]$$

$$+\frac{1}{2}C(I_1,I_k)^2d_Q^2b_I^{(k)}[0]^2+\frac{1}{2}C(I_1,I_k)^2b_I^{(k)}[0]^4+\frac{1}{2}C(I_1,I_k)^2b_I^{(k)}[0]^2b_Q^{(k)}[0]^2\Bigg\}$$

$$+\sum_{\substack{2\le k,l\le K\\ k\neq l}}\left\{C(I_1,I_k)^2b_I^{(k)}[0]^2b_I^{(l)}[0]^2+C(I_1,I_k)^2b_I^{(k)}[0]^2b_Q^{(l)}[0]^2\right\}$$

$$\Upsilon_4=\eta\Psi\sum_{k=1}^{K}C(I_1,Q_k)b_Q^{(k)}[0]\sin\theta^{(k)}$$

$$=\sum_{k=2}^{K}\Bigg\{\frac{3}{2}C(I_1,Q_k)^2d_I^2b_Q^{(k)}[0]^2-C(I_1,Q_k)C(Q_1,I_k)d_Id_Qb_I^{(k)}[0]b_Q^{(k)}[0]$$

$$+\frac{1}{2}C(I_1,Q_k)^2d_Q^2b_Q^{(k)}[0]^2+\frac{1}{2}C(I_1,Q_k)^2b_I^{(k)}[0]^2b_Q^{(k)}[0]^2+\frac{1}{2}C(I_1,Q_k)^2b_Q^{(k)}[0]^4\Bigg\}$$

$$+\sum_{\substack{2\le k,l\le K\\ k\neq l}}\left\{C(I_1,Q_k)^2b_I^{(l)}[0]^2b_Q^{(k)}[0]^2+C(I_1,Q_k)^2b_Q^{(k)}[0]^2b_Q^{(l)}[0]^2\right\}$$

Following the same method, all the α_3^2 terms and cross-corrleation terms can be computed similarly.

9A.3 General Results

By computing the first and second moments of $\mathbf{S_I} + \mathbf{D_I}$, we can get the following results:

$$\begin{aligned}
Y_I &= \alpha_1 d_I + \alpha_3 d_I^3 + \alpha_3 d_I d_Q^2 + \eta\alpha_1 d_Q^3 + \eta\alpha_1 d_I^2 d_Q + \eta\alpha_3 d_Q^5 \\
&\quad + 2\eta\alpha_3 d_I^2 d_Q^3 + \eta\alpha_3 d_I^4 d_Q \\
U_I &= \left(2\alpha_3 d_I + 2\eta\alpha_1 d_Q + 6\eta\alpha_3 d_Q^3 + 6\eta\alpha_3 d_I^2 d_Q\right) \sum (\zeta 20k + \zeta 02k) \\
&\quad + 3\alpha_3 d_Q \sum (\zeta 40k + 2\zeta 22k + \zeta 04k) \\
&\quad + 12\eta\alpha_3 d_Q \sum_{2 \le k < l \le K} (\zeta 20k + \zeta 02k)(\zeta 20l + \zeta 02l) \\
\sigma_I^2 &= \alpha_1^2 (\Upsilon_1 + \Upsilon_2) + 2\alpha_1\alpha_3 (\Upsilon_3 + \Upsilon_4) + \alpha_3^2 \Upsilon_5 - S_I^2 + \sigma_N^2 \\
\sigma_{IQ}^2 &= 2\alpha_1\alpha_3 d_I d_Q \sum R_{1k}^2 (\zeta 20k + \zeta 02k)
\end{aligned}$$

where ζijk, R_{ij}, and Υ_1 to Υ_5 are defined as follows.

$$\begin{aligned}
\zeta ijk &= \frac{\mu_k}{M} \sum_{m=1}^{M} \cos^i \frac{(2m-1)\pi}{M} \sin^j \frac{(2m-1)\pi}{M} (2p)^{i+j} \\
S_I &= \alpha_1 d_I + \alpha_3 d_I^3 + \alpha_3 d_I d_Q^2 + 2\alpha_3 d_I \sum (\zeta 20k + \zeta 02k) \\
\Upsilon_1 &= d_I^2 + \frac{1}{2} \sum R_{1k}^2 \zeta 20k \\
\Upsilon_2 &= \frac{1}{2} \sum R_{1k}^2 \zeta 02k \\
\Upsilon_3 &= d_I^4 + d_I^2 d_Q^2 + \sum \left(2d_I^2 \zeta 20k + 2d_I^2 \zeta 02k + \frac{3}{2} R_{1k}^2 d_I^2 \zeta 20k + \frac{1}{2} R_{1k}^2 d_Q^2 \zeta 20k \right. \\
&\quad \left. + \frac{1}{2} R_{1k}^2 \zeta 40k + \frac{1}{2} R_{1k}^2 \zeta 22k \right) + \sum_{\substack{2 \le k,l \le K \\ k \ne l}} R_{1k}^2 \zeta 20k (\zeta 20l + \zeta 02l) \\
\Upsilon_4 &= \sum R_{1k}^2 \left(\frac{3}{2} d_I^2 \zeta 02k + \frac{1}{2} d_Q^2 \zeta 02k + \frac{1}{2} \zeta 22k + \frac{1}{2} \zeta 04k \right) \\
&\quad + \sum_{\substack{2 \le k,l \le K \\ k \ne l}} R_{1k}^2 \zeta 02l (\zeta 20k + \zeta 02k) \\
\Upsilon_5 &= \Upsilon_{0_1} + \Upsilon_{1_1} + \Upsilon_{1_2} + \Upsilon_{1_3} + \Upsilon_{1_4} + \Upsilon_{2_1_1} + \Upsilon_{2_1_2} \\
&\quad + \Upsilon_{2_2_1} + \Upsilon_{2_2_2} + \Upsilon_{2_3} + \Upsilon_{3_1} + \Upsilon_{3_2} + \Upsilon_{3_3} \\
\Upsilon_{0_1} &= d_I^6 + 2d_I^4 d_Q^2 + d_I^2 d_Q^4
\end{aligned}$$

$$\Upsilon_{1\_1} = \sum \left\{ \frac{1}{2}\left(9d_I^2 + d_Q^2\right)(\zeta 40k + \zeta 04k) + \frac{1}{2}\left(9d_I^4 + d_Q^4\right) R_{1k}^2 (\zeta 20k + \zeta 022k) \right.$$
$$\left. + \frac{1}{2}(\zeta 60k + \zeta 06k + \zeta 24k + \zeta 42k) R_{1k}^2 + 7d_I^2 \zeta 22k \right\}$$

$$\Upsilon_{1\_2} = \sum \left\{ 4\left(d_I^4 + d_I^2 d_Q^2\right)(\zeta 20k + \zeta 02k) + 2d_I^2 \zeta 22k \right.$$
$$\left. + 5d_I^2 d_Q^2 R_{1k}^2 (\zeta 20k + \zeta 02k) + 3d_I^2 R_{1k}^2 (\zeta 40k + 2\zeta 22k + \zeta 04k) \right\}$$

$$\Upsilon_{1\_3} = \sum \left\{ -d_Q^2 \zeta 22k + 2d_Q^2 \zeta 22k \left(R_{1k}^2 + 1\right) + d_Q^2 R_{1k}^2 (\zeta 40k + \zeta 04k) \right\}$$

$$\Upsilon_{1\_4} = \sum R_{1k}^2 (\zeta 42k + \zeta 24k)$$

$$\Upsilon_{2\_1\_1} = \sum_{\substack{2 \le k,l \le K \\ k \ne l}} \left\{ \frac{5}{2} R_{1k}^2 (\zeta 20k + \zeta 02k)(\zeta 40l + \zeta 04l) + d_I^2 \zeta 20k \zeta 02l \left(8 + 10R_{kl}^2\right) \right.$$
$$\left. + 6R_{1k}^2 d_I^2 (\zeta 20k \zeta 20l + \zeta 20k \zeta 02l + \zeta 20l \zeta 02k + \zeta 02k \zeta 02l) \right\}$$

$$\Upsilon_{2\_1\_2} = \sum_{2 \le k < l \le K} d_I^2 (8 + 10R_{kl}^2)(\zeta 20k \zeta 20l + \zeta 02k \zeta 02l)$$

$$\Upsilon_{2\_2\_1} = 2 \sum_{\substack{2 \le k,l \le K \\ k \ne l}} \left\{ d_Q^2 R_{1k}^2 (\zeta 20k + \zeta 02k)(\zeta 20l + \zeta 02l) + d_Q^2 R_{kl}^2 \zeta 20k \zeta 02l \right.$$
$$+ R_{1k}^2 (\zeta 40k + \zeta 04k)(\zeta 20l + \zeta 02l)$$
$$\left. + 2R_{1k}^2 \zeta 22k (\zeta 20l + \zeta 02l) + 2R_{1l}^2 \zeta 22k \zeta 02l \right\}$$

$$\Upsilon_{2\_2\_2} = 2d_Q^2 \sum_{2 \le k < l \le K} (\zeta 02k + \zeta 20k)\, \zeta 02l R_{kl}^2$$

$$\Upsilon_{2\_3} = 3 \sum_{\substack{2 \le k,l \le K \\ k \ne l}} (\zeta 20k + \zeta 02k) \zeta 22l R_{1k}^2$$

$$\Upsilon_{3\_1} = 6 \sum_{2 \le k < l < m \le K} (\zeta 20k \zeta 20l \zeta 20m + \zeta 02k \zeta 02l \zeta 02m)\, R_{1klm}^2$$

$$\Upsilon_{3\_2} = \sum \sum_{\substack{2 \le l < m \le K \\ l \ne k, m \ne k}} \left\{ 6\,(\zeta 20k \zeta 20l \zeta 20m + \zeta 02k \zeta 20l \zeta 20m)\, R_{1klm}^2 \right.$$
$$+ 4\,(\zeta 20k \zeta 02l \zeta 02m + \zeta 20k \zeta 20l \zeta 20m$$
$$\left. + \zeta 02k \zeta 20l \zeta 20m + \zeta 02k \zeta 02l \zeta 02m)\, R_{1k}^2 \right\}$$

$$\Upsilon_{3\_3} = 4 \sum_{\substack{2 \le k,l,m \le K \\ l \ne k, m \ne k}} (\zeta 20k + \zeta 02k)\, \zeta 20l \zeta 02m R_{1k}^2$$

$$R_{ij} = \frac{1}{N} \sum_{n=0}^{N-1} c^{(i)}[n] c^{(j)}[n], \quad i \ne j$$

$$R_{ijkl} = \frac{1}{N} \sum_{n=0}^{N-1} c^{(i)}[n] c^{(j)}[n] c^{(k)}[n] c^{(l)}[n], \quad i \ne j \ne k \ne l$$

9A.4 SER Evaluation for MPSK CDMA

Let the outputs of the two branches of the first correlator be

$$\mathbf{Z_I} = \mathbf{D_I} + \mathbf{U_I} + \mathbf{N_I}$$
$$\mathbf{Z_Q} = \mathbf{D_Q} + \mathbf{U_Q} + \mathbf{N_Q}$$

Let $Z_I = \eta \mathbf{Z_I}$ and $Z_Q = \eta \mathbf{Z_Q}$. The three terms represent the desired signal, other-user interference plus crosstalk, and AWGN. Then for the joint distribution of (z_I, z_Q) we have

$$f(z_I, z_Q) = \frac{1}{2\pi\sigma_I\sigma_Q\sqrt{1-\rho_{I,Q}^2}} \exp\left[-\frac{(z_I - Z_I)^2 + (z_Q - Z_Q)^2 - 2\rho_{I,Q}(z_I - Z_I)(z_Q - Z_Q)}{2(1-\rho_{I,Q}^2)\sigma_I\sigma_Q}\right]$$

The random variables $\mathbf{Z_I}$ and $\mathbf{Z_Q}$ are correlated; we denote their correlation coefficient by

$$f(z_I, z_Q)dz_I dz_Q = \hat{f}(\rho, \phi)\rho d\rho d\phi$$

Joint distribution in polar coordinates can be shown as

$$\begin{aligned}\hat{f}(\rho,\phi) &= \frac{1}{2\pi\sigma_I\sigma_Q\sqrt{1-\rho_{I,Q}^2}}\exp[X_1] = \frac{1}{2\pi\sigma_I\sigma_Q\sqrt{1-\rho_{I,Q}^2}}\exp[X_2] \\ &= \frac{1}{2\pi\sigma_I\sigma_Q\sqrt{1-\rho_{I,Q}^2}}\exp\left\{-\frac{[1-\rho_{I,Q}\sin(2\phi)]\left[\rho-\frac{\cos(\phi-\phi_m)-\rho_{I,Q}\sin(\phi+\phi_m)}{1-\rho_{I,Q}\sin(2\phi)}\right]^2}{2(1-\rho_{I,Q}^2)\sigma_I\sigma_Q}\right\} \\ &\quad\cdot\exp\left\{-\frac{[1-\rho_{I,Q}\sin(2\phi_m)]-\frac{[\cos(\phi-\phi_m)-\rho_{I,Q}\sin(\phi+\phi_m)]^2}{1-\rho_{I,Q}\sin(2\phi)}}{2(1-\rho_{I,Q}^2)\sigma_I\sigma_Q}\right\}\end{aligned}$$

where the exponents X_1 and X_2 are given by

$$X_1 = -\frac{(\rho\cos\phi-\cos\phi_m)^2+(\rho\sin\phi-\sin\phi_m)^2-2\rho_{I,Q}(\rho\cos\phi-\cos\phi_m)(\rho\sin\phi-\sin\phi_m)}{2(1-\rho_{I,Q}^2)\sigma_I\sigma_Q}$$

$$X_2 = -\frac{[1-\rho_{I,Q}\sin(2\phi)]\rho^2-2[\cos(\phi-\phi_m)-\rho_{I,Q}\sin(\phi+\phi_m)]\rho+[1-\rho_{I,Q}\sin(2\phi_m)]}{2(1-\rho_{I,Q}^2)\sigma_I\sigma_Q}$$

Then the probability of error given m is

$$P_{e|m} = 1 - \int_{\phi_m-\frac{\pi}{M}}^{\phi_m+\frac{\pi}{M}}\left[\int_0^\infty \hat{f}(\rho,\phi)\rho d\rho\right]d\phi$$

We first compute the integral with respect to ρ in closed form as

$$\frac{1}{2\pi\sigma_I\sigma_Q\sqrt{1-\rho_{I,Q}^2}}\int_0^\infty \exp\left\{-\frac{[1-\rho_{I,Q}\sin(2\phi)]\left[\rho-\frac{\cos(\phi-\phi_m)-\rho_{I,Q}\sin(\phi+\phi_m)}{1-\rho_{I,Q}\sin(2\phi)}\right]^2}{2\left(1-\rho_{I,Q}^2\right)\sigma_I\sigma_Q}\right\}\rho d\rho$$

$$= \frac{1}{\sqrt{2\pi\sigma_I\sigma_Q}\sqrt{1-\rho_{I,Q}\sin(2\phi)}}\frac{\cos(\phi-\phi_m)-\rho_{I,Q}\sin(\phi+\phi_m)}{1-\rho_{I,Q}\sin(2\phi)}$$

However, the integral with respect to ϕ cannot be put in closed form. Thus, the final result is

$$P_{e|m} = 1 - \int_{\phi_m-\frac{\pi}{M}}^{\phi_m+\frac{\pi}{M}} \bar{f}(\phi,\phi_m)d\phi$$

where

$$\bar{f}(\phi,\phi_m) = \frac{1}{\sqrt{2\pi\sigma_I\sigma_Q}\sqrt{1-\rho_{I,Q}\sin(2\phi)}}\frac{\cos(\phi-\phi_m)-\rho_{I,Q}\sin(\phi+\phi_m)}{1-\rho_{I,Q}\sin(2\phi)}\cdot\exp\{X\}$$

In the above expression the exponent X is given by

$$X = -\frac{[1-\rho_{I,Q}\sin(2\phi_m)] - \frac{[\cos(\phi-\phi_m)-\rho_{I,Q}\sin(\phi+\phi_m)]^2}{1-\rho_{I,Q}\sin(2\phi)}}{2(1-\rho_{I,Q}^2)\sigma_I\sigma_Q}$$

where $\phi_m = (2m-1)\pi/M$ for $m = 1, 2, \ldots, M$. For equi-probable M-ary PSK symbols, the probability of error is thus

$$P_e = \frac{1}{M}\sum_{m=1}^{M} P_{e|m}$$

9A.5 Input Power, Output Power and Input Back-Off

Suppose all users have the same transmitting power, i.e. $X_k \in \left\{\pm\sqrt{2P}\right\}$ for all k.

A.5.1 Input Power

BPSK:

$$Y = \sum_{k=1}^{K} X_k$$

where K is the number of users. The average power at the input of a nonlinear amplifier is thus

$$\eta\frac{1}{T_s}\int_0^{T_s}(Y\cos(\omega_c t))^2\,dt = \frac{\eta Y^2}{2} = KP$$

M-PSK ($M > 2$):

$$Y_I = \sum_{k=1}^{K} X_{k,I}$$

$$Y_Q = \sum_{k=1}^{K} X_{k,Q}$$

where K is the number of users. $X_{k,I} \in \left\{\sqrt{2P}\cos\frac{2m-1}{M}\pi, m = 1, .., M\right\}$, $X_{k,Q} \in \left\{\sqrt{2P}\sin\frac{2m-1}{M}\pi, m = 1, .., M\right\}$. Then the average power at the input of the amplifier is

$$\begin{aligned}
&\eta\frac{1}{T_s}\int_0^{T_s}\left(Y_I\cos(\omega_c t) + Y_Q\sin(\omega_c t)\right)^2 dt \\
&= \frac{1}{T_s}\left(\frac{T_s}{2}\eta Y_I^2 + \frac{T_s}{2}\eta Y_Q^2\right) \\
&= KP
\end{aligned}$$

Therefore, the average power for both BPSK and M-PSK ($M > 2$) is always KP.

9A.5.2 Output Power

BPSK: note that the variance of X_k is $2P$. Since K is large, according to the Central Limit Theorem, Y is Gaussian distributed with mean 0 and variance $2KP$. Let the input power of amplifier be $P_{\text{in}} = KP$, and suppose the output of amplifier is Z. Then $Z = \alpha_1 Y + \alpha_3 Y^3$ for $Z \in [-z_s, z_s]$, $Z = -z_s$ for $Z < -y_s$ and $Z = z_s$ for $Z > y_s$. The corresponding output power is thus

$$\begin{aligned}
P_{\text{out}} &= \eta\frac{1}{T_s}\int_0^{T_s}\left(Z\cos(\omega_c t)\right)^2 dt \\
&= \frac{1}{2}\eta Z^2 \\
&= \frac{1}{2}\left[z_s^2\int_{-\infty}^{-y_s} f_Y(y)dy + \int_{-y_s}^{y_s}\left(\alpha_1 Y + \alpha_3 Y^3\right)^2 f_Y(y)dy + z_s^2\int_{y_s}^{\infty} f_Y(y)dy\right] \\
&= z_s^2 Q\left(\frac{y_s}{\sqrt{2Kp}}\right) + \int_{-y_s}^{y_s}\left(\alpha_1 Y + \alpha_3 Y^3\right)^2 f_Y(y)dy
\end{aligned}$$

where

$$f_Y(y) = \frac{1}{\sqrt{4\pi Kp}}\exp\left(\frac{-y^2}{4Kp}\right)$$

$$Q(y) = \int_y^{\infty}\frac{1}{\sqrt{2\pi}}\exp\left(\frac{-y^2}{2}\right)dy$$

M-PSK ($M > 2$): again the variance of X_k is $2P$. Since K is large, according to the Central Limit Theorem, Y is Rayleigh distributed with parameter $\sigma^2 = KP = P_{\text{in}}$.

Suppose the output of TWT is Z. Then similarly, $Z = \alpha_1 Y + \alpha_3 Y^3$ for $Z \in [-z_s, z_s]$, $Z = -z_s$ for $Z < -y_s$ and $Z = z_s$ for $Z > y_s$. The corresponding output power is thus

$$\begin{aligned}
P_{\text{out}} &= \eta \frac{1}{T_s} \int_0^{T_s} (Z \cos(\omega_c t))^2 \, dt \\
&= \frac{1}{2} \eta Z^2 \\
&= \int_0^{y_s} (\alpha_1 Y + \alpha_3 Y^3)^2 \, g_Y(y) dy + z_s^2 \int_{y_s}^{\infty} g_Y(y) dy \\
&= z_s^2 \exp\left(\frac{-y_s^2}{2Kp}\right) + \int_0^{y_s} (\alpha_1 Y + \alpha_3 Y^3)^2 \, g_Y(y) dy
\end{aligned}$$

where

$$g_Y(y) = \begin{cases} \frac{y}{Kp} \exp\left(\frac{-y^2}{2Kp}\right) & 0 \le y \le \infty \\ 0 & y < 0 \end{cases}$$

The received power for each user is thus P_{out}/K. The amplitude of the received constellation is $\sqrt{2P_{\text{out}}/K}$

A.5.3 Input Back-Off (IBO)

If the input power back-off is l db from the saturation point of the TWT which corresponds to amplitude X_{sat}, then P can be written as

$$\begin{aligned}
20 \log \frac{\sqrt{2KP}}{X_{\text{sat}}} &= -l \\
\Rightarrow P &= \frac{X_{\text{sat}}^2}{2K} 10^{\frac{-l}{10}}
\end{aligned}$$

10

Optimization Techniques for 'Pseudo-Orthogonal' CDMA

10.1 Overview

The CDMA systems presented in the previous chapters were mainly based on the synchronous or orthogonal approach. As we have discussed, orthogonal CDMA achieves maximum capacity, but it requires synchronization of all transmitting users in a multipoint-to-point access network. Such a synchronization, however, may not always be possible in a high mobility environment. In such an enviroment, we use 'Pseudo-Orthogonal' (PO) CDMA. In the PO-CDMA, capacity (users/CDMA-band) is limited by interference resulting from the use of imperfectly orthogonal codes (PN-codes, see Chapter 2) to separate the users. Thus, power 'leakage' occurs between the signals of different users.

In this chapter we present two techniques which are used to optimize the performance or maximize the capacity of a PO-CDMA for terrestrial mobile or satellite networks in uplink transmission. These techniques are (1) adaptive power control, and (2) multi-user detection.

Power control is used to mitigate the 'near-far' problem which appears at the PO-CDMA receiver. That is, the power 'leakage' to the signal of 'far' user from the signal of a 'near' user may be so severe that reception by the far-user may not be possible. A power control mechanism adjusts the transmit power of each user so that the received signal power of each user is approximately the same. Such a power control mechanism is presented in Section 10.2.

Another, more advanced technique that a PO-CDMA receiver may use to optimize performance is interference cancelation or multi-user detection. In Section 10.3 we present a survey of multi-user detection methods that appears in the literature, and we propose a new one based on minimum mean square error estimation and iterative decoding.

10.2 Adaptive Power Control

Power control is vital in pseudo-orthogonal CDMA transmission. It compensates for the effects of 'path-loss' and reduces the Multiple Access Interference (MAI). The power control problem has been investigated extensively. The work given in this section is part of the work that appeared in reference [1]. Previous publications include centralized [2] and distributed [3], [4], power control methods. The distributed algorithms are simpler to implement and will be the focus of this section. Among them, some mainly deal with alleviating the 'path-loss' effects [4], while others deal with the convergence of transmit power level in a static environment [5]. In general, there are two kinds of power control mechanisms, open-loop and closed-loop, which are considered either separately or jointly [6], [7]. Open-loop power control provides an approximate level of the power required for the uplink (or reverse link) transmission based on an estimate of the downlink (or forward link) attenuation of the signal. The downlink transmission, however, may be in another frequency band (if frequency division douplexing is used) which may have different propagation characteristics. Closed-loop power control, on the other hand, uses the measured channel and interference information of the link under consideration to control the transmission power [3], [7]. Therefore, it is more efficient and suitable for any kind of environment, although its performance may be degraded by delays or bit-errors of the feedback channel.

As shown [4], since the power updating command is multiplicative and the path-loss gain is log-normally distributed, the power control error is also (approximately) log-normally distributed with mean target signal-to-interference-plus-noise ratio (SINR) (in dB). The fact that the received SINR cannot be perfectly controlled degrades the average Bit Error Rate (BER) performance. To overcome this situation, a certain power margin proportional to the amount of power control error has to be added in order to meet the BER requirement. For this reason, minimizing power control error is considered necessary in achieving power efficiency.

One practical constraint imposed on closed-loop power control schemes is the limited amount of feedback information. The criterion for a better design therefore aims at achieving the required BER performance with the lowest power consumption given the available feedback bandwidth. This is a classical quantization (of the feedback information) problem, with the cost function defined according to the power efficiency [8]. Given that the power control error is approximately log-normally distributed, the cost function can be deduced to the variance of this distribution. A Minimum Mean Squared Error (MMSE) quantization is therefore our best choice. To combat the mismatching problem between the quantizer and the time-varying error statistics (due to time-varying fading), a power control error measurement can be used to render the *quantizer adaptive.*

In addition to the above, we consider utilizing a loop filter at the transmitter. For one reason, the feedback information is distorted by the quantization and the noisy feedback channel, thus filtering helps in smoothing the feedback and reducing the fluctuation of the received SINR. For the other reason, we have already addressed that power control is never perfect. The power control error gets fed back to the transmitter and affects the next power update. It then can be shown inductively that the feedback (power control error) process will not be memoryless. When we consider quantization of the feedback information, the overload and granularity [9] effects make the time correlation even more evident. We thus conclude that inclusion of a feedback history in the control loop will enhance the power control performance. In other words, the one-tap implementation in references [4], [5] can be improved with higher order filtering. Note that loop filtering is in fact a generalization of the variable power control step size concept.

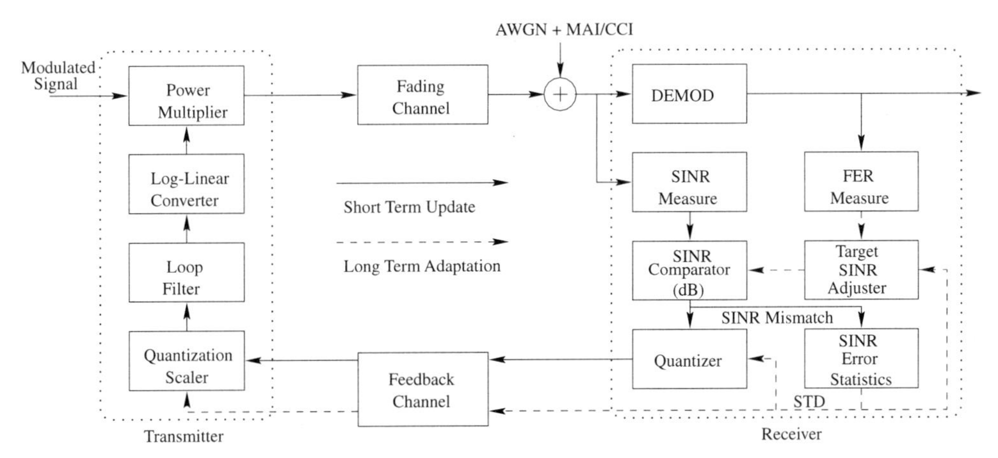

Figure 10.1 Closed-loop power control.

This section is organized as follows. In Section 10.2.2, we present a detail system description of the proposed design. Then we apply this design in a practical example of uplink CDMA transmission in Section 10.2.3. Then, a performance analysis, together with simulation results, are provided. Also, in this subsection we propose the idea of a self-optimizing loop filter.

10.2.1 Power Control System Design

A block diagram of the closed-loop power control system is depicted in Figure 10.1. Before getting into the details, let us adopt the notations from reference [4] and consider the simplified power control loop equation:

$$E(j+1) = E(j) - C\left[\hat{E}(j-k), k = M, M+1, \ldots\right] \\ - [L(j+1) - L(j)] + \delta_c(j+1) \quad \text{(dB)}$$

where $E(j)$ is the average received SINR (in dB) of the jth power updating period, and M is the total number of updating periods needed for the round trip propagation and processing. $C[\cdot]$ is the power multiplier function, depending on the previous SINR error feedbacks, which are derived by comparing the received SINR estimates $(\hat{E}(j-k), k = M, M+1, \ldots)$ with a predefined target. These feedbacks are quantized and subject to the feedback channel distortion. $L(j)$ is the fading loss averaged over the jth updating period, and is typically log-normally distributed.

The above equation differs from a similar one given in reference [4] in a correction term $\delta_c(j+1)$. This correction term is due to the change in the overall noise plus interference power. In the CDMA uplink environment where all users apply power control towards the (same) receiving station, this correction term is very small because of the near-constant interference power spectrum.

The equivalent loop model derived from the above equation is shown in Figure 10.2. Under normal (stable) operation the transmission power $T(j)$ is log-normally distributed (resulting from integration in the transmitter). The slow (shadowing) fading $L(j)$ is log-normally distributed, and the MAI can be approximated as log-

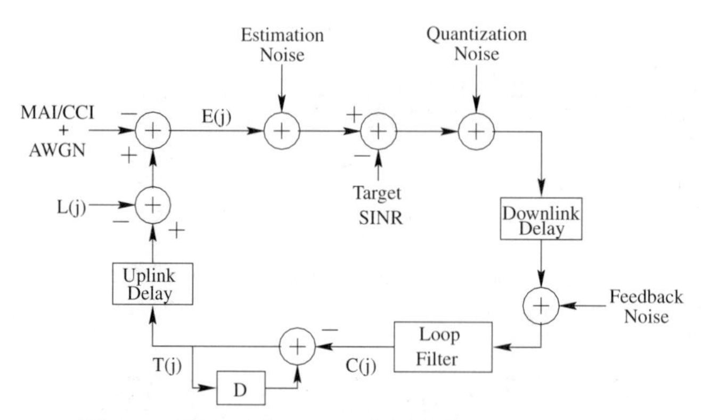

Figure 10.2 Linear model of the control loop.

normal. Given that the dominant interference is MAI, we may conclude that the received SINR $E(j)$ is approximately log-normally distributed.

The components of the entire loop design are shown in Figure 10.1. At the receiver, there are four major blocks pertaining to the power control loop: the SINR measurement, the SINR comparator, the quantizer, and the SINR error statistics producer.

- The SINR measurement block can be any SINR estimation circuitry. The accuracy of the measurements depends on the estimation algorithm; usually a higher accuracy can be obtained with higher computational complexity. The length (in terms of transmission symbols) of the measurement period and the rate of the fast (Rayleigh or Rician) fading also affect the accuracy. In practical situations, locally varying random processes such as the Additive White Gaussian Noise (AWGN) and the fast fading process will be taken care of by Forward Error Control (FEC) coding. The information which is important to the power control loop is the average SINR. Therefore, a longer measurement period and higher mobile speed (hence a higher fading rate) are advantageous for the measurement. However, if the measurement period is too long such that the slow fading process changes significantly during this period, the feedback information will become outdated. A trade-off between the measurement accuracy and feedback effectiveness thus emerges.
- The second block at the receiver is the SINR comparator. This block compares the measured SINR with the target SINR, defined jointly by the Frame Error Rate (FER) statistics and the SINR error statistics. As mentioned before, the SINR error statistic is approximately log-normally distributed. Given the standard deviation of the SINR error statistics, one will be able to estimate how much the target SINR should be shifted so that the BER requirement can be met. The target SINR adjustment is done once for a number of power updating periods.
- The SINR error is computed with high precision and fed into the quantizer and the SINR error statistics producer.
- At the quantizer, an MMSE quantization law (in dB) is used and the quantized SINR error information is sent to the transmitter in bits. The reason why we use an MMSE quantizer is due to the log-normal approximation of the SINR

error distribution. Since the Gaussian process is a second order statistic, we try to minimize the second moment of the SINR error. In this way, the target SINR can be set at the minimum, and the power consumption is reduced. We note that if the feedback channel is noisy, the quantization levels must be optimized with the feedback BER P_b considered [9]. The resulting quantizer will still be MMSE in a quantization/reconstruction sense.

- In order to avoid mismatch between the SINR error distribution and the quantizer, the standard deviation of the SINR error is provided to the quantizer by the SINR error statistics producer. The SINR error statistics producer averages a number of SINR error measurements and produces the standard deviation of the corresponding Gaussian process. This information is used in the target SINR adjustment as well as the quantization. Furthermore, it is sent to the transmitter to adjust the corresponding reconstruction scale. Since we only need to convey the second order statistics, and the adaptation of the system is done less frequently as compared to the power updates, this standard deviation is assumed to be stored with high precision and encoded with FEC. The error probability and the inaccuracy of this information will be ignored. When the fading statistics are slowly varying, this standard deviation can further be differentially encoded to save on feedback bandwidth.

At the transmitter side, there are three main components: the quantization scaler, the loop filter, and the power multiplier:

- The quantization scaler reconstructs the SINR error from the received feedback bits. There is a normalized reconstruction table built in the quantization scaler which is optimized with respect to the SINR error distribution (log-normal) and the feedback channel BER. Since the SINR error statistic is Gaussian in dB, the scale of the reconstruction levels depends only on the standard deviation passed from the receiver.
- The reconstructed SINR error is directed into the loop filter. This is where the history of the feedback gets exploited. The loop filter should be designed so as to maintain the stability of the loop. On the other hand, careful design of this filter can give a minimum power control error (the loop filter design issues will be addressed later). Although the feedback is quantized and has only a few levels, the output of the loop filter does not have this restriction. Computation inside the loop filter is done with a higher precision, as is the power multiplier. In practice, finer output power levels can be achieved with voltage controlled amplifiers. However, if the power level quantization is not fine enough, an additional quantization error should be considered. In this chapter the output of the loop filter as well as the power multiplier will be treated as continuous.

To conclude the system description, we provide some intuitive justifications for our design. The entire design is based on the fact that the received SINR is approximately log-normally distributed. With such a Gaussian distribution in dB, the power consumption and feedback quantization can be optimized with MMSE. The only parameter that needs to be passed around the system for reconfiguration is the second order statistics, therefore adaptation can be achieved with low additional overhead. Target SINR adjustment can also be estimated through this information. Lower power consumption and higher system capacity may thus be obtained.

At the transmitter side, a loop filter is applied to smooth the distorted feedback, enhance the system stability, and exploit the memory of the feedback. The way in which the quantization levels are set also helps in minimizing the steady state SINR variance given fixed feedback bandwidth. The rationale stems from the property of MMSE quantization that there are finer levels in the lower range of SINR error. In the scenario of noncooperative cochannel transmission, once the power vector is close to convergence, resolution of the quantization becomes better and the power vector fluctuation becomes less severe.

10.2.2 Uplink Power Control Performance

In the CDMA uplink scenario, assuming that the user population is large and all users are power controlled, the MAI plus AWGN power is approximately constant, with its strength depending on the number of users. Given a fixed SINR target, the resulting steady state loop model can be simplified from Figure 10.2 to Figure 10.3.

In this model, $\Delta L(j) = L(j)-L(j-1)$, $e(j)$ is the power control error, and ϵ_M, ϵ_Q, ϵ_F are the measurement error, quantization error, and feedback error, respectively. They are all randomly distributed. Among the latter three error terms, the measurement error depends on the channel estimation algorithm and the received SINR. The quantization error depends on $e(j)$ and its standard deviation σ_e. The feedback error is a function of both σ_e and the feedback channel BER P_b. The round trip loop delay is assumed to be M power updating steps, with $M \geq 1$, depending on the application. For example, M can be in the order from tens to hundreds in satellite communication, while it is usually 1 in terrestrial systems. In the loop filter block we consider a filtering function $F(z)$ which needs to be designed to achieve the smallest σ_e while maintaining the loop stability.

It is obvious that the mean of $e(j)$ is zero since all inputs have zero means. In order to derive the steady state standard deviation of $e(j)$, let us first consider the three error terms. In the steady state, the received SINR is distributed around the (fixed) target SINR, so ϵ_M can be treated as a stationary process with its variance depending only on the channel estimation algorithm. For simplicity, we assume that a simple

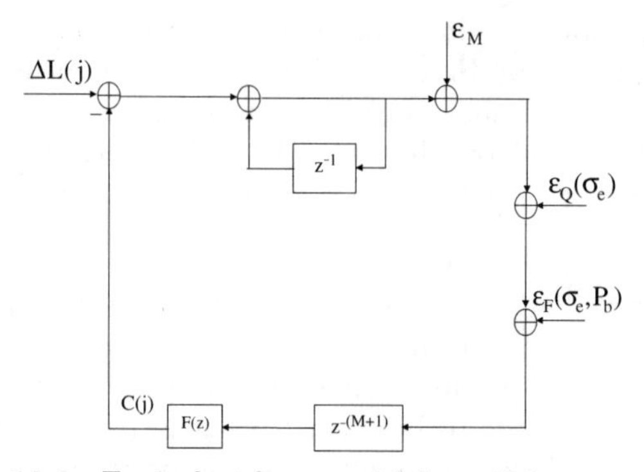

Figure 10.3 Equivalent loop model for uplink power control.

averaging algorithm is used. Since in this case the measurement error is dominated by AWGN, it is reasonable to assume that ϵ_M is independent identically distributed (i.i.d.) with constant variance σ_M^2. We further assume that the feedback BER P_b is fixed, and denote the normalized variances of the quantization error and the feedback error by σ_Q^2 and σ_F^2. These two errors are uncorrelated when a Max-Quantizer is used [9], which is the case we are considering. The variances of ϵ_Q and ϵ_F are then $\sigma_e^2\sigma_Q^2$ and $\sigma_e^2\sigma_F^2$, respectively. According to reference [9], the net result caused by these two errors can further be minimized if the feedback BER P_b is known. The advantage of this kind of re-optimization, however, is not significant when P_b is small ($< 10^{-2}$). Thus, it will not be considered here.

The values of σ_Q^2 can be easily found in a Max-Quantization table. ϵ_Q, however, is correlated with $e(j)$. The feedback error σ_F^2 depends on the feedback bit mapping, and is given by

$$\sum_{k=1}^{L}\sum_{j=1}^{L}(y_k - y_j)^2 P_{kj} P(x \in \mathcal{J}_k)$$

where y_k denotes the reconstruction level and $\mathcal{J}_k$ is the quantization input decision interval; both can be found in a Max-Quantization table. P_{kj} is the conditional probability that y_j will be received when y_k was sent. For memoryless feedback channels, we have

$$P_{kj} = P_b^{D_{kj}}(1 - P_b)^{R-D_{kj}}$$

where R is the number of bits per feedback, and D_{kj} is the Hamming distance between the R-bit codewords representing y_k and y_j. In these circumstances, ϵ_F is i.i.d.

The steady state power control error variance can be upper bounded by assuming i.i.d. ϵ_Q and independent ΔL and ϵ_Q:

$$\sigma_e^2 \leq \frac{1}{2\pi}\left(\sigma_{\Delta L}^2 \int_{-\pi}^{\pi}\left|\frac{S_{\Delta L}(e^{j\omega})}{(1 - e^{-j\omega})(1 + H(e^{j\omega}))}\right|^2 d\omega \right.$$
$$\left. + \left(\sigma_M^2 + \sigma_e^2\sigma_Q^2 + \sigma_e^2\sigma_F^2\right)\int_{-\pi}^{\pi}\left|\frac{H(e^{j\omega})}{1 + H(e^{j\omega})}\right|^2 d\omega\right)$$

where $S_{\Delta L}(e^{j\omega})$ is the normalized spectrum of ΔL and

$$H(e^{j\omega}) = \frac{e^{-j(M+1)\omega}F(e^{j\omega})}{1 - e^{-j\omega}}$$

is the loop gain. This inequality can be rearranged to approximate the steady state power control error variance

$$\sigma_e^2 \approx \frac{\sigma_{\Delta L}^2 \int_{-\pi}^{\pi}\left|\frac{S_{\Delta L}(e^{j\omega})}{(1 - e^{-j\omega})(1 + H(e^{j\omega}))}\right|^2 d\omega + \sigma_M^2 \int_{-\pi}^{\pi}\left|\frac{H(e^{j\omega})}{1 + H(e^{j\omega})}\right|^2 d\omega}{2\pi - \left(\sigma_Q^2 + \sigma_F^2\right)\int_{-\pi}^{\pi}\left|\frac{H(e^{j\omega})}{1+H(e^{j\omega})}\right|^2 d\omega}$$

and find the optimal $F(z)$ minimizing σ_e^2 when a certain filter form is given.

Loop stability is also a major concern. The characteristic function of this loop can be derived from the expression for $H(e^{j\omega})$

$$1 - z^{-1} + z^{-(M+1)}F(z)$$

which can be checked by using the Jury Stability Test [10].

To verify the analysis and illustrate the loop filter design issues, we consider a simple example where a first order loop filter $F(z) = a_0$ is used. Other parameters are: 2-bit power control error quantization; feedback BER $P_b = 10^{-3}$; and the round trip delay $M = 1$. The slow fading model is the same as in reference [4]. That is, the fading process in dB is a Gaussian independent-increment ($S_{\Delta L}(e^{j\omega}) = 1$), with the standard deviation of the increment equal to 1 dB. We assume that the SINR measurement is perfect, so $\sigma_M^2 = 0$. For this particular case, we have from expression of σ_e^2 above

$$\sigma_e^2 = \left(\frac{1}{\chi(a_0)} - a_0^2(\sigma_Q^2 + \sigma_F^2) \right)^{-1}$$

where

$$\chi(a_0) = \frac{1}{2\pi} \int_{-\pi}^{\pi} \left| \frac{1}{(1 - e^{-j\omega})(1 + H(e^{j\omega}))} \right|^2 d\omega$$

The condition of stability for this case is $0 < a_0 < 1$, therefore we plot the standard deviation of the power control error with respect to a_0 in this region in Figure 10.4. From Figure 10.3, it can be seen that σ_e^2 is convex on a_0 and there is a point with minimum σ_e^2. This result is not surprising, since σ_e^2 is infinite on the boundary of the stability region, while it is affected by at most the second order of a_0 within that region. The lowest power control error happens around $a_0 = 0.5$.

In the same figure we also depict the simulation result of the proposed design with its quantizer adaptation period equal to 20 power control iterations. The quantizer adaptation follows reference [9],

$$\beta(n) = \left(\gamma \cdot \beta^2(n-1) + (1-\gamma) \cdot \hat{\sigma}_e^2\right)^{\frac{1}{2}}$$

where β is the quantization/reconstruction scaler, γ is the learning coefficient, and $\hat{\sigma}_e^2$ is the power control error variance estimated via averaging in the $(n-1)^{th}$ interval. In order to reduce the adaptation excess error, we set $\gamma = 0.9$. The two curves in the plot basically follow the same trend except for a small discrepancy. This is due to our assumption of independence between ΔL and ϵ_Q in the analysis. When a_0 is small, the weight of the quantization error σ_Q^2 in the expression for σ_e^2 is small. So the two curves are very close to each other, with the simulation result being higher due to the adaptation excess error. As a_0 increases, the quantization error affects the performance more. The analytical result, as mentioned before, becomes an upper bound. It is also seen from the figure that the adaptive scheme somehow manages to maintain much lower power control error than the upper bound when a_0 is very close to one. The stability range of the adaptive scheme is therefore expected to be wider.

Simulation Results

The simulation results regarding different fading conditions with constant MAI are shown in Figures 10.5 and 10.6. The parameters for this simulation are: feedback BER

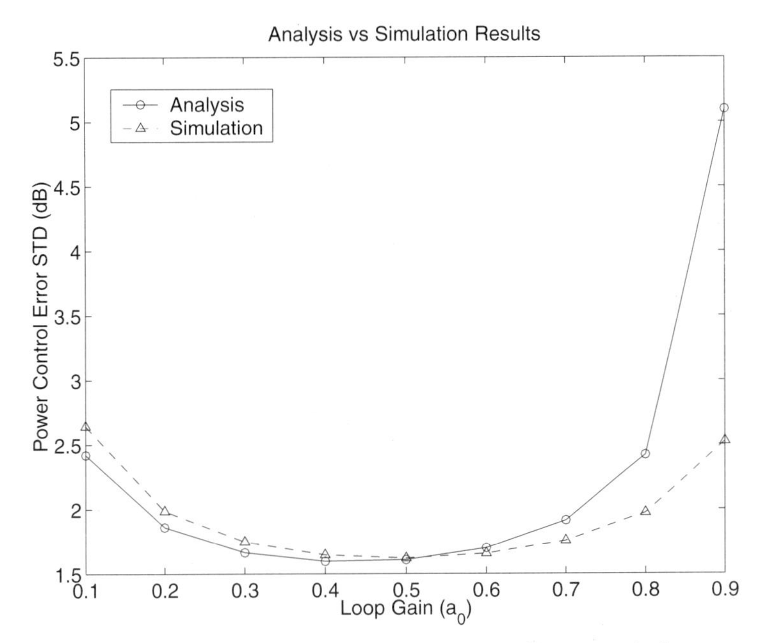

Figure 10.4 Optimization of the loop filter gain (a_0).

$P_b = 10^{-3}$; and round trip delay $M = 1$ (terrestrial). The quantization/reconstruction scale updates once per 20 power control iterations with its learning coefficient $\gamma = 0.8$. The fading process is a Gaussian independent-increment. The standard deviation of the increment ranges from 0.5 dB to 2.0 dB. The SINR measurement is assumed to be perfect. In order to have a common ground for performance comparison, the target SINR is fixed at 8 dB for every simulation. 50 000 power control iterations were simulated for each instance. In Figure 10.5 we first show the received SINR histograms of the proposed schemes. As shown in this figure, the log-normal approximation is quite accurate, therefore use of the MMSE criterion is justified. Through simulation we noticed that the log-normal approximation does not fit well for the fixed schemes when the mismatch between the quantization and fading parameters is large. For this reason, the performance will be compared in terms of the 1% received SINR ($SINR_{1\%}$). In Figure 10.6, the 1% SINR indicates the amount of power needed to shift the target SINR in order to meet the 1% outage probability requirement. In our example, if the demodulator/decoder imposes an SINR requirement $SINR_{req}$ for maintaining a certain BER, then the target SINR will have to be raised by ($SINR_{req} - SINR_{1\%}$) dB, which reflects an increase in the average transmission power (not necessarily ($SINR_{req} - SINR_{1\%}$) dB, since the averaging is done in the linear domain). We have tested five different schemes. For the case with one Power Control Bit (PCB) and fixed quantization, the quantization/reconstruction scaler was 1 while the loop filter gain was set so that each time the transmission power was adjusted ± 0.5 dB. The scheme with two PCBs and fixed quantization took the same quantization/reconstruction scaler and loop filter gain as its 1 PCB counterpart. For the adaptive schemes with constant loop filter (i.e. one tap), the loop filter gain $a_0 = 0.5$, as was determined in the previous optimization. An adaptive scheme with

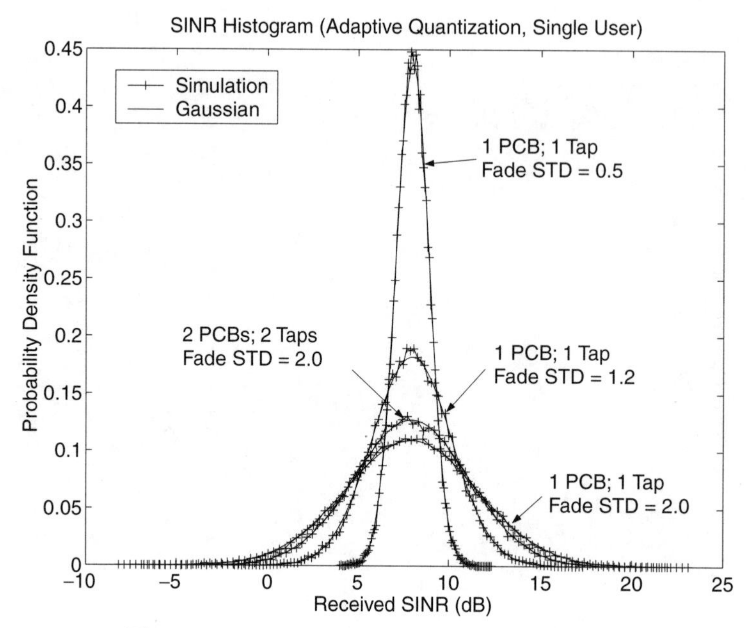

Figure 10.5 Histogram of the received SINR.

two PCBs and a 2-tap filter was also simulated. Its loop filter $F(z) = 0.78 - 0.39z^{-1}$ was obtained through two-dimensional optimization. In the simulations of the adaptive schemes, the quantization/reconstruction scaler was initialized to 1.

From Figure 10.6, it can be shown that the adaptive schemes outperform the fixed schemes except when the fading is mild and the mismatch between the fixed quantization and the fading is small. The performance improvements of the adaptive schemes become larger as the fading gets severer. As expected, the cases with two PCBs have higher $SINR_{1\%}$ than those with one PCB. It is, however, important to note that the gain by using more PCBs decreases as the number of PCBs increases.

In the simulation we assumed that the quantization scaler at the transmitter was updated perfectly. In reality, this long term update requires additional feedback bandwidth. When we compare the fixed scheme with two PCBs and the adaptive scheme with one PCB, it is immediately seen that the adaptive scheme is allowed 20 bits per quantization scaler feedback. This guarantees high precision even when a rate of 1/2 FEC is applied. The use of the adaptive scheme (with one PCB) subject to limited feedback bandwidth, however, is preferred only when the fading increment standard deviation is larger than 1.5 dB.

Finally, the performance when using a 2-tap loop filter is also compared. Due to the assumption of independent-increment fading, the power control error process is almost i.i.d., so the improvement by using a 2-tap loop filter is very limited. When the fading increment is correlated, the benefit of 2-tap filtering is expected to be more visible (see Figure 10.9).

Figure 10.7 shows the impact of the MAI intensity on the CDMA uplink scenario. The same independent-increment fading model and power control parameters as in Figure 10.6 were used. The fading increment processes for different users were assumed independent but with the same statistics (standard deviation = 1.5 dB). In addition,

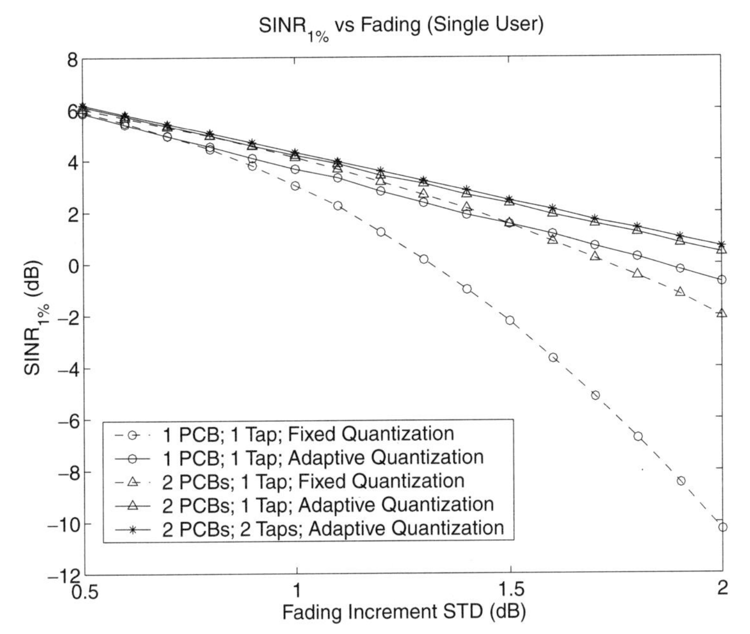

Figure 10.6 1% SINR vs. fading strength.

the CDMA processing gain was 64, and the modulation was BPSK. From this figure, it can be seen that the 1% SINR decreases very slowly with the number of users in the range we simulated. Outside this range, the CDMA network was simply not able to be supported. The relation between the performances of different power control schemes, in the meanwhile, remains similar to before.

The effect of a long propagation delay was examined by applying the proposed design to a GEO satellite communication system. For this example, the satellite was used as a bend pipe, so the round trip propagation delay for power control was about 0.5 sec. The power control updates happened every 50 ms. Hence, including the time required for measurement and processing, the total delay was $M = 11$ power updates. Due to such a long delay, the stability condition becomes very restrictive. For a first order loop filter, the stability condition is $0 < a_0 < 0.1365$. Evaluations similar to Figure 10.4 were carried out to obtain the optimal loop filters. The resulting first and second order loop filters were $F(z) = 0.08$ and $F(z) = 0.124 - 0.062z^{-1}$, respectively.

The simulation results of the geostationary satellite applications are shown in Figure 10.8. In this figure, except for the long delay $M = 11$ and different loop filters for the adaptive schemes, the other parameters are the same as in Figure 10.6. Note that the independent-increment fading process (with a time unit equal to 50 ms), which was chosen to simplify the model and be consistent with the previous examples, may be pessimistic. As shown in Figure 10.8, the adaptive schemes basically follow the same trend as in Figure 10.6. The fixed schemes, however, perform very differently. To explain the behaviors of the fixed schemes, we first note that their loop filter is $F(z) = 0.6266$, which is not in the stability region. These schemes, as we have mentioned previously, are always stable, for their transmission power adjustments are limited. In other words, fixing the dynamic range of the transmission power adjustment

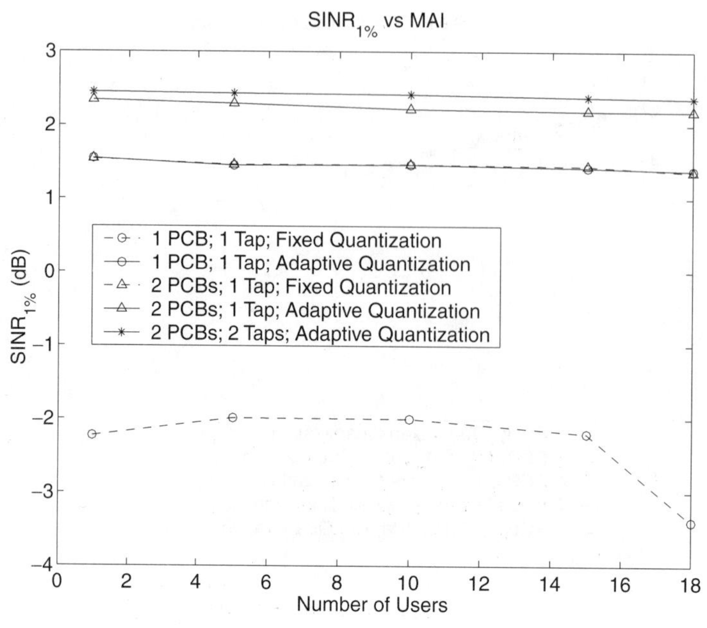

Figure 10.7 1% SINR vs. MAI.

is equivalent to decreasing the effective loop filter gain as the power control error increases. Once the effective loop filter gain touches the boundary of the stability region, the power control error will stop growing; and the steady state power control performance depends on the dynamic range of the power adjustment. The scheme with one PCB outperforms the scheme with two PCBs because it has smaller dynamic range when the two schemes have the same quantization/reconstruction scaler. As the fading increment (hence the power control error) increases, the aforementioned effective loop filter gain may fall inside the stability region from the beginning, and the performance is again dominated by how well the transmission power adjustment can track the fading process. In this situation, the fixed 2-PCB performance becomes better than the 1-PCB one. Figure 10.8 also shows that there is a region where the fixed 1-PCB quantization is close to the optimum. In this region, the adaptive 1-PCB scheme is slightly worse than the fixed one due to the adaptation excess error. The utilization of a 2-tap loop filter is, again, not necessary under such a fading model.

Self-Optimizing Loop Filter

In the above application, the selection of loop filter relies on either numerical analysis or simulation. These approaches impose extra computation on the system design, and may not give exact optimization, since it is very difficult to consider all the error processes, not to mention that the fading statistics are time-varying. Fortunately, as Figure 10.4 shows, the power control error is a convex function of the loop filter gain within the stability region. This suggests that the loop filter may also be adjusted adaptively. In that case, the loop filter works like a channel identifier. As we can observe from Figure 10.2, the construction we have now differs from an ordinary system identification model in that our prediction of the channel is an accumulated

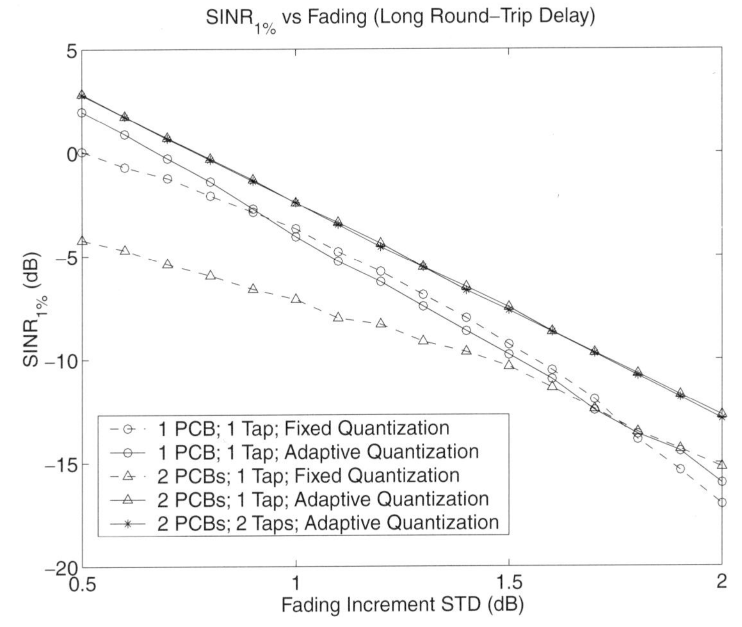

Figure 10.8 1% SINR with long round-trip delay ($M = 11$).

version of the filter output. The feedback is distorted and delayed and the driving process to the filter is the feedback itself. Despite these, the system identification principle remains the same.

To see how the loop filter adaptation can be implemented, we adopt the constant MAI analysis for simplicity, and drop the error processes. The steady state power control error variance can be written as

$$\sigma_e^2 = \frac{\sigma_{\Delta L}^2}{2\pi} \int_{-\pi}^{\pi} \left| \frac{S_{\Delta L}(e^{j\omega})}{1 - e^{-j\omega} + e^{-j(M+1)\omega} F(e^{j\omega})} \right|^2 d\omega$$

and the characteristic function is still $1 - z^{-1} + z^{-(M+1)} F(z)$.

Now if we consider a transversal loop filter $F(z)$ with its tap-weights denoted by $a_k, k = 0, 1, \ldots, K-1$, where K is the order of the filter, a characteristic polynomial can be obtained. The system stability is maintained if all zeros of the characteristic polynomial are within the unit circle. Let us denote by $Z_1, Z_2, \ldots, Z_{M+K}$ and $b_0, b_1, \ldots, b_{M+K}$, $(b_0 = 1)$, the zeros and the coefficients of the polynomial. The characteristic polynomial can be seen as a continuous mapping between the $\mathbf{C}^{M+K}$ domain of zeros and the $\mathbf{R}^{M+K}$ domain of coefficients excluding b_0. The stability condition is $|Z_i| < 1, 1 \leq i \leq M + K$, which is a connected region in $\mathbf{C}^{M+K}$. This means that the stability region of the coefficients is also connected. In other words, there is only one stability region of the filter tap-weights. Within this region, if we fix all tap-weights except for one which is left variable, the above equation for σ_e^2 can be used to compute the power control error variance as a function of this coefficient. The shape of this function, as intuition suggests, is convex, at least for the low order (≤ 3) filters we have evaluated. This result guarantees the validity of gradient search for the global minimum when the loop filter is first order. For higher order filters we have

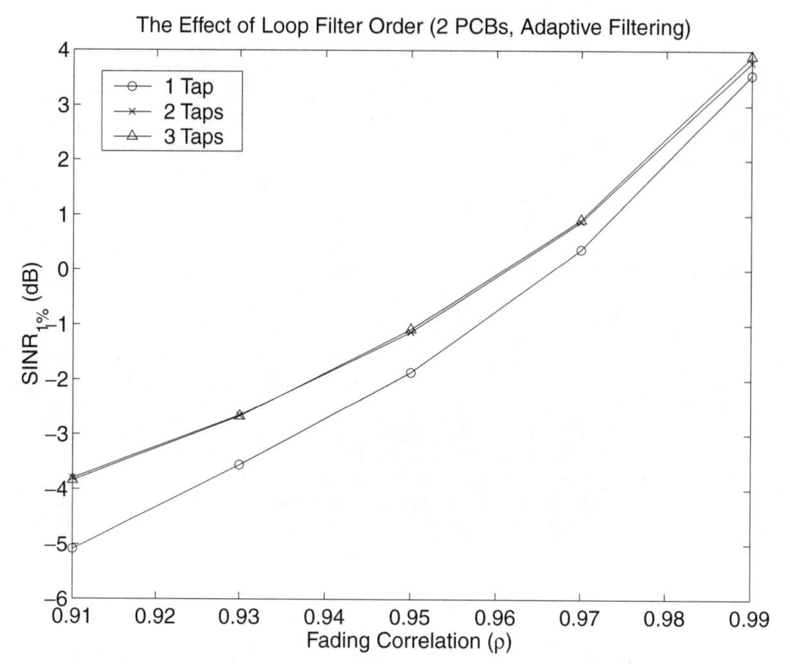

Figure 10.9 1% SINR for two PCB adaptive filtering.

no proof at this moment whether this result implies a single minimum. Simulation results, however, seem to suggest so.

To implement the gradient method, standard adaptive filtering techniques can be applied. We use in the following the Least-Mean-Square (LMS) [11] algorithm as an example. For this setup, the loop filter tap-weights a_k are adapted with

$$a_k(j+1) = a_k(j) + \mu \cdot e\,(j - 2(M+1) - k)\, e(j - M - 1)$$

where μ is the step-size parameter. When we take into consideration all different kinds of noises, the LMS algorithm suffers from a severe gradient noise. This problem can be solved by averaging the tap-weight increments, thus we replace the above equation by

$$a_k(n+1) = a_k(n) + \mu \Delta \hat{a}_k(n)$$

where $\Delta \hat{a}_k(n)$ is the average of the negative gradient $e\,(j - 2(M+1) - k)\, e(j - M - 1)$ over a number of power updating steps, and n is the time index of the adaptation.

Figure 10.9 shows the simulation results regarding the filter order and adaptation. In this figure, only the cases with two PCBs are shown. The systems utilized, in addition to adaptive quantization, the above LMS algorithm with step size $\mu = 0.005$. The step size can be chosen to increase the convergence speed of the filter tap-weights or reduce the steady state error. More importantly, it must not destroy the system stability. The stability condition of the step size depends on the fading model, the quantization, the filter order, the averaging period of the negative gradient, etc., and is difficult to determine analytically. We used simulation to search for the stability region. The step size used here is more on the fast convergence side. For every simulation, the loop filter was initialized with all zero taps, then 50 000 power control iterations were simulated. Both adaptations (quantization and filter) happened once every 20 power control steps. Figure 10.9 shows that, with such an exponentially decorrelating fading, using more than second order filtering does not improve further the performance.

10.3 Multi-User Detection

As we mentioned above, the capabilities of a pseudo-orthogonal or Asynchronous CDMA (A-CDMA) are limited by the near-far effect, and in general by the Multiple Access Interference (MAI). The near-far problem is mitigated with tight power control, as we have discussed in the previous section. The MAI may also be reduced or canceled with MAI cancelers or multi-user detectors. The basic idea behind a multi-user detector comes from the fact that the MAI has an inherent structure, which can be exploited by the detector to increase capacity and improve the performance. However, the computational complexity of the optimum multi-user detector, measured in terms of the number of arithmetic operations per modulated symbol, grows exponentially with the number of users, and is thus impractical unless the number of active users is quite small. Therefore, over the last decade, most research has focused on finding suboptimal multi-user detectors which provide near-optimal performance without incurring the cost of exponential complexity. Among the numerous suboptimal schemes, we pay more attention on the following: *Linear decorrelating detector, Minimum Mean Squared Error (MMSE) detector, Multistage detector, Decision feedback detector, Successive Interference Canceler (SIC)*, and *Joint design of multiuser detection and decoding.*

Many of the aforementioned multi-user detection schemes require exact knowledge of one or several system parameters, such as received powers, phases and propagation delays for all of the users. Exact knowledge of these parameters is unrealistic, and the parameters need to be estimated in real systems. If the receivers are operating in a near-far environment, the parameter estimators must be near-far resistant as well. Analysis of the effects of channel mismatch on the performance of multiuser detectors shows that the detectors are sensitive to parameter estimation errors, and hence lose the desired near-far resistance. Therefore, accurate and efficient channel estimation schemes are essential to the validity of multiuser detectors.

In Section 10.3.1 we present a survey of multiuser detection methods, while in Section 10.3.2 we present a novel method for multiuser detection based on interative decoding.

10.3.1 Methods of Multiuser Detection

In this section, for the sake of simplicity of presentation, we focus our discussion on coherent multiuser detectors. The underlying assumption is that the receiver is able to estimate and track the phase of each active user in the CDMA scenario. However, the reverse link of a cellular CDMA system may employs noncoherent reception. In this case, various noncoherent multiuser detection schemes have been proposed. The basic idea is to pass the received signal through the in-phase and quadrature branches, so the phase information can be preserved without explicitly being tracked. The research results show that the performance loss compared to coherent reception is the same as in a single user detection case.

For a complete up-to-date survey of various multiuser schemes, the interested readers should refer to Verdu [12].

Maximum Likelihood Sequence Estimation (MLSE) Detector

The objective of an MLSE detector [13] is to find the input sequence which maximizes the conditional probability or maximum likelihood of the given output sequence. Let us consider the case of synchronous transmission first; then we address asynchronous transmission. On the synchronous channel, the desired user symbol is interfered exactly by one symbol from other users. In the Additive White Gaussian Noise (AWGN) channel, it is sufficient to consider the signal received in one symbol period, and determine the corresponding detected symbol. For the total number of user K, there are 2^K possible choice of bits in the information sequence. The MLSE detector computes the correlation metrics for each possible sequence, and selects the sequence that has the largest correlation metric. Obviously, the complexity of the MLSE grows exponentially with the number of users K. In the asynchronous case, there are exactly two consecutive symbols from each interferer that overlap a desired symbol. This situation is similar to the single user channel corrupted by InterSymbol Interference (ISI), hence the Viterbi algorithm can be applied to reduced the complexity. Unfortunately, the computational complexity of the Viterbi algorithm is still exponential in terms of the number of users.

Despite the great performance and capacity gains over the conventional detection, such a high complexity makes the MLSE detector impractical. Another disadvantage of the MLSE detector is that it requires knowledge of received amplitudes, phases and propagation delays. These parameters are not available in the receiver *a priori*, and must be estimated accurately.

Linear Multiuser Detectors

An important group of multiuser detectors are linear multiuser detectors. These apply a linear mapping to the soft outputs of the matched filter bank to reduce the MAI seen by each user. In this subsection, we briefly review the two most popular linear multiuser detector, the decorrelating detector and the MMSE detector.

Decorrelating Detector

The decorrelating detector [14], [15] applies the inverse of the correlation matrix to the outputs of the matched filter bank. Thus, the detector completely decorrelates the multiuser interference and results in elimination of the MAI. No knowledge of the received amplitude is required, hence it reduces the burden of the channel estimator. The computational complexity is reduced to linear in terms of the number of users, which is significantly lower than that of the MLSE detector. Lupas and Verdú also proved that the decorrelating detector yields the optimal near-far resistance. A disadvantage of this detector is that it causes noise enhancement, since the power associated with the noise term at the output of the decorrelating detector is always greater than or equal to the noise power at the output of the matched filter bank. Another disadvantage of the decorrelating detector is that the computations needed to invert the correlation matrix are difficult to perform in real time. For synchronous systems, this problem is alleviated because we can decorrelate one symbol at a time. For asynchronous systems, only a finite-length window is used for the decorrelating operation with correction of the edge effects. Therefore, the use of codes that repeat each symbol (short codes) is generally assumed so that the correlation matrix are the same for each symbol. This minimizes the need for recomputation of the correlation matrix inverse from one symbol to the next. Many novel schemes for simplifying the necessary computations in the time-varying (long code) environment have been devised, but the implementability of these schemes is still a question.

MMSE Multiuser Detector

The Minimum Mean Square Error (MMSE) detector [16], [17] is a linear detector which takes into consideration the background noise and MAI at the same time. The detector implements the linear mapping which minimizes the mean-squared error between the actual data and the weighted soft output of the matched filter bank. The solution to this optimization problem shows that the MMSE detector implements a partial or modified inverse of the correlation matrix. Because it takes the background noise into account, the MMSE detector generally (but not in all cases) provides a better probability of error performance than the decorrelating detector. As the background noise goes to zero, the performance of the MMSE detector converges to the decorrelating detector. On the other hand, as the MAI goes to zero, the performance of the MMSE detector approaches to the conventional single user detector. A disadvantage of the MMSE detector is that, unlike the decorrelating detector, it requires the training sequence and estimation of the received amplitude. Another disadvantage is that the performance depends on the energies of the interfering users, which causes some loss of the near-far resistance.

Multistage Multiuser Detector

The multistage detector [18] is analogous to the parallel interference canceler. Each stage takes as its input the data estimates of the previous stage, and produces a new set of estimate at its output. Due to the delay constraint, it is desirable to limit the number of stages to two or three. It was shown that in a system with well designed code waveforms, the performance of the multistage detector closely tracks that of the optimum receiver. As in the case of the MLSE detector, the multistage detector requires a knowledge of the signal amplitude and code timing. The computational complexity of this algorithm is linear in terms of the number of users K. There is one more thing we must note, which is that the performance of the multistage detector depends heavily on the initial data estimates. Too many incorrect initial data estimates may cause the performance to degrade relative to the conventional detector. Therefore, using a decorrelating detector at the first stage significantly improves the performance of the detector and simplifies the analysis of error probability.

Successive Interference Cancelation (SIC) Detector

The main idea of the SIC [19], [20] is to consider what would be the simplest augmentation to the conventional detector which would provide some benefits of multiuser detection. That is how to improve the traditional detector such that it performs reasonably well in a near-far environment. This goal can be achieved by successively canceling the interference generated from the other users. Bearing this concept in mind, it is important to cancel the strongest signal before detecting the other signals because it has the most negative effect. Also, the best estimate of the signal is from the strongest, since the strongest signal has the minimum MAI. Therefore, there are two reasons for doing successive cancelation in order of signal strength. First, it is easiest to achieve acquisition and demodulation on the strongest user. Secondly, the removal of the strongest signal gives the most benefit for the remaining users. The successive cancelation must operate fast enough to keep with the bit rate and not introduce intolerable delay. For this reason, it will be necessary to limit the number of cancelations. The disadvantage of the SIC is that an accurate channel estimate is required for successively canceling the interference from the received signal. Another potential problem with the SIC detector occurs if the initial data estimates are not reliable. Thus, a certain minimum performance level of the conventional detector is required to yield the improvements.

Decision Feedback (DF) Multiuser Detector

The decision feedback detector [21], [22] can be viewed as a combination of a decorrelating detector and a SIC detector. The linear operation of a decorrelating detector partially decorrelates the users without enhancing the noise, then the SIC operation decisions subtract the interference from one additional user at a time, in descending order of signal strength. An important difficulty with the DF detector is the need to compute the Cholesky decomposition of the correlation matrix and the whitening filter. Like the other nonlinear detectors, the DF detector has the disadvantage of requiring channel parameter estimation. If the channel estimates are more reliable than those produced by the decorrelating detector, the DF detector performs better than the decorrelating detector. If the estimates are less reliable, the performance will degrade greatly.

Joint Multiuser Detection and Decoding

The joint design of multiuser detection and decoding for convolutionally coded asynchronous CDMA systems was first considered by Giallorenzi and Wilson [23], who showed that the MLSE is optimal in the sense of minimizing the Bit Error Rate (BER). However, the computational complexity of the MLSE, measured in terms of the number of arithmetic operations per information bit, grows exponentially with the sum of the number of users and the constraint length of the convolutional code, and is thus impractical unless the number of users and the constraint length of the convolutional code are small. Therefore, many researchers have focused on looking for suboptimal receiver designs which provide near-optimal performance without having the cost of exponential complexity [24]. One class of the suboptimal schemes of joint multiuser detection and decoding is the iterative (turbo) multiuser detector which utilizes the turbo processing principle appeared in 1993.

The joint multiuser detection and decoding can be formulated as an iterative decoding process of a serial concatenated code with computational complexity grows exponentially in terms of the number of users. To reduce the high computational complexity, Wang and Poor [25] proposed a low-complexity iterative multiuser detection scheme which consists of soft interference cancelation and MMSE filtering. At each iteration, extrinsic information extracted from the detection and decoding stages can be used as *a priori* information in the next stage. El Gamal and Geraniotis [26] also presented a similar scheme where the Pilot Symbol Aided Modulation (PSAM) was employed to obtain the reliable channel estimate, and this scheme was jointly considered with the iterative decoding process. This approach will also be presented in the following subsection.

10.3.2 An Interative MMSE Multi-User Detector

Now we present a novel iterative receiver for joint detection and decoding of CDMA signals. This scheme is applicable in two situations: (a) when the receiver is capable of decoding the signals from all users, and (b) when the receiver is only capable of decoding the signals from a subset of users, either due to limited processing power or the unavailability of information about some of the users.

The proposed iterative receiver is different from the MMSE receiver [16] in two major aspects. Firstly, in reference [16], the transmitted symbols are assumed to have a uniform distribution. In the proposed algorithm, this assumption is only valid in the first iteration. In the subsequent iterations, the decoder's soft outputs are used to generate the *a priori* probabilities necessary to find the optimum filter coefficients. Secondly, the MMSE filter in reference [16] is a feed-forward filter. This comes as a result from the uniform distribution assumption. While in the proposed algorithm, the filter has both feed-forward and feed-back coefficients. The feed-back connections represent the subtractive interference cancelation part of the receiver.

The direct implementation of the proposed algorithm requires a complexity of polynomial order in terms of the number of users. However, we believe that an adaptive version of the algorithm similar to structures in reference [25] can be developed with far less complexity.

One of the major disturbances that affects the transmission of digital information over land-mobile links is fading, and the most challenging task is the estimation of the time varying fading parameters (i.e. fading phase and amplitude). In slow fading channels, the demodulation process can be enhanced by inserting some known symbols in the bit stream, and using them at the receiver to estimate the complex fading gains. This scheme, called Symbol Aided Demodulation (SAD), has been presented in Chapter 8.

Now we propose a modification to the SAD technique. This modification allows the channel estimator to use, in addition to the known symbols, the soft information from the previous decoding iteration to obtain better channel estimates. The amplitudes and phases of the user signals thus obtained are used in the multi-user detector module to assist during the interference cancelation. Two iterative soft-input channel estimation algorithms are proposed: the first is based on the MMSE criterion; and the second is a lower-complexity approximation of the first. The multi-user detection and channel estimation schemes of this chapter are suitable for both terrestrial wireless (cellular and PCS) as well as for satellite communications.

The Multi-user Detection Model

First, we consider the case of the AWGN channel. Let us assume that K users are sharing the channel. Each one of the K users encodes the binary information sequence using a rate $1/n$ binary convolutional code. Each user independently interleaves the encoded sequence (the necessity of interleaving will be clarified later). A different spreading sequence of length N-chips is used by each user to modulate the encoded symbols. For simplicity of notation, only the synchronous case is considered. However, it can be easily shown that the extension to slotted asynchronous systems (where synchronization is only performed at the frame level) is straightforward. Using the argument in reference [16], it is easy to show that a slotted asynchronous system is equivalent to a synchronous system with twice the number of users. The modulation scheme is Binary Phase Shift Keying (BPSK), and demodulation is assumed to be done coherently. The baseband output of the chip matched filter bank, in the ith bit duration, is given by

$$\mathbf{r}_i = S\mathbf{b}_i + \mathbf{n}_i$$

where $\mathbf{r}_i$ is the $[N \times 1]$ chip matched filter bank output vector; $\mathbf{b}_i$ is the $[K \times 1]$ vector of the transmitted symbols by the K users; S is the $[N \times K]$ signature matrix where the kth column is the signature sequence of the kth user; $\mathbf{n}_i$ is a $[N \times 1]$ white Gaussian noise vector. The different user amplitudes are included in the signature matrix S.

Before we present the new scheme, we will review briefly the two previously proposed iterative receivers. In the maximum *a posteriori* (MAP) receiver (see reference [27]), without loss of generality, the input to the first decoder at time t is calculated as follows:

$$L_t^{(1)} = \log \left[\frac{E_{b^{(2)},....,b^{(k)}} \left\{ p(\mathbf{r}_t | b_t^{(1)} = 1, b^{(2)}...b^{(k)}) \right\}}{E_{b^{(2)},....,b^{(k)}} \left\{ p(\mathbf{r}_t | b_t^{(1)} = -1, b^{(2)}...b^{(k)}) \right\}} \right]$$

where $L_t^{(1)}$ is the log likelihood ratio, and $E_{b^{(2)},....,b^{(k)}}$ is the expectation with respect to the transmitted symbols form the other users. The *a priori* probabilities used to evaluate this expectation are obtained from the previous decoding iteration soft outputs [27]. This expectation is the sum of $2^{(K-1)}$ terms corresponding to all combinations of transmitted symbols. Therefore, this receiver suffers from a complexity of exponential order in the number of interfering users $(K-1)$.

To solve the complexity problem, the following suboptimal approximation was proposed in reference [28]:

$$L_t^{(1)} = \log \left[\frac{\left\{ p\left(\mathbf{r}_t | b_t^{(1)} = 1, E(b^{(2)})...E(b^{(k)}) \right) \right\}}{\left\{ p\left(\mathbf{r}_t | b_t^{(1)} = -1, E(b^{(2)})...E(b^{(k)}) \right) \right\}} \right]$$

where

$$E(b_t^{(k)}) = \frac{e^{L_t^{(k)}} - 1}{e^{L_t^{(k)}} + 1}$$

$L_t^{(k)}$ is the previous iteration soft output, in the log domain, of the kth decoder at time t. Note also that in the above expression, $E(b_t^{(k)})$ was evaluated using the *a priori* probability obtained from the previous decoding iteration. Based on the fact that the chip matched filter bank output vector has a multi-variate Gaussian distribution, each iteration can be viewed as a soft interference cancelation operation. The previous iteration soft outputs are used to calculate estimates of the transmitted symbols. These estimates are then remodulated, by the corresponding spreading codes, and subtracted from the chip matched filter bank output vector to form the next decoding iteration input vector. The complexity of this algorithm is a linear function of the number of interfering users. Compared with the conventional decision feedback multi-user detector, the iterative soft interference cancelation receiver attempts to reduce the probability of error propagations by feeding back soft information instead of hard decisions.

The Multi-User Receiver for AWGN Channels

The main difference in the proposed scheme, compared with the previously proposed iterative receivers, is the design of the multi-user detection module based on the MMSE criterion. After each decoding iteration, the soft outputs are used to update the *a priori* probabilities of the transmitted symbols. These updated probabilities are then used to calculate the MMSE filter feed-forward and feed-back weights. Two scenarios will be considered in this section. First, we consider the scenario where joint decoding of all users is possible. Then, we outline the necessary modifications for the case of joint decoding of only a subset of users.

Joint Decoding of all Users

Without loss of generality, we derive a set of equations describing the filter coefficients used for demodulating the ith transmitted binary symbol from the kth user. The input y_k to the kth user decoder at time i is given by

$$y_i^{(k)} = \mathbf{c}_{fi}^{(k)^T} \mathbf{r}_i + \mathbf{c}_{bi}^{(k)^T} \hat{\mathbf{b}}^{(K/k)}$$

where $\mathbf{c}_{fi}^{(k)}$ is the $[N \times 1]$ optimized feed-forward coefficients vector; $\mathbf{c}_{bi}^{(k)}$, $\hat{\mathbf{b}}^{(K/k)}$ are the $[K-1 \times 1]$ vectors of the optimized soft feed-back weights, and hard decisions, respectively. Note that, since the feed-back coefficients appear only through their sum, we can assume, without loss of degrees of freedom, that

$$c_{bi}^{(k)} = \mathbf{c}_{bi}^{(k)^T} \hat{\mathbf{b}}^{(K/k)}$$

where $c_{bi}^{(k)}$ is a single coefficient that represents the sum of the feed-back terms. $\mathbf{c}_{fi}^{(k)}$, $c_{bi}^{(k)}$ are obtained through minimizing the mean square value of the error (e) between the data symbol and its estimate, given by

$$\begin{aligned}
e &= E\left[\left(y_i^{(k)} - b_i^{(k)}\right)^2\right] \\
&= E\left[\left(\mathbf{c}_{fi}^{(k)^T}\mathbf{r}_i + c_{bi}^{(k)} - b_i^{(k)}\right)^2\right] \\
&= E\left[\left(\mathbf{c}_{fi}^{(k)^T}\left\{\mathbf{S}^{(k)}b_i^{(k)} + S^{(K/k)}\mathbf{b}_i^{(K/k)} + \mathbf{n}_i\right\} + c_{bi}^{(k)} - b_i^{(k)}\right)^2\right]
\end{aligned}$$

where $\mathbf{S}^{(k)}$ is the $[N \times 1]$ signature vector of the kth user; $S^{(K/k)}$ is the $[N \times K-1]$ matrix composed of the signature vectors of the other $K-1$ users; $\mathbf{b}_i^{(K/k)}$ is the $[K-1 \times 1]$ transmitted data vector form the other $K-1$ users. Using standard minimization techniques, it is easily shown that the MMSE solutions for $\mathbf{c}_{fi}^{(k)}$ and $c_{bi}^{(k)}$ have to satisfy the following relations:

$$E\left[\mathbf{b}_i^{(K/k)}\right]^T S^{(K/k)^T}\mathbf{c}_{fi}^{(k)} + c_{bi}^{(k)} = 0 \qquad \text{(a)}$$

$$\begin{aligned}
&\left\{\mathbf{S}^{(k)}\mathbf{S}^{(k)^T} + S^{(K/k)}E\left[\mathbf{b}_i^{(K/k)}\mathbf{b}_i^{(K/k)^T}\right]S^{(K/k)^T} + E\left[\mathbf{n}_i\mathbf{n}_i^T\right]\right\}\mathbf{c}_{fi}^{(k)} \\
&\quad + S^{(K/k)}E\left[\mathbf{b}_i^{(K/k)}\right]c_{bi}^{(k)} = \mathbf{S}^{(k)} \qquad \text{(b)}
\end{aligned}$$

where

$$\begin{aligned}
E\left[\mathbf{n}_i\mathbf{n}_i^T\right] &= \sigma_n^2 I_{N\times N} \\
E\left[\mathbf{b}_i^{(K/k)}\right] &= \mathbf{E}_b^{(K/k)}
\end{aligned}$$

$$E\left[\mathbf{b}_i^{(K/k)}\mathbf{b}_i^{(K/k)^T}\right] = I_{(K-1)\times(K-1)} - Diag\left(\mathbf{E}_b^{(K/k)}\mathbf{E}_b^{(K/k)^T}\right) + \mathbf{E}_b^{(K/k)}\mathbf{E}_b^{(K/k)^T}$$

σ_n^2 is the white noise variance; $I_{[N\times N]}$ is the identity matrix of order N; $\mathbf{E}_b^{(K/k)}$ is the $[K-1 \times 1]$ vector of the expected values of the transmitted symbols from the other $K-1$ users. The *a priori* probabilities used to evaluate the expectations are obtained from the previous decoding iteration soft outputs, through the following component-wise relation:

$$P(b_t^{(k)} = 1) = 1 - P(b_t^{(k)} = -1) = \frac{e^{L_t^{(k)}}}{1 + e^{L_t^{(k)}}}$$

Note that $E\left[\mathbf{b}_i^{(K/k)}\mathbf{b}_i^{(K/k)^T}\right]$ above is obtained by assuming that the different users soft outputs are independent. This assumption is justified through the different, and independent, interleaving used by each user. To simplify notation, we define the following:

$$A = \mathbf{S}^{(k)}\mathbf{S}^{(k)^T}$$

$$B = S^{(K/k)} \left[I_{(K-1)\times(K-1)} - Diag\left(\mathbf{E}_b^{(K/k)} \mathbf{E}_b^{(K/k)^T} \right) + \mathbf{E}_b^{(K/k)} \mathbf{E}_b^{(K/k)^T} \right] S^{(K/k)^T}$$

$$F = S^{(K/k)} \mathbf{E}_b^{(K/k)}$$

$$R_n = \sigma_n^2 I_{N\times N}$$

Solving equations (a) and (b) above, we obtain the following results for the optimum filter feed-forward and feed-back coefficients:

$$\mathbf{c}_{fi}^{(k)} = \left(A + B + R_n - FF^T\right)^{-1} \mathbf{S}^{(k)}$$

$$c_{bi}^{(k)} = -F^T \mathbf{c}_{fi}^{(k)}$$

In the first decoding iteration, we assume that the transmitted symbols have a uniform distribution, and hence, $\mathbf{E}_b^{(K/k)} = \mathbf{0}$. The feed-forward filter coefficients vector, $\mathbf{c}_{fi}^{(k)}$, in this iteration is given by the MMSE equations derived in reference [16], and the feedback coefficient $c_{bi}^{(k)} = 0$. After each iteration, $\mathbf{E}_b^{(K/k)}$ are recalculated using the decoders soft outputs. $\mathbf{E}_b^{(K/k)}$ are then used to generate the new set of filter coefficients as described. In the asymptotic case when $|\mathbf{E}_b^{(K/k)}| = \mathbf{1}$, the receiver is equivalent to the subtractive interference canceler. This is expected, since $|\mathbf{E}_b^{(K/k)}| = \mathbf{1}$ means that the previous iteration decisions, for the other users, are error free. Under this assumption, the subtractive interference canceler becomes the optimum solution.

Joint Decoding of a Subset of Users

One major drawback of the two previously proposed iterative receivers is the necessary assumption of joint decoding of all users at the receiver. Any undecoded user is treated as white Gaussian noise. This imposes a significant limitation on the receiver performance in the presence of undecoded users with relatively high transmission powers. In the new algorithm, this problem is solved naturally through the use of the MMSE filter as a front-end in the receiver. The MMSE filter asymptotically eliminates the interference coming from the undecoded users through the optimization of the feed-forward coefficients [16], [29]. We assume that the receiver has prior knowledge of the undecoded users spreading codes. However, we believe that this assumption can be relaxed through the use of an adaptive architecture similar to references [29] or [25].

In this scenario, the feed-forward and feed-back filter coefficients, $\mathbf{c}_{fi}^{(k)}$, $c_{bi}^{(k)}$ are still given by the above equations. However, for the undecoded users, the expected values of the transmitted symbols are $E_b(j) = 0$ and do not change with iterations. In closed form, Let K_1 be the number of undecoded users, and K_2 be the number of decoded users where $K - 1 = K_1 + K_2$. $S^{(K_1)}$, $S^{(K_2)}$ are the $[N \times K_1]$ and $[N \times K_2]$ signature matrices of the undecoded and decoded users, respectively. Now, to calculate the feed-forward filter coefficients, B should be evaluated from the following:

$$B = S^{(K_1)} S^{(K_1)^T} + S^{(K_2)} \left[I_{(K_2)\times(K_2)} - Diag\left(\mathbf{E}_b^{(K_2)} \mathbf{E}_b^{(K_2)^T} \right) + \mathbf{E}_b^{(K_2)} \mathbf{E}_b^{(K_2)^T} \right] S^{(K_2)^T}$$

and the feed-back coefficient can be obtained from

$$c_{bi}^{(k)} = -\mathbf{E}_b^{(K_2)^T} S^{(K_2)^T} \mathbf{c}_{fi}^{(k)}$$

Near-Far Resistance

The near-far resistance characterizes the performance of the multi-user detector(s) in the presence of high power interferers. In the case of joint decoding of only a subset of users, the two previously proposed iterative receivers [27], [28] treat the undecoded signals as white Gaussian noise. This means that, in such a scenario, the near-far resistance of these receivers is equal to zero, similar to the simple matched filter receiver.

Due to the soft information feedback in this class of receivers, the near-far resistance, as defined in references [14], [15] cannot be obtained in a closed form. However, for the iterative MMSE algorithm, unlike the other two algorithms, a lower bound on the performance can be easily obtained by analyzing the single sweep receiver. It can be easily seen that the single sweep receiver is equivalent to the MMSE receiver in reference [16]. Therefore, the near-far resistance of this single sweep receiver is equal to the near-far resistance of the decorrelator receiver given in references [14], [15]. It is also clear that this lower bound, on the near far resistance holds in the case of joint decoding of only a subset of users. This supports our claim of the iterative MMSE receiver superiority, compared with the other iterative receivers, in the case where only joint decoding of a subset of users is possible. This argument will be validated by simulation (see performance results).

The Multi-User Receiver for Fading Channels

In the development of the iterative multi-user detection algorithm for the AWGN channel, we assumed prior knowledge of each signal amplitude and phase. This assumption is not valid in fading channels due to the multiplication of each signal by a complex fading parameter. Accordingly, the iterative algorithm has to be modified to account for the unknown amplitudes and phases. Note also that the two previously proposed iterative receivers [27], [28], [30] were designed for AWGN channels and assumed known amplitudes and phases at the receiver. In the following discussion, we will restrict ourselves to Rayleigh fading channels; the extension to Rician fading channels is straightforward.

First, the AWGN model $\mathbf{r}_i = S\mathbf{b}_i + \mathbf{n}_i$ has to be modified to account for the multiplication by the complex fading amplitudes. The baseband output vector of the chip matched filter bank, in the ith bit duration, is now given by

$$\mathbf{r}_i = SF_i\mathbf{b}_i + \mathbf{n}_i$$

where F_i is a $[K \times K]$ diagonal matrix of the complex fading amplitudes (i.e. $F_i(k,g) = f_i(k)\delta_{k-g}$). The fading process is assumed to be frequency nonselective, and the complex fading amplitude is assumed to remain constant over one symbol interval. The *a priori* fading correlation sequences are organized as the diagonal matrices $\{V(n)\}$: $V(n) = E\left[\mathbf{f}_{n_1}^{*}\mathbf{f}_{n_1+n}{}^{T}\right]$; $\mathbf{f}_{n_1}$ is the $[K \times 1]$ complex fading vector at time n_1. It is also assumed that the fading processes of the different users are independent. In Rayleigh fading channels, $\mathbf{f}_{n_1}$ is a complex Gaussian random vector with zero mean. Consequently, $\mathbf{r}_i$ is characterized by a multi-variate Gaussian distribution, when conditioned on the transmitted data vector. Based on this fact, it is straightforward to develop a soft-input-soft-output MAP multi-user detector similar to that proposed for the AWGN in references [27], [30]. However, the receiver thus obtained will have a complexity of exponential order in the product of the number of users and the channel estimation filter length. The channel estimation filter length is a design parameter which should be chosen based on the fading bandwidth and the available processing power. The exponential complexity of the MAP approach makes it impractical, and hence the need arises for lower complexity architectures.

In reference [31], it was shown that under the assumption of uncorrelated errors in the different users, fading parameter estimates, the channel estimation and the multi-user detection can be done separately. It was also shown, through simulation and analytical bounds, that this 'Near-Optimum' detector achieves much better performance than the conventional matched filter receiver for uncoded systems. Accordingly, we will restrict ourselves to the canonical receiver architecture shown in Figure 10.10. In this architecture, the channel estimation and the multi-user detection operations are performed in different modules. However, the loss in performance due to this separation is minimized through the soft information passing between the different modules and the iterative architecture of the receiver.

For the detail receiver description (i.e. the channel estimation and multi-user detection modules) as well as its performance in fading channels the interested readers may should refer to reference [26].

Multi-User Detector Performance Results

In this section we assume that the number of users is $K = 7$, and the spreading gain is $N = 7$ which correspond to a fully loaded system. These low values of K and N were used for the sake of computational speed (short running times) in the multi-user detector simulation. The short spreading codes assigned to different users were chosen randomly.

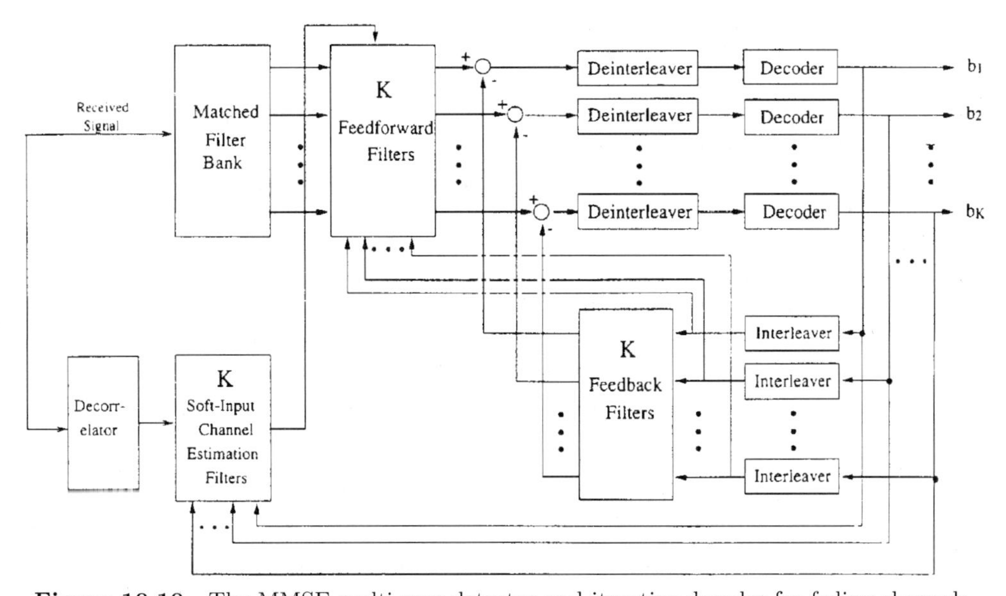

Figure 10.10 The MMSE multi-user detector and iterative decoder for fading channels.

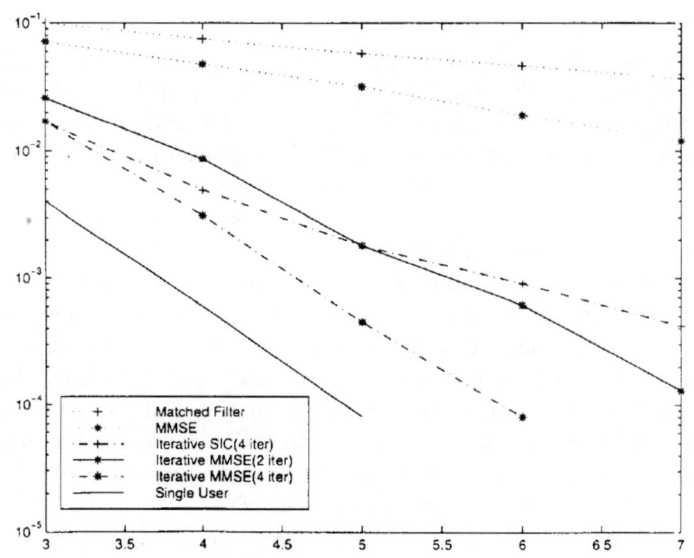

Figure 10.11 Performance comparison of the different CDMA receivers in AWGN channels.

The random phase offsets effect was not accounted for in the simulations (i.e. all users were assumed to have zero phase offset). The FEC codes used are rate 1/2, four states convolutional codes with generator polynomial $(5_8, 7_8)$. The results were obtained by averaging the bit error rates of all decoded users. The channel decoder is based on the Soft Output Viterbi Algorithm (SOVA). The performance results presented here are only consider AWGN channels.

Figure 10.11 compares the performance of the different receiver architectures in AWGN channels when joint decoding of all users is possible. All users were assumed to have equal powers. In the figure 'Iterative SIC' refers to the iterative soft interference cancelation scheme proposed in reference [28]. The iterative MMSE receiver is referred to as 'Iterative MMSE' in the figure. We have also included the performance of the MMSE receiver proposed in reference [16], the conventional matched filter receiver, and the single user system. Note that we have not included the iterative MAP receiver proposed in references [27] and [30] because of its prohibitive exponential complexity. It is clear that the proposed iterative MMSE receiver achieves superior performance to the other receivers in this scenario. It is also shown that the performance gap between the proposed algorithm and the soft interference cancelation increases with the signal-to-noise ratio. We believe that the difference in the performance of the iterative soft interference cancelation receiver reported in the figure and that reported in reference [28] can be attributed to the different conditions in each scenario. In reference [28], the users were assigned long and random spreading codes in an asynchronous multipath channel. Rake receivers that assumed perfect channel knowledge were used. We believe that the additional randomness introduced by the long spreading codes, the asynchronous operation, and the multi-path propagation allowed the soft interference canceler to converge to the single user performance. This may not be the case in practical situations where channel and timing estimation errors may exist. Also, in some practical applications short spreading codes are used because of their desirable

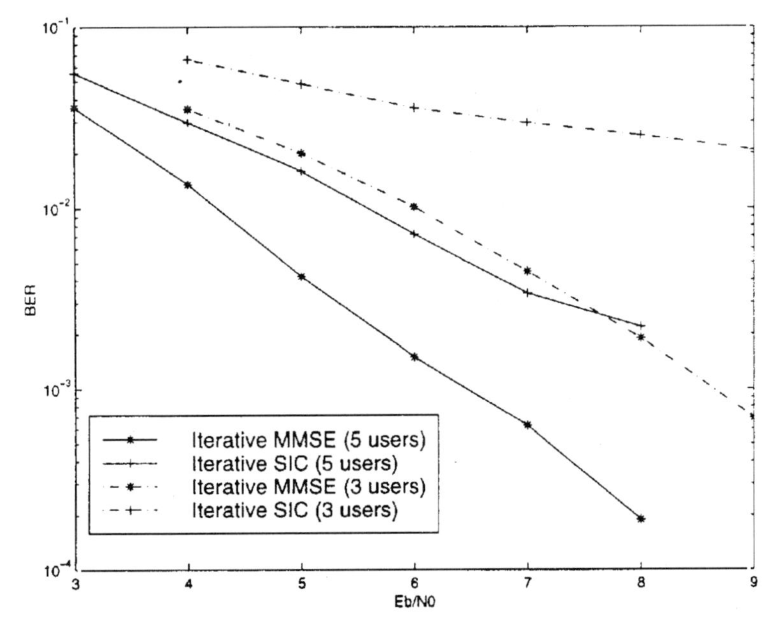

Figure 10.12 Joint decoding of a subset of users in AWGN channels.

synchronization properties. This non-convergence phenomenon will generally become more apparent as the number of simultaneous users sharing the channel increases. While Figure 10.11 does not represent a very practical setting, it indicates that in cases where the soft interference canceler is unable to converge to the single user performance, the iterative MMSE receiver would provide better performance at the expense of added complexity.

Figures 10.12 and 10.13 elaborate further on the comparison between the proposed scheme and the soft interference canceler in terms of the near-far resistance. Four iterations were performed at the receiver in each case. In Figure 10.12, the performance is compared assuming the decoding of three and five users out of a total of seven. It is clear that the performance of the soft interference cancelation scheme is limited by an error floor at moderate to high signal-to-noise ratios. The bit error rate at which the floor begins depends on how many users are decoded at the receiver (i.e. the floor begins at higher bit error rate when decoding smaller number of users). On the other hand, it is shown that the performance of the proposed scheme does not suffer from the error floor phenomenon, as predicted by the discussion in the subsection on 'Near-Far Resistance' in this section.

Figure 10.13 compares the performance of the two schemes in the case of unequal power levels. Four users were assumed to transmit at 3 dBs higher than the nominal E_b/N_0. The receiver is assumed to have prior knowledge of the transmitted power levels. Quite interestingly, it is noted that the performance of both schemes is better than the equal power levels case for small signal-to-noise ratios. This phenomenon is similar to what was observed in the successive interference cancelation detector. At small signal-to-noise ratios, the unbalance in power levels increases the correct decoding probability for the high power users. Canceling those correctly decoded signals in the subsequent iterations increases the correct decoding probability for the

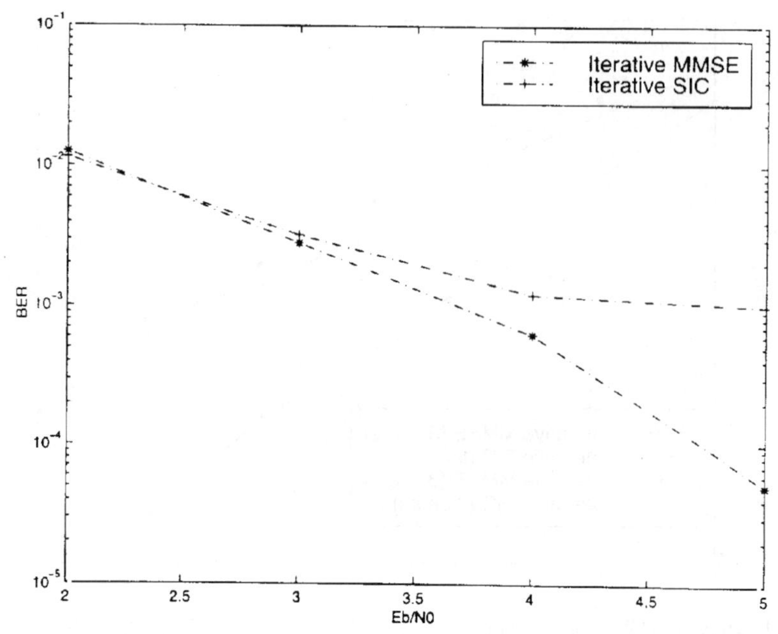

Figure 10.13 Performance in the case of unequal power levels in AWGN channels.

low power users. For the soft interference canceler, the performance is again limited by the error floor phenomenon. As expected, the performance gap between the two algorithms increases with the signal-to-noise ratio.

10.4 Conclusions

In this chapter we have examined two different techniques for optimizing pseudo-orthogonal CDMA performance; namely closed loop power control and multi-user detection.

In Section 10.2 we identified the closed-loop power control problem as an instance of the classical quantization problem. We then proposed a simple adaptive closed-loop power control design based on the statistical characteristics of the fading channel and the MAI process. It was shown that the log-normal approximation of these processes renders the power control error process log-normal, which reduces a great deal of the system complexity and the feedback overhead. In order to smooth the quantized feedback information and exploit its memory, a loop filter was further introduced. Optimization of this filter, according to our analysis and simulation, can be achieved on-line in a self design fashion. Extra computation is therefore held at minimum.

In Section 10.3, we presented a survey of multi-user detection methods, and proposed a novel method for multi-user detection based on an iterative MMSE algorithm for joint decoding and multi-user detection. The performance of the proposed algorithm was compared with the iterative soft interference canceler [28] in AWGN channels. It was shown that the proposed algorithm achieves superior performance under different conditions. The superiority of the proposed algorithm, with respect to the near-far resistance, was demonstrated in the case of joint decoding of a subset of users and the case of unequal power levels.

References

[1] H.-J. Su and E. Geraniotis 'Adaptive Closed-Loop Power Control with Quantized Feedback and Loop Filtering' Electrical Engineering Dept., University of Maryland College Park, MD 20742.

[2] J. Zander 'Performance of Optimum Transmitter Power Control in Cellular Radio Systems' *IEEE Trans. on Vehicular Technology*, Vol. 41, No. 1, February 1992, pp. 57–62.

[3] G. J. Foschini and Z. Miljanic 'A Simple Distributed Autonomous Power Control Algorithm and its Convergence' *IEEE Trans. on Vehicular Technology*, Vol. 42, No. 4, November 1993, pp. 641–646.

[4] A. J. Viterbi, A. M. Viterbi and E. Zehavi 'Performance of Power-Controlled Wideband Terrestrial Digital Communication' *IEEE Trans. on Comm.* Vol. 41, No. 4, April 1993, pp. 559–569.

[5] R. D. Yates and C.-Y. Huang 'Integrated Power Control and Base Station Assignment' *IEEE Trans. on Vehicular Technology*, Vol. 44, No. 3, August 1995, pp. 638–644.

[6] A. M. Monk and L. B. Milstein 'Open Loop Power Control Error in a Land Mobile Satellite System' *IEEE Journal on Selected Areas in Comm.*, Vol. 13, No. 2, February 1995, pp. 205–212.

[7] A. J. Viterbi *CDMA: Principles of Spread-Spectrum Communications*, Addison-Wesley, Massachusetts, 1995.

[8] J. Proakis '*Digital Communications*' *3rd Ed.*, McGraw-Hill, 1995.

[9] N. S. Jayant and P. Noll *Digital Coding of Waveforms*, Prentice Hall, New Jersey, 1984.

[10] G. F. Franklin, J. D. Powell and M. L. Workman *Digital Control of Dynamic Systems, 2nd Ed.* Addison-Wesley, Massachusetts, 1990.

[11] S. Haykin '*Adaptive Filter Theory*' *3rd Ed.*, Prentice Hall, New Jersey, 1996.

[12] S. Verdu *Multiuser Detection*, Cambridge University Press, 1998.

[13] S. Verdu, 'Minimum probability of error for asynchronous Gaussian multiple-access channel' *IEEE Trans. Inform. Theory*, Vol. 32, Janunary 1986, pp. 85–96.

[14] R. Lupas and S. Verdu 'Linear multiuser detectors for synchronous code-division multiple-access channels' *IEEE Trans. Inform. Theory*, Vol. 35, January 1989, pp. 123–136.

[15] R. Lupas and S. Verdu 'Near-far resistance of multiuser detectors in asynchronous channels' *IEEE Trans. Comm.*, Vol. 38, April 1990, pp. 496–508.

[16] U. Madhow and M. L. Honig 'MMSE interference suppression for direct-sequence spread spectrum CDMA' *IEEE Trans. Comm.*, Vol. 42, December 1994, pp. 3178–3188.

[17] Z. Xie, R. Short and C. Rushforth 'A family of suboptimal detectors for coherent multiuser communications' *IEEE Journal on Selected Areas in Comm.*, Vol. 8, May 1990, pp. 683–690.

[18] M. K. Varanasi and B. Aazhang 'Multistage detection in asynchronous code-division multiple-access communications' *IEEE Trans. Comm.*, Vol. 38, April 1990, pp. 509–519.

[19] R. Khono 'Pseudo-noise sequences and interference cancelation techniques for spread spectrum systems-spread spectrum theory and techniques in Japan' *IEICE Trans.*, Vol. E.74, May 1991, pp. 1083–1092.

[20] P. Patel and J. Holtzman 'Analysis of a simple successive interference cancelation scheme in DS/CDMA system' *IEEE Journal on Selected Areas in Comm.*, Vol. 12, June 1994, pp. 796–807.

[21] A. Duel-Hallen 'A family of multiuser decision-feedback detectors for asynchronous code-division multiple-access channels' *IEEE Trans. Comm.*, Vol. 43, Feb./Mar./Apr. 1995, pp. 421–434.

[22] M. K. Varanasi 'Decision feedback multiuser detection: a systematic approach' *IEEE Trans. on Inform. Theory.*, Vol. 45, January 1999, pp. 219–240.

[23] T. R. Giallorenzi and S. G. Wilson 'Multiuser ML sequence estimator for convolutionally coded asynchronous DS-CDMA systems' *IEEE Trans. Comm.*, Vol. 44, August 1996, pp. 997–1008.

[24] T. R. Giallorenzi and S. G. Wilson 'Suboptimal multiuser receivers for convolutionally coded asynchronous DS-CDMA systems' *IEEE Trans. Commun.*, Vol. 44, September 1996, pp. 1183–1196.

[25] X. Wang and H. V. Poor 'Iterative (Turbo) soft interference cancelation and decoding for coded CDMA' *IEEE Trans. Comm.*, Vol. 47, July 1999, pp. 1046–1061.

[26] H. El Gamal and E. Geraniotis 'Iterative multiuser detection for coded CDMA signals in AWGN and fading channels' *IEEE Journal on Selected Areas in Comm.*, Vol. 18, January 2000, pp. 30–41.

[27] M. C. Reed, C. B. Schlegel, P. D. Alexander and J. A. Asenstorfer 'Interative multiuser detection for DS-CDMA with FEC' *Proc. International Symposium on Turbo Codes and Related Topics*, Brest, France, September 1997, pp. 162–165.

[28] P. D. Alexander, A. J. Grant and M. C. Reed 'Interative detection in code division multiple access with error control coding' *European Trans. on Telecommunications*, Vol. 9, September/October 1998, pp. 419–425.

[29] M. Honig, U. Madhow and S. Verdu 'Blind adaptive multiuser detection' *IEEE Trans. on Inform. Theory*, Vol. 41, July 1995, pp. 944–960.

[30] M. Moher 'An iterative multiuser decoder for near-capacity communications' *IEEE Trans. on Comm.*, Vol. 46, July 1998, pp. 870–880.

[31] S. Vsudevan and M. Varanasi 'Achieving near-optimum asymptotic efficiency and fading resistance for time-varying Rayleigh fading CDMA channel' *IEEE Trans. on Comm.*, Vol. 44, September 1996, pp. 1130–1143.

Index